The Amateur Astronomer's Introduction to the Celes

This introduction to the night sky is for amateur astronomers who desire a deeper understanding of the principles and observations of naked-eye-astronomy. It covers topics such as terrestrial and astronomical coordinate systems, stars, and constellations; the relative motions of the sky, Sun, Moon, and Earth; leading to an understanding of the seasons, phases of the Moon, and eclipses. Topics are discussed and compared for observers located in both the northern and southern hemispheres.

Written in a conversational style, only addition and subtraction are needed to understand the basic principles, with a more advanced mathematical treatment available in the appendices. Each chapter contains a set of review questions and simple exercises to reinforce the reader's understanding of the material. The last chapter is a set of self-contained observation projects to get readers started with making observations related to the concepts they have learned.

WILLIAM MILLAR is Professor of Astronomy at Grand Rapids Community College. He gained his B.S. in physics from Calvin College and his M.A. in physics from Western Michigan University, and he is currently working on a doctorate in astronomy at James Cook University in Australia. He has taught astronomy for almost 20 years and has also taught courses in physics, mathematics, electronics, and acoustics and is author of the astronomy textbook *Descriptive Astronomy* (Harcourt Brace, 1997). He is very involved in amateur astronomy groups and serves on the board of directors for, and is a past vice-president of, the Grand Rapids Amateur Astronomy Association. He is a member of the Planetary Society and a Technical Member of the Astronomical Society of the Pacific.

The Amateur Astronomer's Introduction to the Celestial Sphere

WILLIAM MILLAR
Grand Rapids Community College

CAMBRIDGE UNIVERSITY PRESS
Cambridge, New York, Melbourne, Madrid, Cape Town, Singapore, São Paulo

Cambridge University Press
The Edinburgh Building, Cambridge CB2 2RU, UK

Published in the United States of America by Cambridge University Press, New York

www.cambridge.org
Information on this title: www.cambridge.org/9780521671231

First published 2006

Printed in the United Kingdom at the University Press, Cambridge

A catalog record for this publication is available from the British Library

ISBN-13 978-0-521-67123-1 paperback
ISBN-10 0-521-67123-X paperback

For Larry Oppliger, Jerry Hardie, Michitoshi Soga, and Leo Parpart
My teachers and my friends

Contents

Preface

This is the first of a series of five introductory astronomy books written specifically for amateur astronomers. They are the result of many years of teaching college-level introductory astronomy courses and working with amateur astronomers. Amateur astronomers have a wide variety of backgrounds and levels of current knowledge. As a group, this description also fits the students attending a typical community college.

There are many introductory texts on the market geared for one-semester courses with plenty of full-color photos and drawings. All of them are well-written books covering the many aspects of astronomy. However, I have found that amateur astronomers have a natural desire for a deeper understanding of the sky than the majority of the students for whom these books are written. Therefore I chose to write this series of introductory books with the depth and detail that my interactions with amateurs have led me to believe they desire.

All five volumes in this series use a minimum of mathematics. There is a set of appendices giving greater mathematical details for those readers wishing to learn them, but ignoring the mathematics should not detract from your understanding of the material. The text is written in a conversational style without side-bars, footnotes, or end notes, which, in my opinion, only act to distract the reader. The pages are not a collage of pictures and text boxes. You can start at page one and simply read straight through the story.

The Amateur Astronomer's Introduction to the Celestial Sphere describes the general characteristics of the sky for both northern and southern hemisphere observers. It covers stars and constellations (Chapter 3), and the motion of the sky (Chapter 4). As part of understanding this motion the reader is introduced to the terrestrial and celestial coordinate systems used by astronomers (Chapter 2). The sky maps included with this volume are used for examples of simple conversions between the coordinate systems: larger, printable copies are available from the book's website, which may be visited at http://www.cambridge.org/9780521671231. This volume then goes on to discuss the motion of the Sun and the measure of time (Chapter 4). The cause of the seasons is covered in Chapter 5. The motions and phases of the Moon are discussed in Chapter 6. Solar and lunar eclipses are covered in Chapter 7. Chapter 8 contains a set of sky observation projects to get you started on the types of observations emphasizing the topics discussed in the book.

There are some paragraphs set-off from the text by a little extra space and a line along the left side of the paragraph. I call these, "concept paragraphs."

| *This is a concept paragraph.*

Their purpose is to make sure you see the major points of the discussion. There are also some places where I want to make sure you understand the arithmetic in the discussion and so I've included examples. They look like this:

Example: This is an example of an example.

Each chapter has a set of review questions and short review problems to reinforce your understanding of the material. Make sure you can answer these questions and problems before moving on to the next chapter. The answers are available as a .pdf type file on the book's associated website.

The main difficulty in writing astronomy books is the deep connections between the topics. These connections make hard work out of deciding on the order of topics and deciding how deeply to cover each topic. The deeper you go, the more connections there are to be seen. In certain places within this volume I was forced to drop the discussion of a topic in further detail and leave it to a later volume in the series. This was necessary in part because there is material needed for that detail which is not discussed until the later volume and in part because I needed to keep the number of pages in this volume reasonable.

Most of us are aware that astronomy grew out of astrology. The history of astrology is discussed in the context of the roots of modern astronomy in *The Amateur Astronomer's Introduction to the History of Astronomy* – the next volume of this series. In that discussion, I present the basic ideas of astrology and present both astrologers' and astronomers' points of view in the discussion of its validity. The second volume also covers archeoastronomy, pre-telescope discoveries, and the work of the major Renaissance astronomers in far greater detail than the astronomy textbooks. It discusses astronomical discovery through the twentieth century.

The third volume, *The Amateur Astronomer's Introduction to the Solar System*, discusses the Sun and the planets. It covers observations of the Sun, the planets' and moons' orbital motions, naked-eye and telescopic appearance, geology and environments – all of which are based on our observations and explorations of the planets. The fourth volume is *The Amateur Astronomer's Introduction to Deep Space*, covering the characteristics of the stars and the galaxies they inhabit. This includes the stellar and galactic classification schemes based on spectroscopy and observation. The fifth volume is *The Amateur Astronomer's Introduction to Cosmology*, which discusses galactic redshift and its implications for cosmology, and the various cosmological models developed through the ages.

I hope that you find these books both useful and interesting. Please enjoy and learn!

Acknowledgments

Any book project requires the work of many people to complete. I must first thank Dr. Larry Oppliger and Leo Parpart of Western Michigan University for changing a flame of interest in astronomy into a bonfire. Thanks to Dr. Kirk Korista also of Western for encouragement and gently nudging me into doing this project after reviewing an ancient version of the manuscript.

James R. Henningsen provided important aid with some of the photographs. Kevin S. Jung of the Grand Rapids Amateur Astronomical Association (GRAAA) provided important photographs and proof-read the entire manuscript, offering valuable suggestions. Kevin Kozarski of the GRAAA also proof-read the text. I also want to thank the members of the GRAAA and the Kalamazoo Astronomical Society (Kalamazoo, Michigan) for their support.

The Linda Hall Library of Science and Technology provided the photograph from Bayer's *Uranometria* and the United States Naval Observatory provided the clock photos. I had a rather eventful trip to Bolivia with Herb and Shirley DeVries to see the November 1994 total solar eclipse. Thanks go to Herb for allowing me to use his photo of the Diamond Ring in this and other of my publications. The three of us also witnessed the May 1994 annular solar eclipse.

Thanks to my friends and colleagues at Grand Rapids Community College for their help and support – especially Dr. Tom Neils, Bob Cebelak, and Elaine Kampmueller. Tom read the entire manuscript offering many helpful suggestions for clarity and readability. Bob checked my physics and Elaine helped with geology.

Many thanks go to all the people at Cambridge University Press, particularly to Jacqueline Garget for believing the project to be worth the effort and putting in the work to get it approved by the CUP Syndicate. I believe her patience with my many questions to be inexhaustible. The CUP referees' suggestions made this a far better book than it started out to be.

I also thank my friends and family for putting up with me while I worked on "the book."

The book's website may be visited at http://www.cambridge.org/9780521671231.

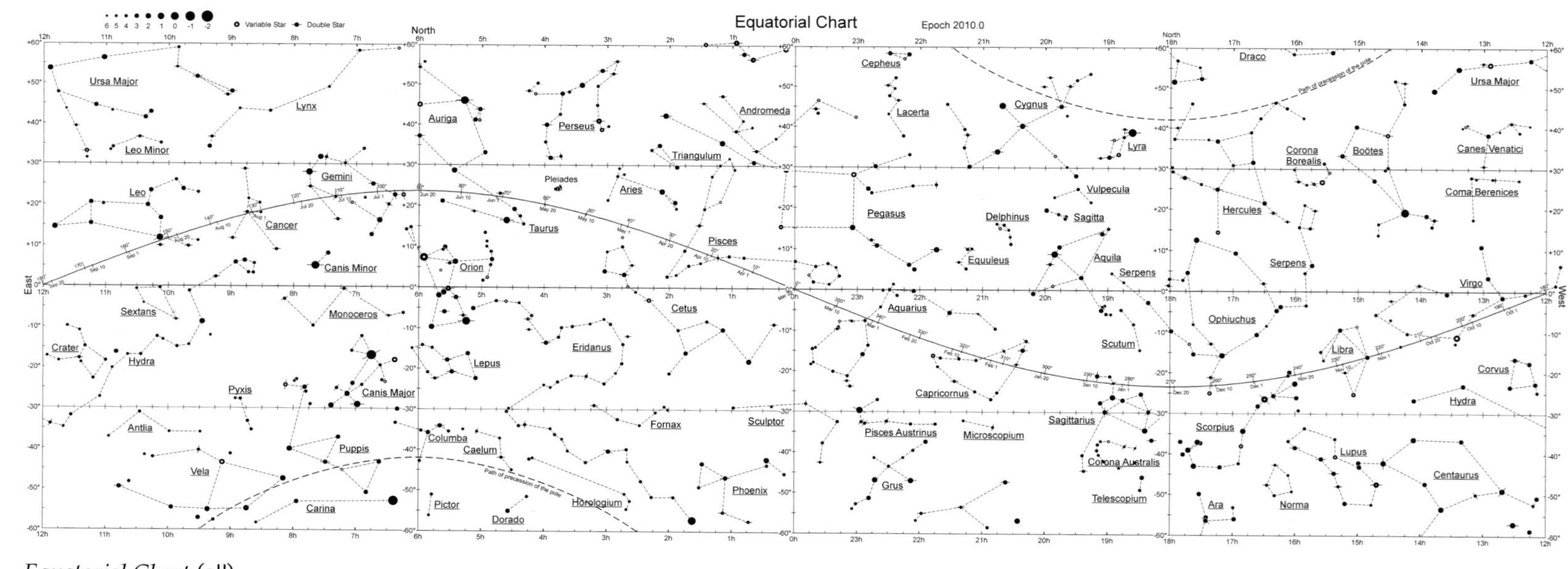

Equatorial Chart (all)

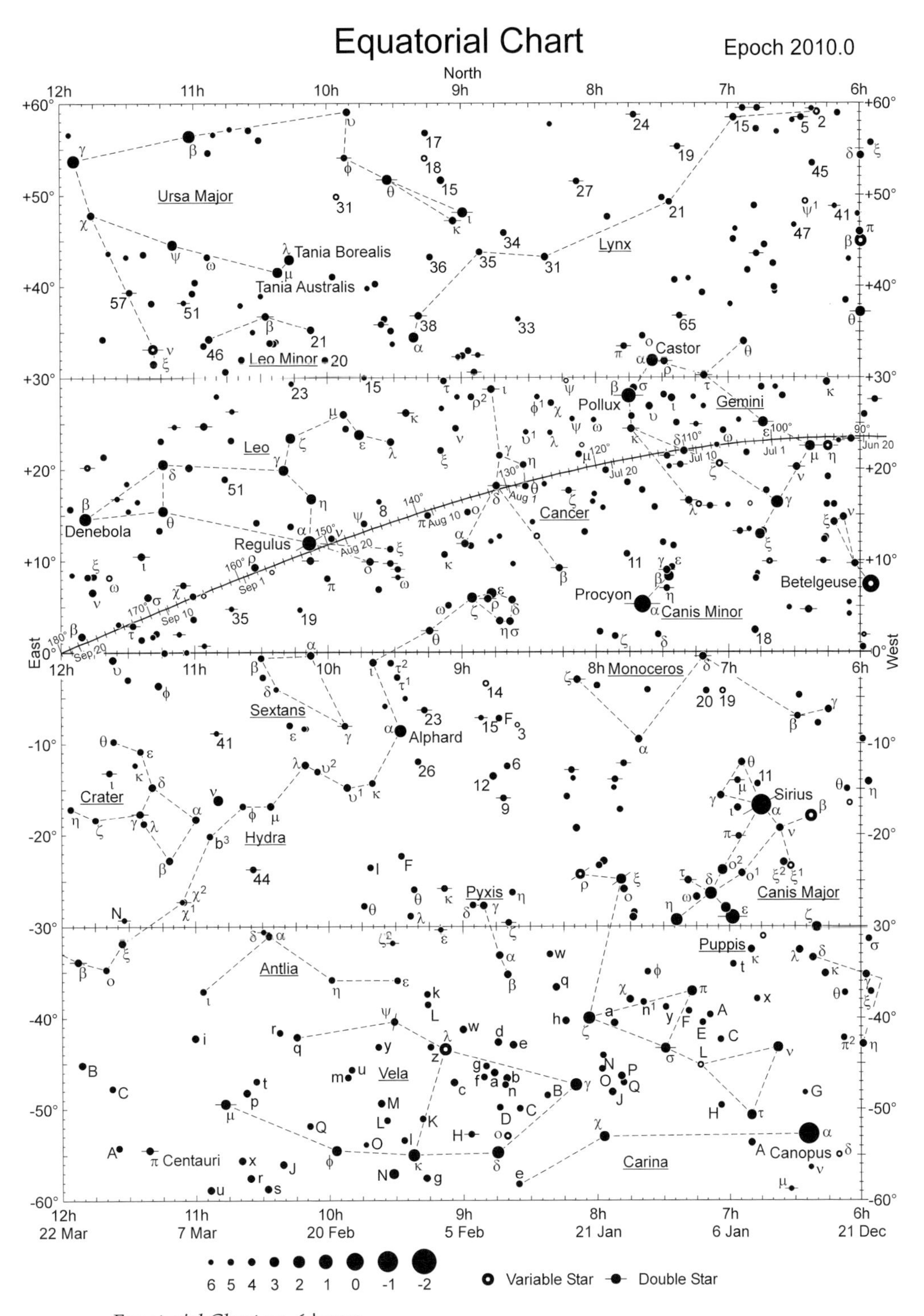

Equatorial Chart 12–6 hours

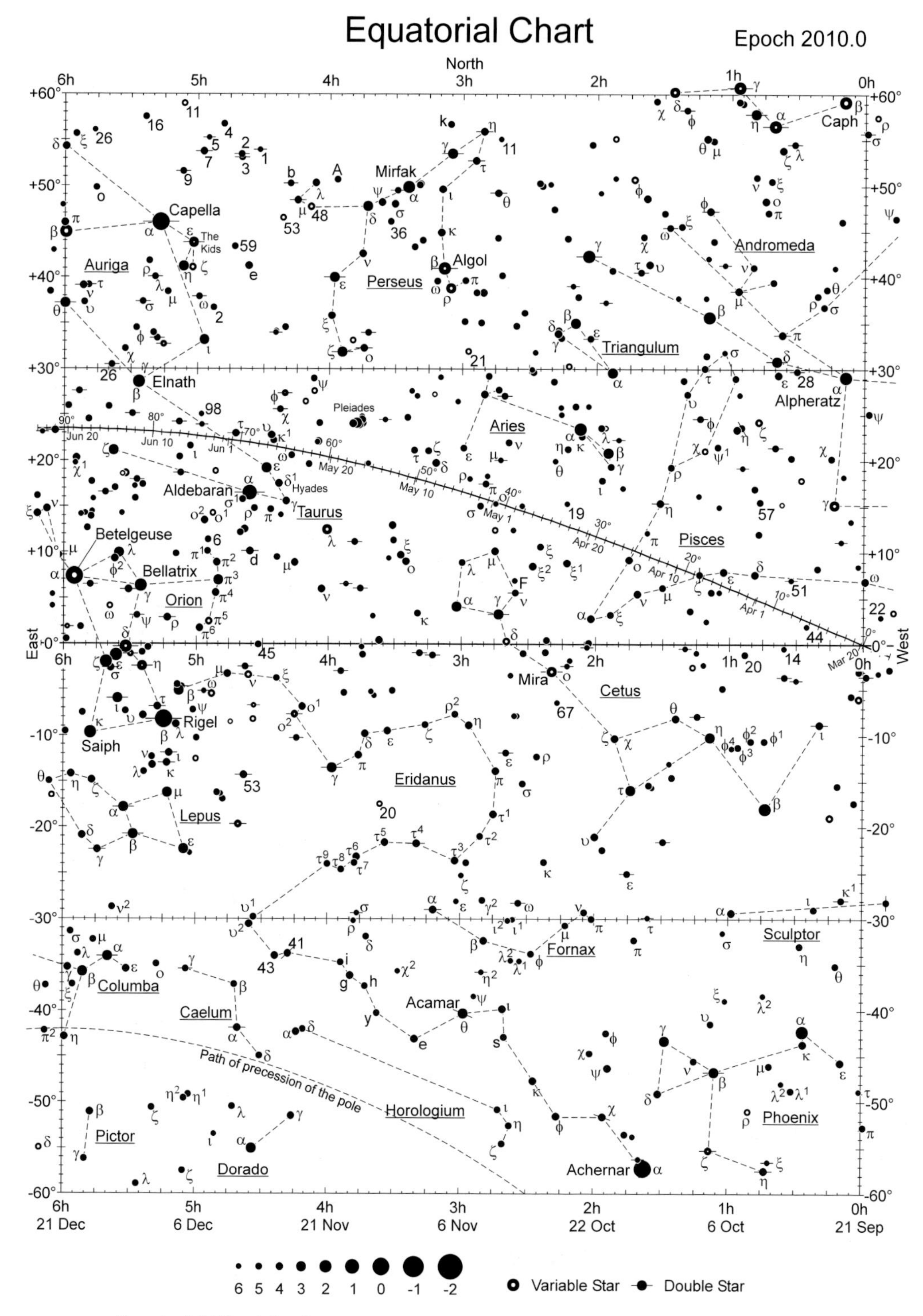

Equatorial Chart 6–o hours

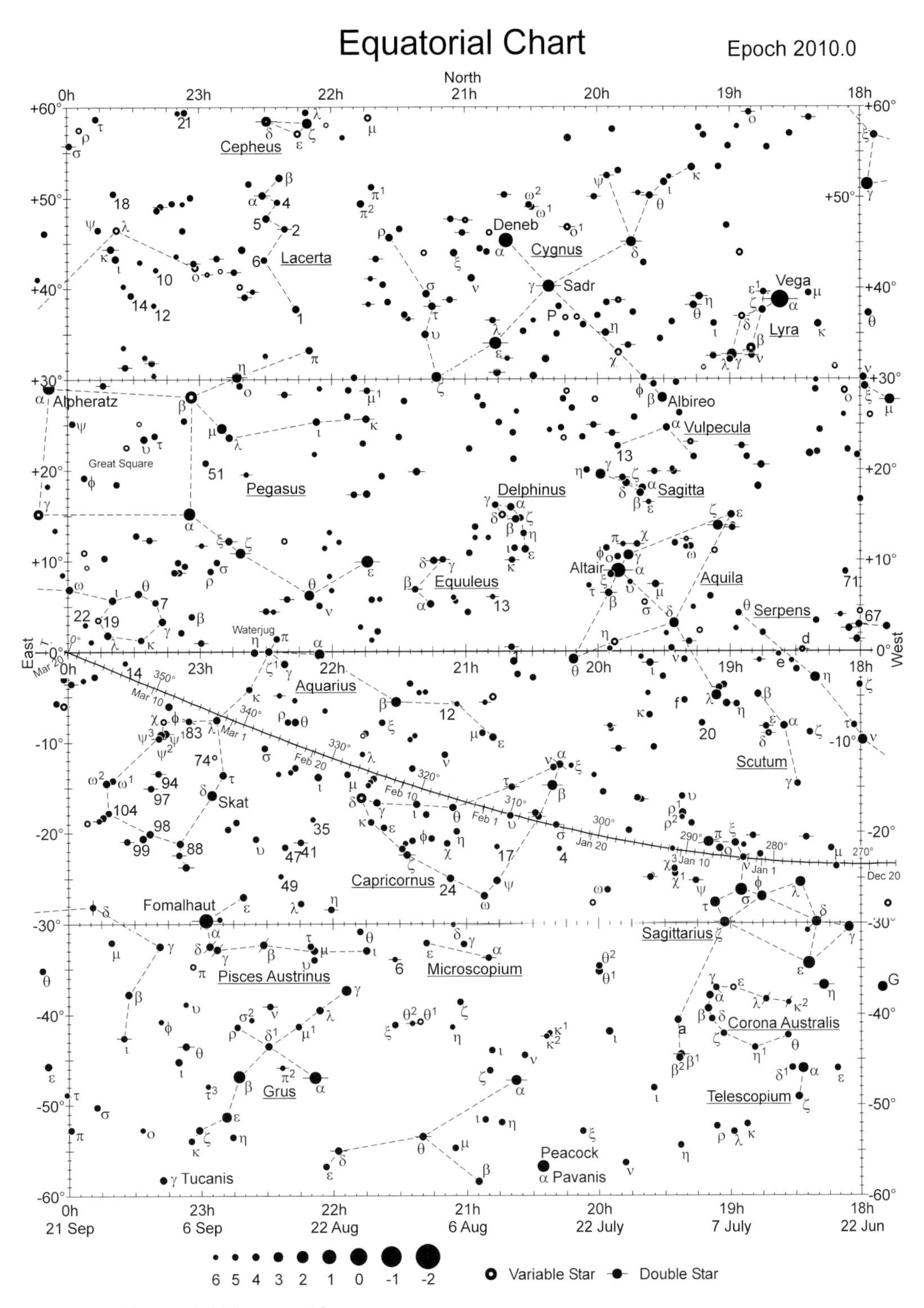

Equatorial Chart 0–18 hours

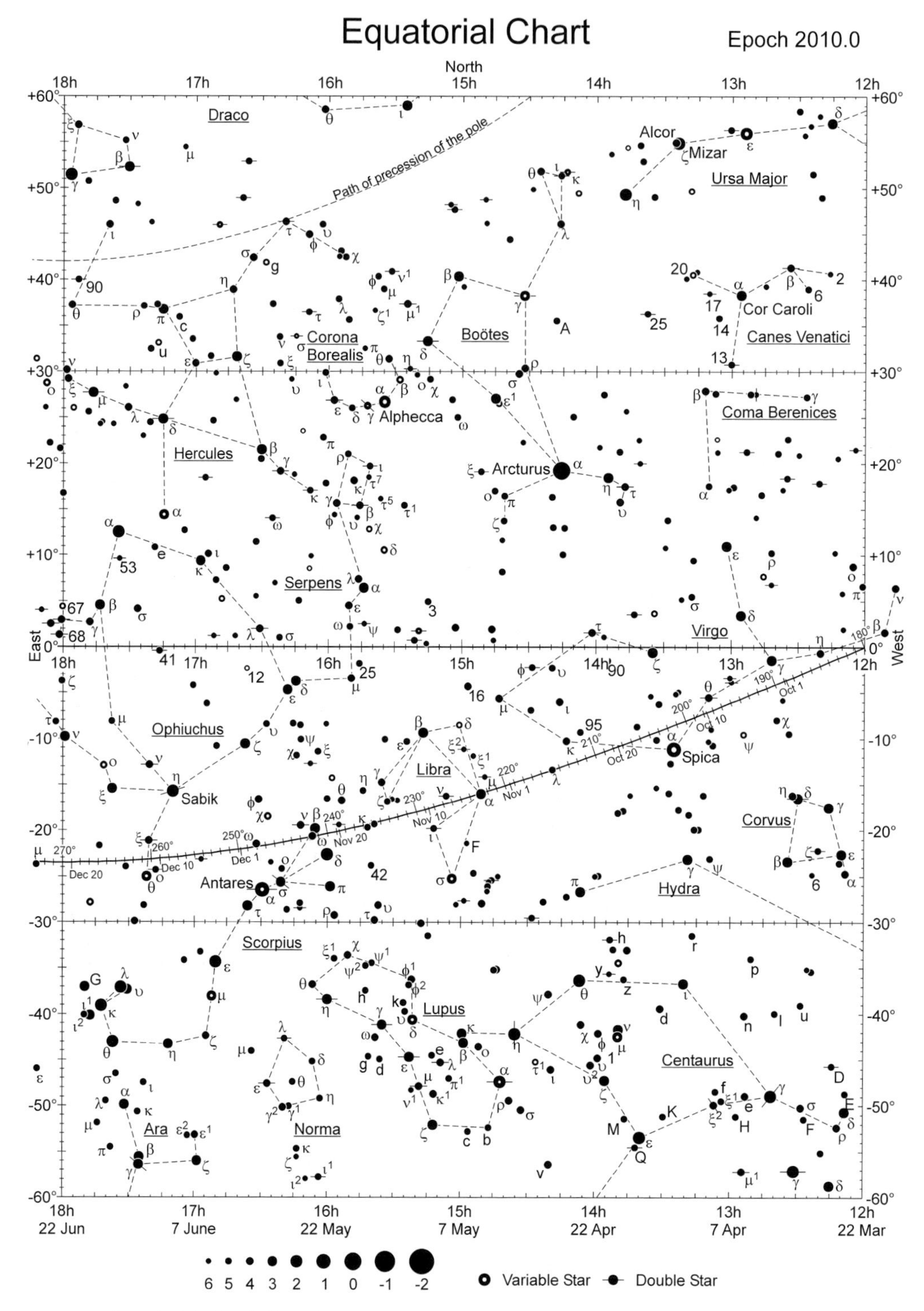

Equatorial Chart 18–12 hours

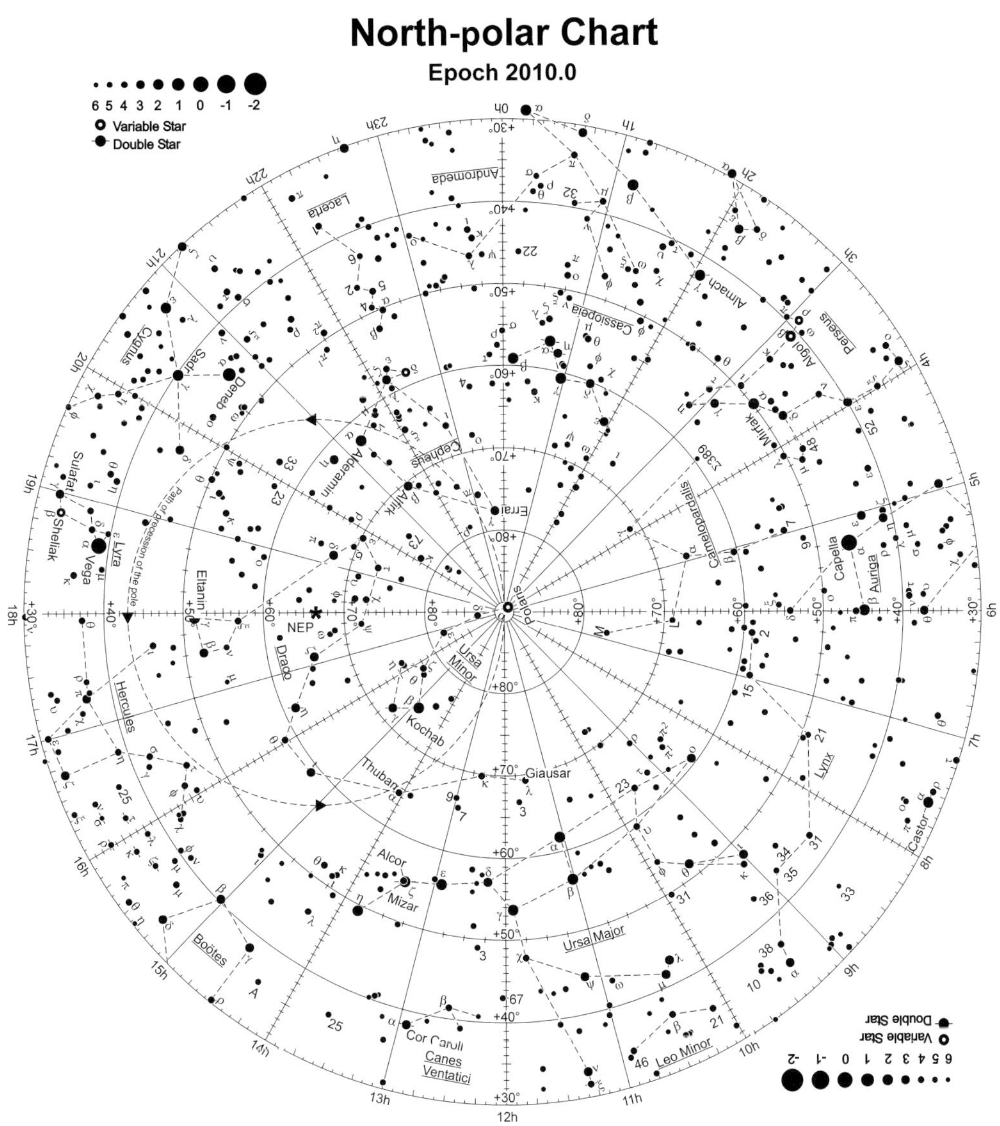

North-polar Chart

South-polar Chart

Epoch 2010.0

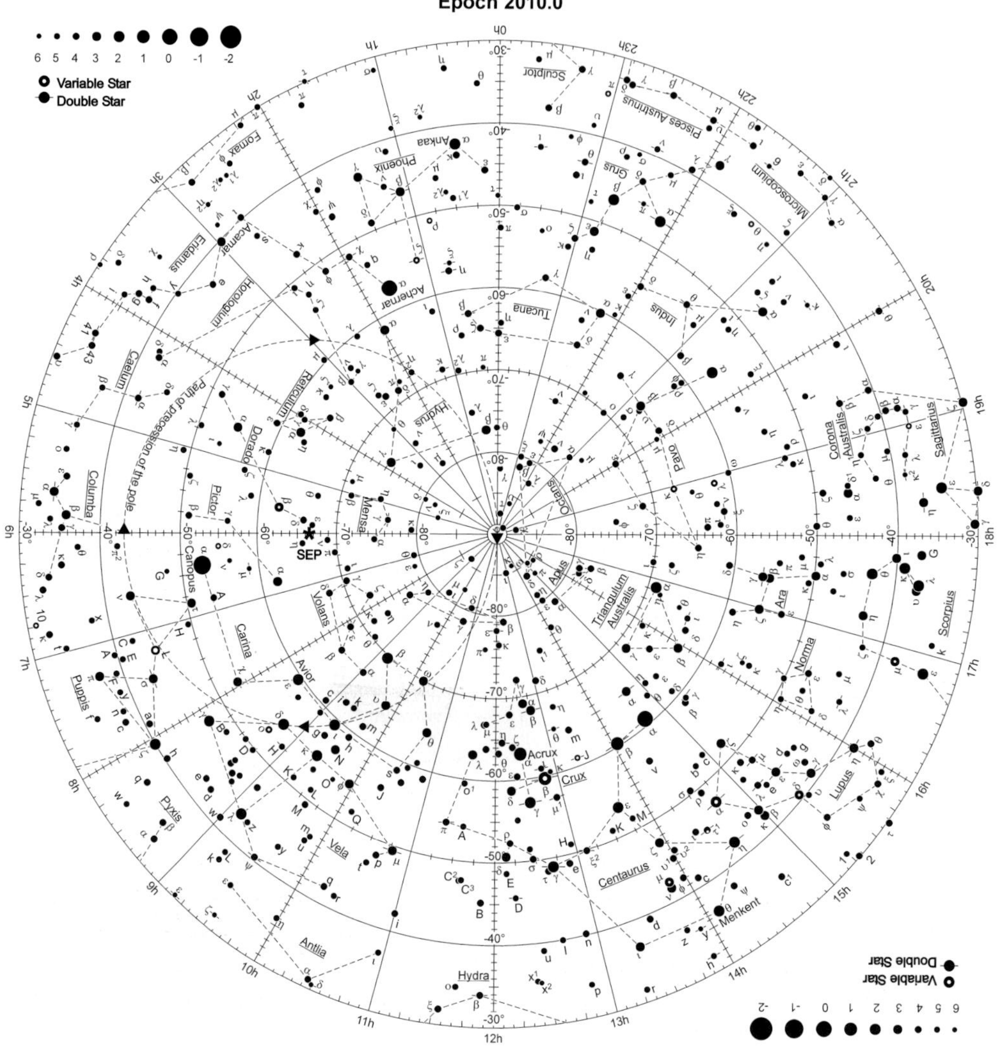

South-polar Chart

1
The lure of the sky

A dark, clear evening sky is one of the most alluring sights – inciting wonder and sparking the imagination. It beckons us to jump to hyperspace or travel a wormhole to distant suns and worlds – to the possibility of alien civilizations. Images of worlds with multiple moons or a twin sunrise shimmer in the mind's eye. Because we now know of over 100 planets circling other suns, there can be little doubt that such worlds exist somewhere, along with other worlds we have not yet even imagined.

Many of us who have adopted the hobby of astronomy were drawn to it by television shows such as *Star Trek* or movies such as *Star Wars*. Others may have been fortunate enough to have parents who introduced them to the stars and constellations. Each of us has a personal story behind how we became interested in the glorious patterns of twinkling lights and the beautifully colored objects photographed through a telescope. We might feel like secret agents seeking the deepest inner workings of the Universe.

For children, the sky can be fascinating or frightening – full of wonder or full of confusion. I seriously doubt that anyone can forget their first sight of Saturn through a telescope. I also doubt that anyone can identify every constellation on the first attempt. Learning the sky, like anything else, takes time. As adults, I believe we should continue to learn more about the world and the Universe in which we live. I hope with increased knowledge, any and all confusion will turn to wonder and fascination.

Through the centuries humans have asked many questions about the sky. What *are* those lights in the sky? How many are there? Where did they come from? Why is the sky blue? Why are sunsets red? Why do stars twinkle? What is a planet? Why does the Moon change its shape? What makes the Sun shine? The history of astronomy is filled with stories of our quest to find the answer to these questions and many others.

We estimate that there are 100 billion galaxies in the Universe. Our home galaxy, the Milky Way, contains an estimated 200 billion stars. How could we possibly count so many galaxies and stars? There are more galaxies in the Universe than there are people who have ever lived on Earth (estimated as 37 billion). There are more stars in the Universe than all the grains of sand on all the beaches of the world. How does someone figure these things out? (The "grains of sand" estimate is worked out on p. 219.)

The star nearest to ours, α Centauri (Alpha Centauri), is known to be at least a binary star system. If the small, red-dwarf star known as Proxima Centauri, which lies between us and α Centauri, is not gravitationally bound to α Centauri A and B (this has not been established) then Proxima Centauri is the nearest star. At this time, we assume Proxima Centauri is part of the α Centauri system (making it a ternary system) and thus the α Centauri system is the nearest star – at 4.3 light years away. How far is a light year? How far is 4.3 of them?

Alpha Centauri is approximately 40 trillion (40 000 000 000 000) kilometers (25 trillion miles) away, as measured by an old surveyor's trick, known to astronomers as *stellar parallax*. This is the *nearest* star and we are already looking at mind-boggling sized numbers. (Hence the phrase "astronomical.") When we speak of the distance to the stars, we must comprehend such vast distances that spaceships – as we know them – would take many human generations to travel to even the nearest star. The fastest spacecraft we have built (the two *Voyager* spacecraft) travel at about 110 000 kilometers (70 000 miles) per hour. A ship traveling this fast would have to support 1950 human generations, requiring about 39 000 years to reach the nearest star. (This claim is calculated on p. 219.) All of written human history is about 6000 years, or 300 generations. A ship may start with us, but it would arrive carrying our great-great . . . great-grandchildren.

Even traveling to our neighboring planets requires years by unmanned spacecraft. The reason I explicitly stated unmanned spacecraft is because we can accelerate these crafts at a higher rate than manned spacecraft. Humans cannot tolerate high accelerations, and these accelerations are needed to produce the velocities to get the crafts to their destinations in a reasonable amount of time. The *Voyager* missions were launched in 1974. They reached Jupiter in 1978, Saturn in 1982, Uranus in 1986, and Neptune in 1989. They are now on their way to the stars, at about six billion miles from Earth. Six billion miles in roughly 30 years. They have not yet even left the Solar System. The latest mission to Saturn, the *Cassini* mission, required seven years to travel to Saturn.

The *Apollo* astronauts took three days to reach the Moon. (*Voyager* made it in ten hours.) Each mission spent a little more time on the surface than the previous. They brought back nearly 382 kilograms (843 pounds) of Moon rocks. Analysis of the rocks has allowed us to come closer to understanding how the Moon formed. This has increased our understanding of how the entire Solar System may have formed. The study of the formation of our Solar System and of planetary systems, in general, is an extremely active field of astronomical research.

Human beings have a strong, built-in drive to explore the world around them. This is demonstrated by the fact that we now occupy nearly every corner of this planet. Our need to explore naturally extends to the worlds beyond our own. The exploration of Mars, which is possibly the next target of our manned explorations, has been discussed in both science and science fiction for over a 100 years. For many years people believed Mars to be inhabited by beings much like ourselves – or perhaps not like ourselves – but far more advanced than us. So far, we have put several landing craft on Mars but none of them have detected any sign of Martians, even of microbial size. In all likelihood, as Dr. Carl Sagan put it, "we will be the Martians" (Sagan, 1980).

Two of the first civilizations to take serious interest in the sky (as indicated by written records) were the Egyptians and the Babylonians. The Egyptians used stars and constellations to create "star clocks" that allowed them to mark the passage of time. The Babylonians created the houses of the zodiac, through which the seven planets (at that time, the Sun and Moon were considered planets) moved – the basis of modern Western astrology. These zodiac signs were later adopted by the Greeks and associated with nearby star patterns. The Greeks added many new constellations related to or based on the mythology surrounding their gods. Other civilizations around the world established their own versions of astrology. One of the more popular forms still practiced is Chinese astrology. Undoubtedly, all of these versions of astrology were first built upon sky observations that were noticeably associated with natural events on Earth.

The Egyptians were the first to realize that there are 365.25 days in a year. They noticed that every four years the helical rising of the bright star Sothis (Sirius) came a day late. With this they created the "Calendar of Sothis," containing 365 days for three years and 366 days during the fourth. The helical rising of Sothis was important to the Egyptians because it signaled the start of the annual flooding of the Nile.

Coincidences such as this and direct relationships such as the motion of the Sun and the cycle of the seasons, and the motion of the Moon and its phases, are probably what led the ancients to connect sky events to human concerns. The Babylonians kept extensive records (*ephemerides*) of the planetary positions for hundreds of years. These records became the basis of their numerical model of the world. This model was used to predict the positions of the planets along the zodiac and thus predict events on Earth. This is the foundation of modern astrology.

Astronomy is generally considered a science. Dick Teresi writes an interesting discussion on the definition of *science* (Teresi, 2002, pp. 15–16). If defined too strictly, astronomy is excluded from the sciences and if defined too loosely, astrology is included. I prefer a definition which includes modern astronomy and excludes astrology, ancient or modern. Thus I am guilty of holding the same, possibly (but not probably) biased, opinion held by most modern astronomers.

Kenneth Davis writes that the word **astronomy** is made from the Greek words *astron*, meaning "star," and *nomos*, meaning "law" (Davis, 2001, p. 9). The earliest known use is about 2500 years ago. Its original purpose was the ordering or classification of the stars. That is, astronomy in astrology was the equivalent of taxonomy in biology. Astrology is the study of the motion of the planets and their affect on our lives. In ancient times, if one studied the sky, one studied astrology. Astronomy was simply a branch of astrology. Astrology and astronomy began to take on their separate, modern meanings with the work of Johannes Kepler in the early 1600s. It was Kepler who realized that we need to use precise observations rather than imprecise philosophies or mysticism to understand the motions of the planets.

Our fascination with the sky is ancient and rooted deep in our collective psyche. What ever the reason you have been lured to the sky, there is much to learn. There's no better time to start than now.

2
Location and coordinates

Telling someone to look for an object within a constellation is one way of expressing the location of the object. However, telling someone NGC 2566 is in the constellation Puppis conveys as much information as telling them the Barringer Meteor Crater is in the state of Arizona. They now know the general region but may still have trouble finding it. To specify the location of an object more precisely, for instance when we want to use a telescope, we need to use a coordinate system. If we want to send a letter to the Visitors' Center at Meteor Crater, we need its exact location – its address. If we want to point a telescope at a particular galaxy, we need its exact location – its coordinates.

Somewhere in our education we have all been introduced to coordinate systems. Usually this occurs in a mathematics class. The objective of all coordinate systems is to specify a point (location) accurately in space. While a point is itself dimensionless (it has no size), the space containing it may be of any number of dimensions, from one to infinity. Probably, the most commonly used coordinate system is the postal address system.

2.1 Coordinate systems

The simplest coordinate system is a number line, shown in Figure 2.1. A line is a one-dimensional space and therefore it takes only one number to specify any point-location in this space. This single number is the **coordinate** and its value specifies the distance between the point-location and the *origin*. In this case, the zero mark on the scale is used as the origin of the coordinate system. The **origin** is the *starting point* of the coordinate system – the point from which all distance measurements begin. The location of the origin within the space is arbitrary so any point-location on the line can be called the origin. Usually the zero mark on the system's scale is placed at the origin, which generally makes the system easier to use but this is not a requirement. Any number on the scale

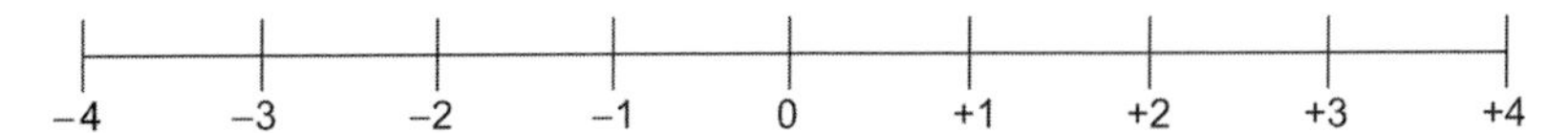

Figure 2.1 A simple number line acts as a one-dimensional coordinate system. Each number describes the distance of a particular point-location from the origin, or zero. The unit of measure for the distance is arbitrary.

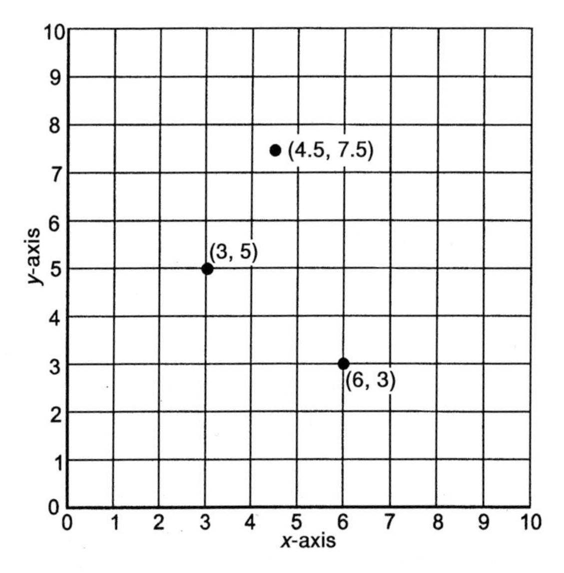

Figure 2.2 Here is a simple x–y-coordinate system with the coordinates of a few points shown as examples. Coordinates are given as an ordered pair of numbers, with the x-coordinate first. The origin is the lower left, at (0, 0). The reference lines are the x- and y-axes. All grid lines are drawn parallel to the two reference lines. The purpose of the grid is to make it easier to make distance measurements between a location-point and the reference lines.

can be placed at the origin. The most common example of a one-dimensional coordinate system is a simple meter stick.

We can also specify which side of the origin a point-location is found by using positive and negative numbers. In the example of Figure 2.1, positive is directed to the right of the origin. Determining the direction of positive coordinate values is referred to as setting the *sense* (as in *sense of direction*) of the coordinate system. The sense is also arbitrary. The positive coordinates can be sent along either direction of the line.

Rectangular coordinates

In most cases, our introduction to two-dimensional coordinate systems is a rectangular grid of fine lines where the bottom edge is the x-axis and the left edge is the y-axis. Although some may refer to this as a graph, this is in fact a coordinate system (Figure 2.2). The x- and y-axes become the **reference lines** for the system.

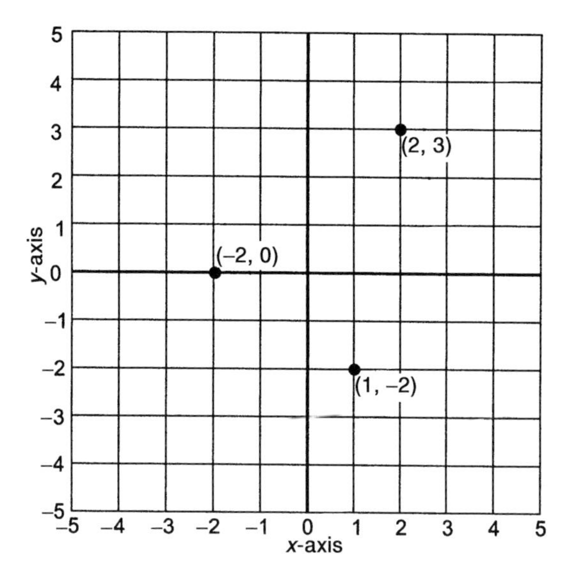

Figure 2.3 A more general coordinate grid places the origin (0, 0) at the center of the grid. The coordinates may have either positive or negative values. The sign merely indicates whether the point is left or right (x), or above or below (y), the axis.

Reference lines are used for making distance measurements from the origin to the location-point of interest.

Because a surface is a two-dimensional space it requires two coordinates to specify any location-point. The x- and y-values are the coordinates of a point within the grid, drawn on the surface. The type of coordinate system shown in Figures 2.2 through 2.4 is called a **rectangular** or **Cartesian coordinate system**. The French mathematician René Descartes is the creator of *analytic geometry*. He used a rectangular coordinate system to translate between algebraic equations and geometric curves in space. For rectangular systems, the reference lines are straight and at right angles to each other.

> *A coordinate system consists of an arbitrarily located origin and arbitrarily positioned, mutually perpendicular reference lines, intersecting at the origin. The reference lines use a scale in units of distance and the zero mark of each scale is usually located at the origin.*

For the simple x–y-coordinates of Figure 2.2 the origin is the lower left corner, where $x = 0$ and $y = 0$. The reference lines are the x- and y-axes. Generally, we locate the origin at a point which makes the most sense for the solution to a problem or where it makes it easy to use the coordinate system. The origin of the rectangular system could just as well be placed at the center of the grid, as shown in Figure 2.3. The same logic is used to position and set the sense of the reference lines. Coordinate systems where the reference lines are perpendicular (cross at 90°) to each other are called *orthogonal coordinate systems*. Non-orthogonal systems have their uses, but they are rare.

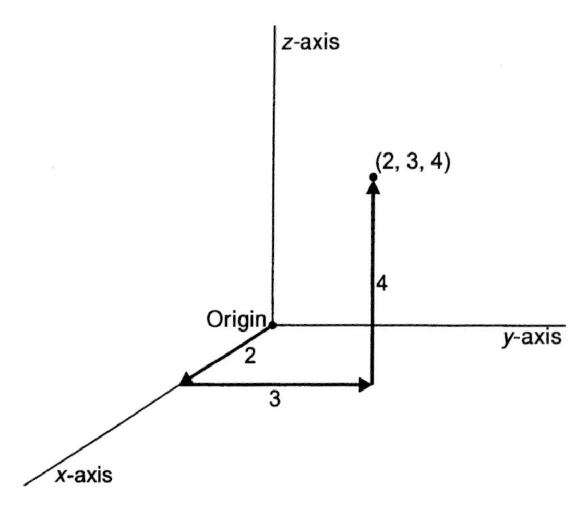

Figure 2.4 A three-dimensional space, such as a room, requires a three-dimensional coordinate system. Each point-location is specified by an ordered set of measurements of distances from the origin made along or parallel to a reference line.

With the origin at the center of the grid, the coordinates can take on negative values. Just like in a one-dimensional system, positive and negative values are simply an indication of direction along the reference lines. Negative y-values are locations (usually) below the x-axis. Any point on the surface, mapped out by the grid, can be specified as exactly as we need by describing a location as a pair of numbers. This number-pair is the coordinates of the location-point on the grid. Figures 2.2 and 2.3 show some example number-pair coordinates. The number-pair is sometimes called an *ordered pair*. The first number is the x- (or horizontal) coordinate and the second is the y- (or vertical) coordinate. (Note that the x- and y-axes do not have to be horizontal and vertical, they can be at any angle, or even swapped. It depends on how the coordinate system is being used.)

> *A point's coordinates are distance measurements made between the point and each reference line. Each distance measurement is made along a grid line that is parallel to a reference line.*

Coordinates and spatial dimensions

The number of coordinates required for the system depends on the number of spatial dimensions the system is describing. We live and think in three dimensions – up/down, left/right, and forward/backward. It requires three coordinates to specify any point-location in a three-dimensional space, as shown in Figure 2.4. If you wrap your fingers around the z-axis with your thumb extended, pointing in the positive-z sense, and your fingers point from the positive x-axis to the positive y-axis through the smallest angle between them, then you are using a *right-handed coordinate system*. It is of course, possible to define and use a left-handed system, but its use is fairly rare.

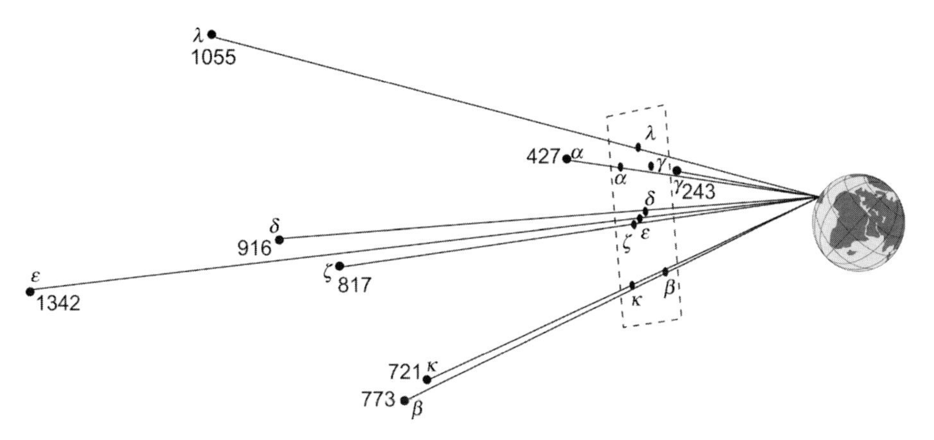

Figure 2.5 The stars of Orion are distributed through three-dimensional space but our view of those stars does not suggest any radial distance, except by the stars' apparent brightness. Apparent brightness, however, is not a valid indicator of true distance. The Greek letters are the Bayer designations for each star (p. 45). The numbers are the actual distance to the stars, in light years, as given by the *Hipparcos Catalog*.

Geometry and mathematics put no limit on the number of dimensions of space and so there is no limit on the number of coordinates of a possible coordinate system. For instance, our current understanding of gravity, general relativity, uses a four-dimensional space – three spatial dimensions and one corresponding to time (this four-dimensional space is usually called *space-time*).

Some of the more recent physical theories use anywhere from a 9-dimensional to a 24-dimensional space. These are the theories that attempt to meld modern quantum physics with the classical physics of general relativity – our attempts to create a theory of quantum gravity. These theories go by names such as "string theory," "M theory," "super gravity," or "loop quantum gravity." A working theory of quantum gravity is needed to understand further the earliest stages of the Universe (cosmology) and our understanding of black holes.

Fortunately, the coordinate systems we use in astronomy treat the sky as a surface, a two-dimensional space. Yes, it is true that space is three dimensional. However, when we are looking at something in the sky – or more importantly, pointing our telescopes to look at something – we can ignore the radial distance between us and the object (except in terms of the effect this distance has on brightness, of course). This allows us to treat the sky as two dimensional (Figure 2.5).

Because the distance to a celestial object plays no role in the line of sight to the object, we need only two coordinates to describe the observed location of an object. That is, from our Earth-bound point of view, we can treat the sky as a surface.

A coordinate system's reference and grid lines need not be straight – they could also be simple curves or arcs – and positions can be measured with angles rather than distances.

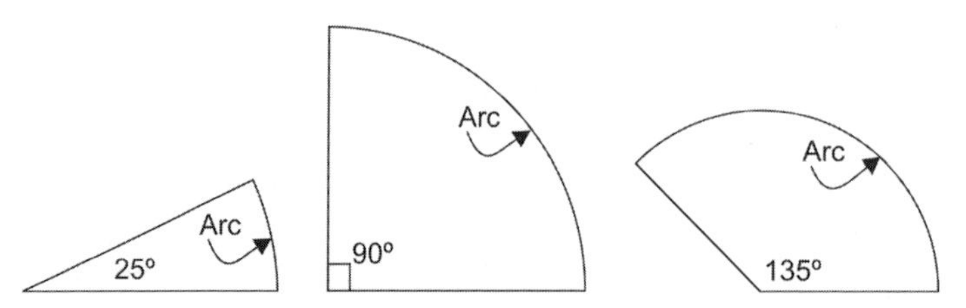

Figure 2.6 A few angles are shown with an arc line between the lines creating the angle. The 90° angle is also known as a "right angle." Such angles are shown with a small square at the vertex.

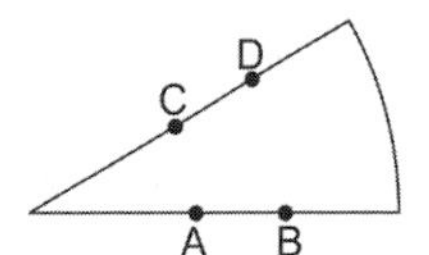

Figure 2.7 The angle between two points is independent of the distance of the points from the vertex of the angle. The angle between points A and C is the same as between B and D. Notice also, that this same angle occurs between points B and C.

Angles

We can build other types of coordinate systems completely equivalent to the above rectangular system but using curved reference lines and angles. An **angle** is formed by the intersection of two lines. Even if the lines are curved we can measure an angle between them by measuring the angle between tangent lines (one for each curve) located at the intersection of the two curves.

Figure 2.6 shows a few example angles. Angles are measured (for most astronomical purposes) in degrees (symbol ^{d} or $^\circ$), minutes (symbol ^{m} or $'$), and seconds (symbol ^{s} or $''$). There are 360 degrees (360°) in a circle, 60 minutes (60′) in a degree and 60 seconds (60″) in a minute. Thus, there are 3600 seconds in one degree, 21 600 minutes in a circle, and, also, 1 296 000 seconds in a circle. As can be seen, one second is a very small angle. This method of measuring angles was created by the Babylonians, whose numbering system was based on 60.

The angle is measured at its **vertex** – the point where the two lines forming the angle meet. We can also see from Figure 2.7 that the angle between two points is independent of the distance of the points from the vertex. It is this property of angles that allows us to treat the sky as a surface. The portion of a circle drawn between the two lines (Figure 2.6) is referred to as an **arc**. Therefore, angles are sometimes quoted as "degrees of arc" or "seconds of arc," or also "arc seconds." This type of phrasing helps us to distinguish between minutes or seconds of angle and minutes or seconds of time, particularly in cases where the two might be confused.

Astronomers have a different way of writing decimal degrees. Instead of writing "twenty-five point five degrees" as 25.5°, it is written as 25°.5 with the degree symbol above the decimal point. Decimal minutes and seconds are

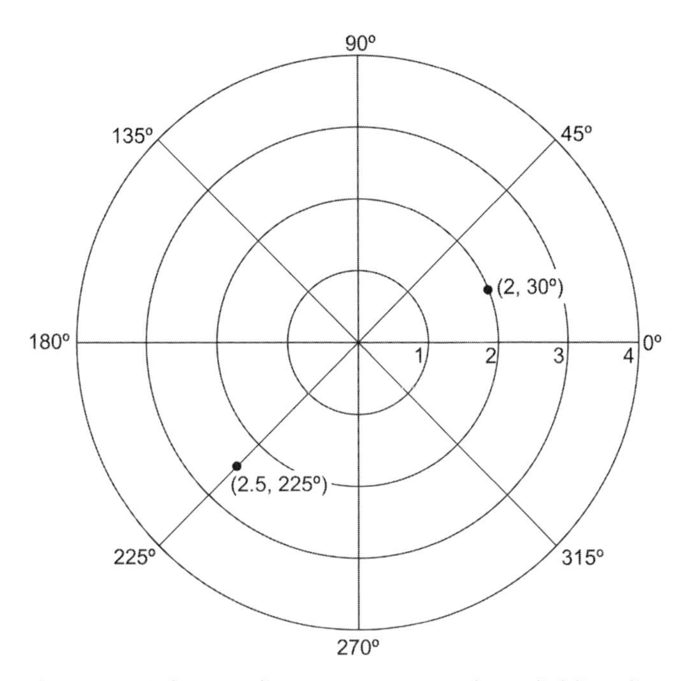

Figure 2.8 Polar coordinate systems use the radial line distance from the origin to a location-point, and the angle between the location-point's radial line and a reference line. The origin is sometimes called "the pole" because the angle coordinate has no meaning if the radial distance is zero. In this system, generally, there are no negative coordinate values. However, we could have the angle coordinate run from -180° to $+180^\circ$ rather than from 0° to 360°. This may have advantages in some situations.

written similarly. This technique allows us to mark a decimal position much more clearly, helping to prevent the mis-location of a decimal point or confusion about the decimal point caused by extraneous marks on the paper.

Polar coordinates

A **polar coordinate system** has its origin located in the center of a grid made from radial lines emanating from the origin and circles which are concentric on the origin (Figure 2.8). In this system the coordinates of a location-point are specified by a distance measurement and an angle measurement. The distance measurement is made along a radial line between the origin and the location-point. The angle measurement is made between this same radial line and the angle reference line, which is usually in the same position as the positive x-axis is generally drawn (horizontal and to the right of the origin). Of course, the position of the angle reference line is arbitrary.

The *Equatorial Chart* included with this book is an example of a rectangular coordinate system. The two polar charts are examples of a polar coordinate system. These maps show special views of a coordinate system drawn on the sky – a celestial coordinate system. ("Celestial" comes from the Latin word

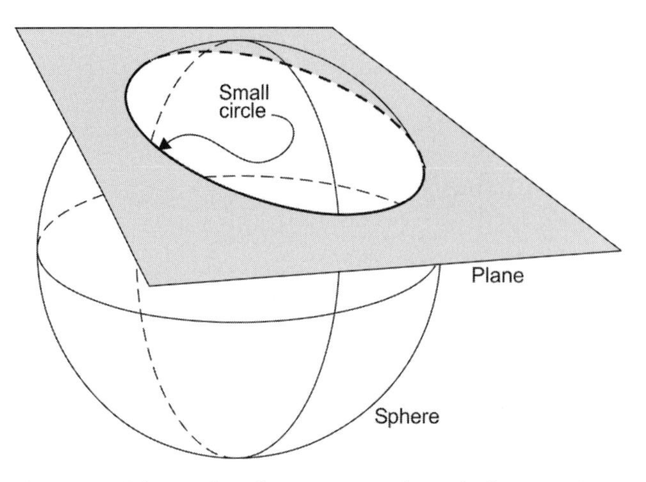

Figure 2.9 When a plane intersects a sphere the intersection creates a circle on the surface of the sphere and on the plane.

caelestis for "belonging to heaven.") When we look at the night sky, we perceive it as a sphere or, at least, as a hemisphere. ("Hemisphere" comes from the Greek words *ἥμι, hemi* for "half," and *σφαιρα, sphaira* for "ball.") Because it seems we are looking at the inside of a giant ball or inverted bowl, we sometimes refer to the night sky as "the bowl of night." Celestial coordinate systems are mapped out on a sphere. Thus, we must also understand spherical coordinate systems.

Coordinates on a sphere

When creating a coordinate system, we do not want to use just any random lines for our system's references. We must use lines that not only make sense but are easily reproducible. The simplest lines on a sphere are created by the intersection of a plane with the sphere's surface. These lines are perfect circles on the sphere and on the plane. Circles generated by planes that do not pass through the center of the sphere are called **small circles** (Figure 2.9). The line created in the special case where the plane does pass through the center of the sphere is called a **great circle** (Figure 2.10).

A plane also defines a line in space. This line is the unique line that is perpendicular to the plane. (Actually, mathematicians consider this in the opposite sense – a line defines a plane.) With a rotating sphere like the Earth, the two points defined by the passage of the body's rotational axis through its surface are called the **poles**. The great circle created on the sphere by the plane defined by (perpendicular to) this axial-polar line is called the **equator** (Figure 2.10).

> *On a rotating sphere, the equator is a great circle found 90° away from the poles. The position of the poles and therefore the equator is dictated by the sphere's rotational axis.*

With the poles and the equator now defined, we can draw a set of great circles, each containing the two poles and intersecting the equator at 90°. The planes

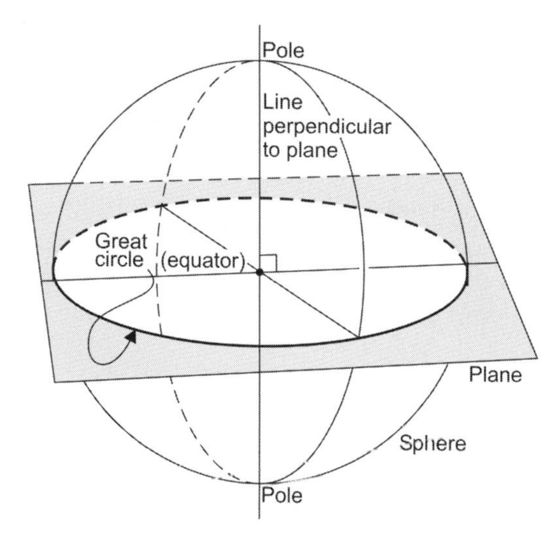

Figure 2.10 A plane passing through the center of the sphere creates a great circle on the sphere. When a line matching the sphere's rotational axis is drawn perpendicular to the plane, the line pierces the sphere at its poles. In this case the great circle is called the sphere's equator.

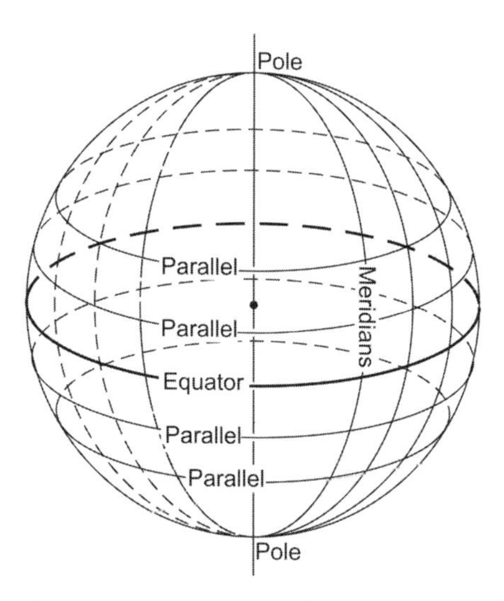

Figure 2.11 Planes, parallel to the plane creating the equator, create small circles that are parallel to the equator. On Earth, the lines are called *parallels*. Great circles created by planes containing the polar axis are called *meridians*.

creating these great circles all contain (and intersect at) the polar or rotational axis line. The circles are called **meridians** (Figure 2.11). They are part of our coordinate grid on the sphere's surface. We must choose one of them to be the reference line for measuring distance around the sphere's equator. The choice is arbitrary and, once made, the chosen line is called the **prime meridian**. The intersection of the equator and the prime meridian is the origin of this *spherical coordinate system*.

To create the second set of lines for the grid we use planes parallel to the equatorial plane. The lines these planes make are small circles because the planes do not pass through the sphere's center. These lines allow us to measure distance between the equator and the poles. They are shown in Figure 2.11. Because these small circles are parallel to the equatorial circle they are usually called **parallels**. The planes that create the parallels are positioned at intervals along the polar axis such that the angles between the parallels and the equator are consistent multiples.

Distances on the surface of a sphere are measured by angles with the vertex of all angular measure located at the center of the sphere. Because we are interested only in the surface of the sphere, we need only two (angular) coordinates. For the two coordinates, one angle is measured along the equator starting at the prime meridian. The second angle is measured from the equator along a meridian line to the location-point of interest.

2.2 The terrestrial coordinate system

The most common application of a spherical coordinate system is the **terrestrial coordinate system**. We need to be familiar with the terrestrial system not only as an example of a spherical coordinate system but also for transforming coordinates between the celestial systems. The angle measured along the equator is known as **longitude**. The angle measured from the equator along a meridian to the location-point of interest is called **latitude**. The vertex of both these angles is at the center of the Earth. Both angles are measured in degrees, minutes, and seconds.

Latitude

We can draw guidelines of equal latitude on the surface of the Earth. When looking at a globe, we can see these equal latitude lines running east–west, parallel to the equator. These are the parallels as described above. For example, the "45th parallel" is halfway between the equator and the pole. These east–west lines measure the latitude (north–south) angle between themselves and the equator.

The latitude of the equator is 0°. The latitude of the North Pole is 90° N and the South Pole's latitude is 90° S. Latitude angles are never greater than 90°. Never forget to specify whether the angle is measured north or south of the equator.

On the surface of the Earth, a difference of one degree, one minute, and one second of latitude is about 111 kilometers (69 miles), 1855 meters (1.2 miles), and 30.9 meters (102 ft) respectively. The distance corresponding to one degree of longitude, depends on latitude. This can be seen in an observation of longitude

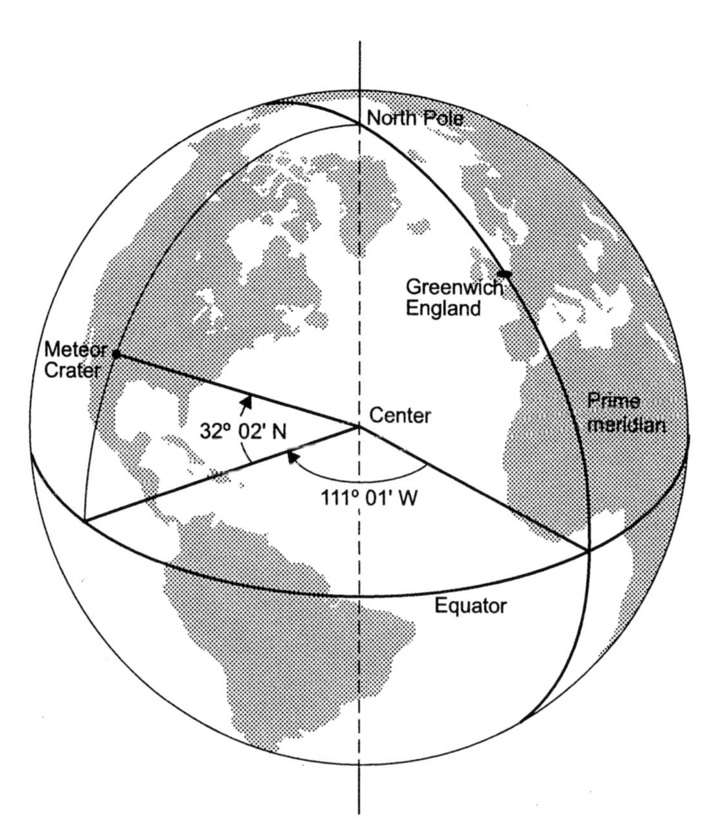

Figure 2.12 This is an illustration of the terrestrial or latitude–longitude system. The location of the Barringer Meteor Crater, Arizona, is 111° 01′ W longitude, 32° 02′ N latitude. The angles are measured at the center of the Earth.

on a world globe by seeing how the meridian lines on the globe get closer to each other as they approach the pole. One degree of longitude gives a maximum distance at the Earth's equator (0° latitude). As latitude increases the distance corresponding to one degree of longitude grows steadily smaller.

When standing on a pole (north or south) longitude no longer has any meaning. When standing on the North Pole the only direction you can face is south. When standing on the South Pole the only direction you can face is north. In either case, longitude no longer has any meaning.

Example: Let us use the latitude of Barringer Meteor Crater, Arizona. The latitude of Meteor Crater is 32° 02′ N. Here is one way to picture the creation of this angle. Draw a line from Meteor Crater to the center of the Earth. Draw another line from the center of the Earth to a point on the equator directly south of Meteor Crater. These two lines are both contained in the plane creating the meridian line for Meteor Crater. The angle between these two lines is 32° 02′ (Figure 2.12).

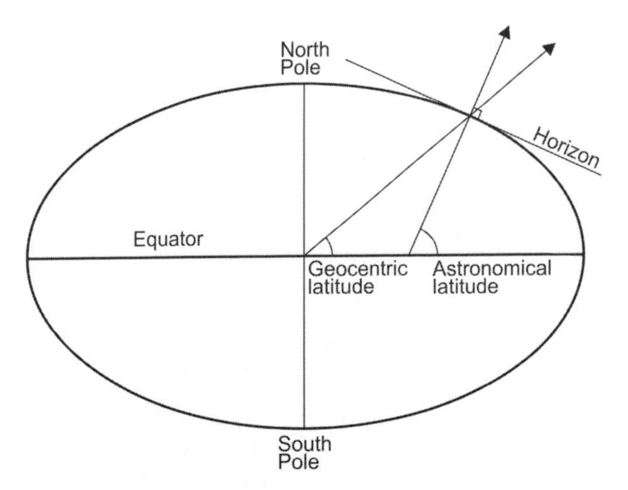

Figure 2.13 Two definitions of latitude are caused by the Earth's polar flattening, which in turn is caused by its rotational motion. The flattening shown here is drastically exaggerated to make the difference clearer.

Geocentric and astronomical latitude The Earth rotates on its axis and because it is not solid, this rotational motion causes the polar diameter to be slightly less than the equatorial diameter. The Earth's polar flattening causes a dual definition of latitude (Figure 2.13). We have defined latitude as an angle measured at the center of the Earth. This is called *geocentric latitude*. We can also define latitude relative to the horizon. This is called *astronomical latitude*. Because the polar flattening is small, the difference is small. However, because the horizon is the reference most accessible to observers, measured latitudes are actually astronomical latitudes, rather than geocentric latitudes.

Longitude

Longitude is a little more difficult than latitude but only because the Earth has no *natural* dividing line between east and west – like the equator divides north and south. We had to agree upon an east–west dividing line. At one time, French sailors divided east and west by the meridian passing through Paris, and English sailors used London. The fifth Astronomer Royal, Nevil Maskelyne, moved the English prime meridian from London to the Greenwich Observatory while he lived at the observatory – from 1765 to 1811. Maskelyne published 49 issues of the *Nautical Almanac*, where he listed all lunar–solar and lunar–stellar distances as measured from the new Greenwich meridian. (Lunar comes from *Luna*, the Latin name for the Moon.) These measurements were so precise they became the unofficial standard for navigation at sea (using the lunar method) and for making maps of uncharted territories for many countries. The International Meridian Conference was held in Washington, DC, in 1884. There, 26 countries adopted the Greenwich Observatory meridian as the prime meridian of the world (Sobel and Andrewes, 1995; Figure 2.14). The French held out until 1911.

Figure 2.14 This is the "transit house" through which the terrestrial coordinate system's prime meridian passes, at the Old Royal Observatory in Greenwich, England. The line on the ground in front of the house is the prime meridian.

> *The meridian line drawn from the North Pole, through the Old Royal Observatory in Greenwich, England, through the equator to the South Pole, is the terrestrial system's prime meridian.*

Greenwich, England, is on the Thames River, about 8 km (5 miles) east of London.

Longitude measures the distance along the equator, from its intersection with the prime meridian, to the equatorial intercept with the location's meridian (Figure 2.12). We call all locations west of the prime meridian "west longitude," and all locations east, "east longitude." Barringer Meteor Crater is at 111° 01′ W longitude. An Earth globe also has longitudinal (meridian) lines running from the North Pole to the South Pole. Longitude runs from 0° to 180° W, and from 0° to 180° E. The meridian line on the opposite side of the Earth from the prime meridian is the *International Date Line*. When it is high noon in Greenwich, England, it is midnight on the International Date Line and, thus, the calendar date is changing. The International Date Line is discussed further on p. 163.

2.3 Celestial coordinate systems

There are four coordinate systems used in astronomy. They are the *horizon system*, the *equatorial system*, the *ecliptic system*, and the *galactic system*. Each has its uses. The first two are the most useful for amateur astronomy. The ecliptic system is used for Solar System objects and is used to a limited extent in amateur astronomy. The galactic system (which has two versions) is used for study of the Milky Way.

We might imagine drawing guidelines in the sky to act like the guidelines we drew on the Earth globe for latitude and longitude. To do this we use the concept of the **celestial sphere**. Some ancient peoples believed an opaque celestial sphere about the size of the orbit of the Moon – or in some cases just at the top of the atmosphere – was pierced with "pin holes," allowing the light of Heaven to shine through, thus creating the stars at night. The celestial sphere is also an important part of the Copernican system of the world.

The celestial sphere

As explained earlier, we can treat the sky as a two-dimensional surface – the celestial sphere. The Earth is concentric with this sphere. However, the Earth is small enough compared to the sphere, that an observer on the Earth's surface can be considered to be at the sphere's center. That is, the distance between the center of the Earth (the true center of the celestial sphere) and the observer standing on the surface, is negligible. When this distance is not negligible, we have to contend with observational effects such as *geocentric parallax*. This effect becomes important when we discuss the required conditions for a solar eclipse in Section 7.5.

The celestial sphere has two important points (Figure 2.15) – the **north celestial pole** (NCP) and the **south celestial pole** (SCP). The Earth rotates on an axis. If we run a thin rod (an axle) through the Earth such that it would rotate about this axle, then the axle also passes directly through the celestial sphere at its north and south poles. Instead of rotating the Earth, we could allow the celestial sphere to rotate (in the opposite direction) on this axle. As seen from the Earth, the resulting sky motions are the same.

> *The NCP is the point directly above the Earth's North Pole. The SCP is the point directly above the Earth's South Pole. They represent the points where the Earth's rotational axis passes through the celestial sphere.*

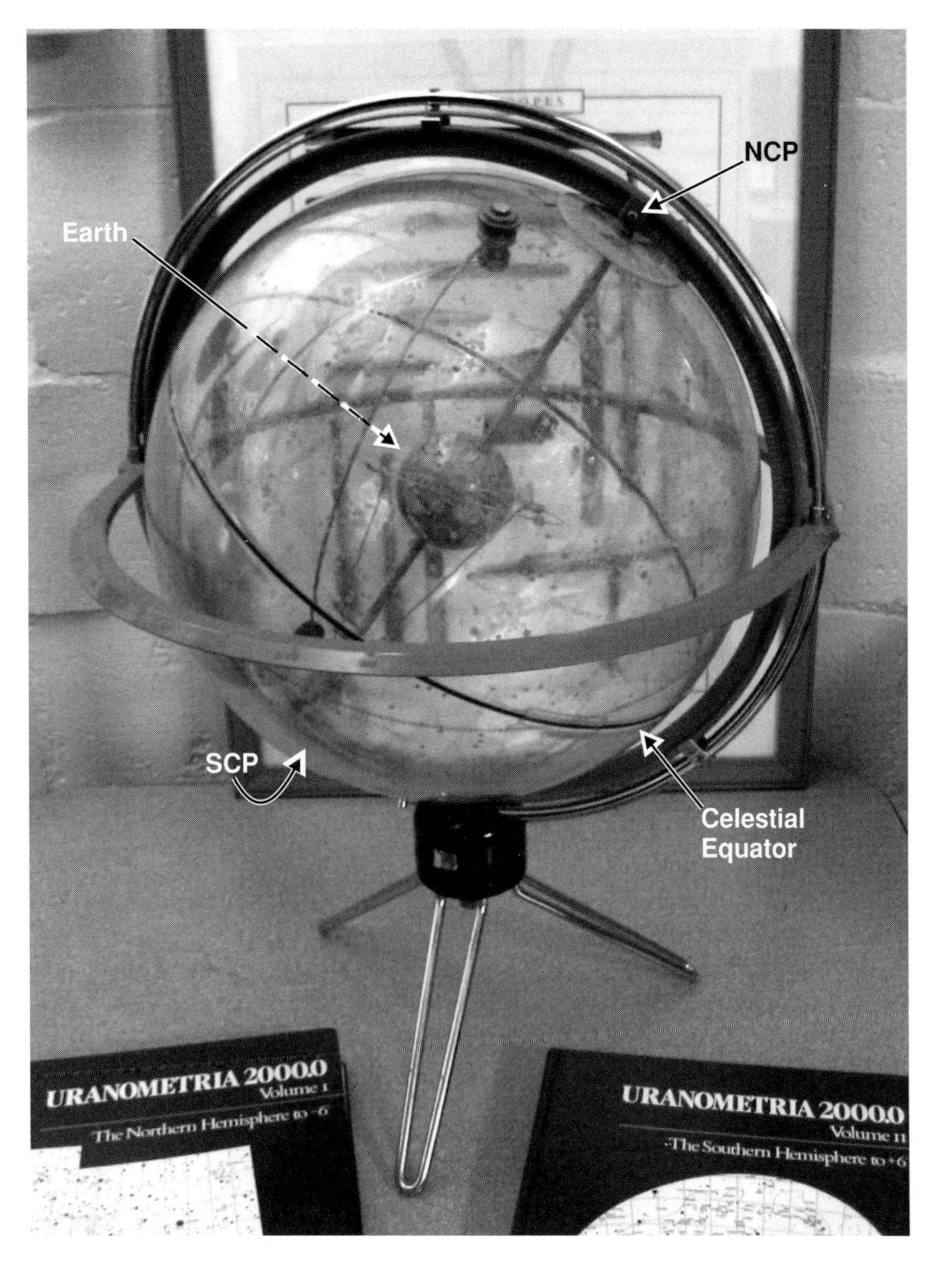

Figure 2.15 The celestial sphere is an imaginary sphere, centered on the Earth. The celestial equator is directly above the Earth's equator and the celestial poles are directly above the Earth's poles. They share the axis of rotation. (Thanks to Kirk Korista for providing the sphere shown in this picture.)

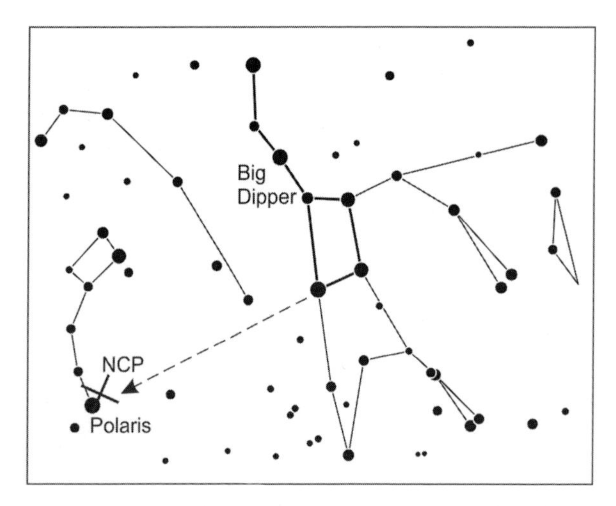

Figure 2.16 Use the "pointer stars" at the end of the Big Dipper bowl to find the North Star (Polaris). The north celestial pole (NCP) is next to Polaris.

The celestial sphere has an equator directly above the Earth's equator. Like the Earth's equator, the **celestial equator** is a line located half way between (90° from) the NCP and SCP. For every observer (except when on one of the Earth's poles), the celestial equator traverses the sky from due east to due west. In fact, the compass directions of east and west are *defined* by the intersection of the celestial equator with the horizon (at sea level).

Because the Earth and the celestial sphere share an axis of rotation the celestial equator is directly above the Earth's equator.

Locating the north and south celestial poles

In the northern hemisphere there is an easy way to determine your latitude. Polaris is directly (almost) above the North Pole. This is why Polaris is often referred to as the *North Star*. The distance between Polaris and the north celestial pole is actually about one-half degree, or about one Moon diameter. This is discussed in more detail on p. 105. Granting that there is this half-degree of difference, we can make the following statement:

At all points in the northern hemisphere, the altitude (p. 21) of the North Star (Polaris) is equal to the observer's latitude.

At 40° N, the altitude of Polaris is 40°. Polaris makes it easy to find the north celestial pole, as shown in Figure 2.16.

There is no pole star for the southern hemisphere and finding the south celestial pole is a little trickier. The only naked-eye star close to the SCP is a dim star labeled, σ (sigma) Octantis. (See p. 45 on Bayer designations for the meaning of the Greek letter.) Sigma Octantis is 1°.06 from the SCP. Figure 2.17 shows one method for locating the south celestial pole.

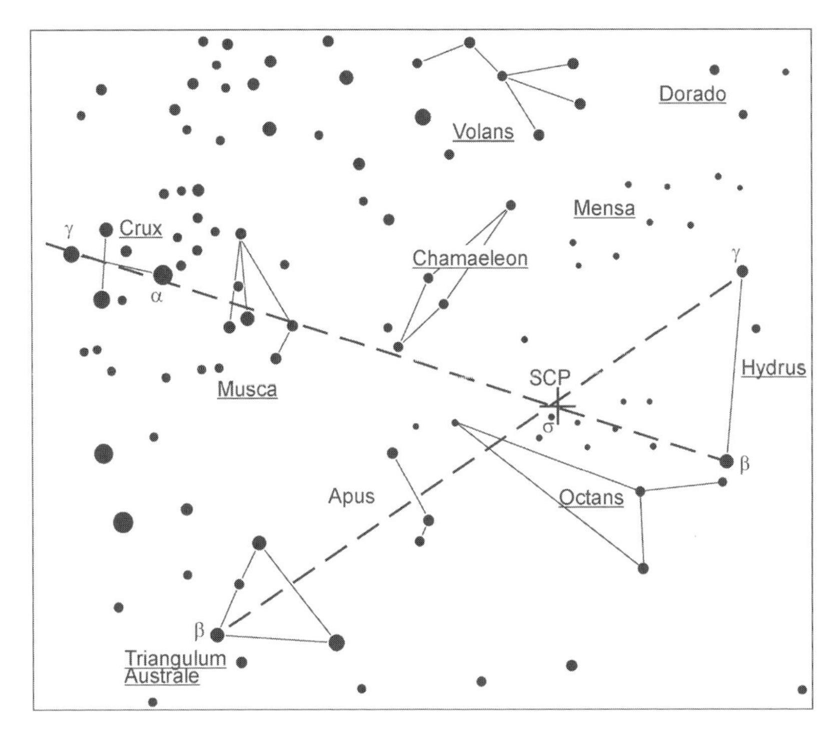

Figure 2.17 To find the south celestial pole, draw a line through γ and α Crucis (Crux) toward β Hydri (Hydrus), then draw a line from β Trianguli Australis to γ Hydri. The intersection of these two lines is near the SCP. Also, the dim star σ Octantis is near the SCP. (The Greek letters used here are the Bayer designations for these stars. Bayer designations are introduced on p. 45.)

2.4 The horizon coordinate system

The **horizon coordinate system** uses *altitude* and *azimuth* to specify the location of an object. Because these coordinates are on the surface of a sphere (the "bowl of night") both measurements are angles, measured in degrees, minutes, and seconds.

Altitude

The **altitude** of an object is an angle measured from the horizon to the object, along a meridian line (Figure 2.18). The **horizon** represents a circle on the celestial sphere and its altitude is zero degrees (0°). Technically speaking, this circle is not a great circle because it does not pass through the center of the Earth – it is tangent to the surface at the observer's location. This is ultimately the cause of geocentric parallax. For observations where geocentric parallax can be ignored (objects outside the Solar System), the horizon can be treated as a great circle. In practical observation terms, the horizon is, of course, simply the line where sky meets land.

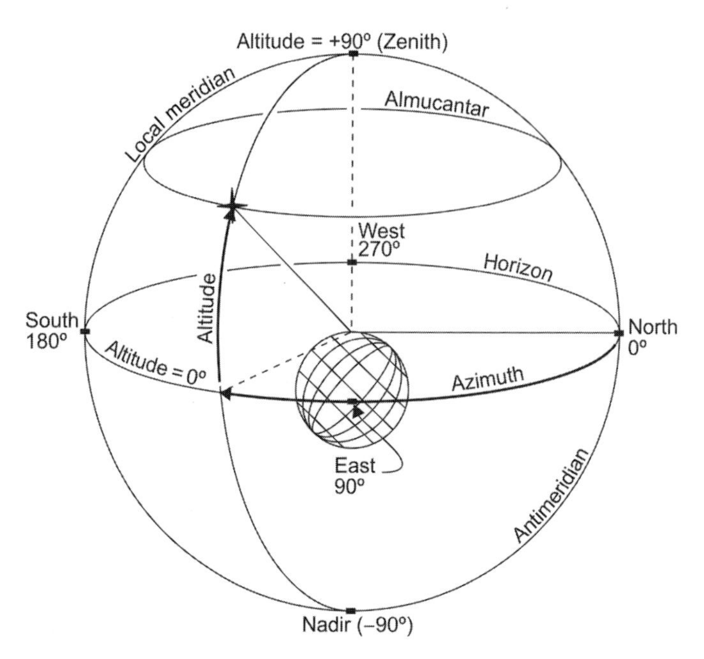

Figure 2.18 This is the basic structure of the horizon coordinate system. Azimuth is measured from 0° to 360° along the horizon, starting from due north. Altitude is measured from 0° to 90° starting from the horizon. Negative altitudes are positions below the horizon.

The sky point straight over your head is called the **zenith** and its altitude is +90°. The sky point directly below your feet (on the celestial sphere, on the other side of the Earth) is called the **nadir**. A carpenter's plumb line held out in your hand shows the line between the zenith and the nadir. Every observer has their own zenith and nadir. The horizon is then defined by the plane perpendicular to the plumb line. This plane is called the **horizon plane** and it is the plane tangent to the Earth's surface at the observer's position. The horizon line is the intersection of the horizon plane with the celestial sphere. The meridian line along which we measure an object's altitude, is a great circle running from the zenith to the nadir and its plane contains the zenith–nadir plumb line. Because the meridian passes through the zenith, it intersects the horizon at a right angle.

Altitude angles for all objects above the horizon run from 0° to +90° and are not allowed to be greater than +90°. (If the altitude becomes greater than +90°, it is easier to turn around than to bend over backwards.) If the altitude of an object is *negative*, the object is *below* the observer's horizon. Certainly an observer will never measure a negative altitude, but negative altitudes can arise from coordinate system transformations. When this happens, it merely informs us that the object is not visible at the observer's location. The full range of altitude is then, −90° (the nadir) to +90° (the zenith). The **almucantar** is a small circle of equal altitude.

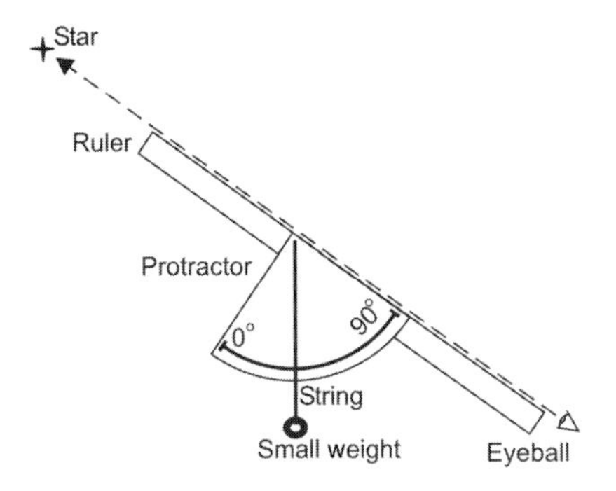

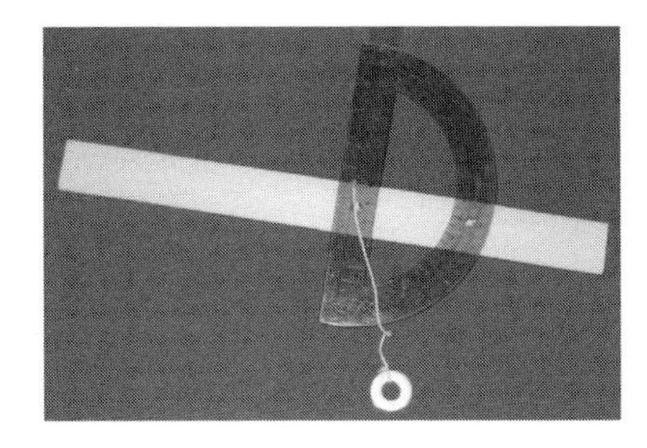

Figure 2.19 Here is a simple, inexpensive quadrant. Measure a star's altitude by sighting the star along the top edge of the ruler. Pinch the string against the protractor to read the altitude angle.

Altitude is measured from the horizon to the object along a meridian line passing through the zenith. It ranges from −90° (the nadir) to 0° (the horizon) to +90° (the zenith). Objects below the observer's horizon have negative altitude.

Altitude can be measured with a **quadrant**. Figure 2.19 shows an inexpensive quadrant made with a ruler, a protractor, a piece of string and a small weight (such as a machine nut or washer). Section 8.2 describes the construction of this quadrant. You can use a quadrant like this to measure altitudes during your sky observations (with a reasonable error for such informal observations as those in Chapter 8). In the northern hemisphere, you can check the accuracy of your quadrant by measuring the altitude of Polaris. If your quadrant is well made, Polaris' altitude should match your latitude. In the southern hemisphere, look for σ Octantis and compare its altitude with your latitude.

In either hemisphere, the altitude of the celestial pole is equal to the observer's latitude. At 40° N latitude the altitude of the NCP is +40°, while the SCP is at −40°. At 40° S latitude the altitude of the SCP is +40°, while the altitude of the NCP is −40°.

If a quadrant is not available, an observer can use an outstretched hand to obtain a rough measure of altitudes (with a slightly larger error). If the index finger is held at arm's length, it is about 1° across. A fist is about 10° across and when the fingers are spread out, there is about 20° between the finger tips. Remember to hold your hand out at arm's length. Using your fists, hold the first fist so its bottom is at eye-level, then alternately stack them one atop the other until reaching the sky object. You may have to interpret a smaller angle with the last fist in the stack. This gives a rough estimate of the object's altitude (Figure 2.20).

Azimuth

Because azimuth is the coordinate measured along the horizon, it is necessarily related to the compass directions. The four basic directions, *north*, *south*, *east*, and *west*, are called the **cardinal points** of the compass.

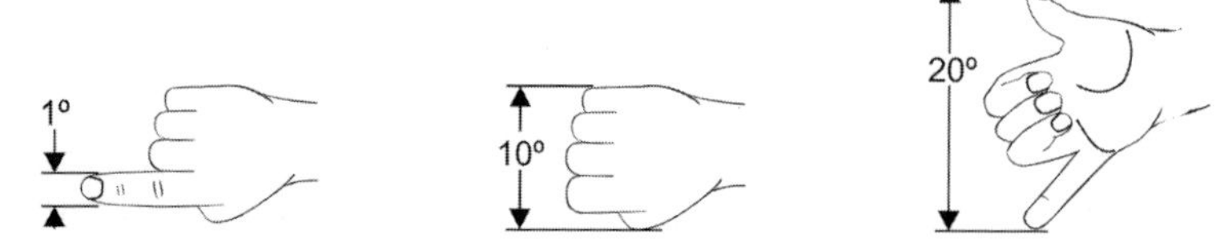

Figure 2.20 Your hand can be used to measure altitudes and other angles when a quadrant is not available. Hold your hand at arm's length and the equivalent angles are shown above.

> *The* **azimuth** *coordinate specifies the location along the horizon from which the altitude of the object is measured. In other words, it specifies the location of the intersection of the object's altitude meridian line with the horizon. Like altitude, azimuth is also measured as an angle. Following the custom from navigation, due north is 0° azimuth, east is 90°, south is 180°, and west is 270° azimuth. Thus, azimuth runs from 0° to 359°.*

In some cases, due south is taken as 0° azimuth. Remember, the position of the coordinate system origin is arbitrary and chosen for ease of use or convenience for solving a problem.

Look again at Figure 2.18. If we draw a sky line from the northern horizon (0° azimuth) through the zenith to the southern horizon (180° azimuth), we have drawn the **local meridian**. The local meridian is half of a great circle. Because every observer has their own zenith, every observer also has their own local meridian. In the northern hemisphere, the local meridian also contains the north celestial pole. In the southern hemisphere, it contains the south celestial pole. The line drawn from the north through the nadir to the south (the other half of the great circle) is called the **antimeridian**. The antimeridian contains your nadir and the opposite celestial pole for your hemisphere. The local meridian and the antimeridian together form the great circle simply called the *meridian*.

> *Every observer has their own local meridian, which is an imaginary line drawn from due north, through the zenith, to due south. The antimeridian is the continuation of this line through the nadir. The local meridian and the antimeridian combined is called the meridian.*

Problems with the horizon system

The problems with the horizon coordinate system are that it depends on the time and date of the observation, and on the location (latitude and longitude) of the observer. These problems are created by the motion of the sky (actually, the Earth is moving, but at this point it makes no difference) and the fact that every Earth-bound observer has their own horizon. Because the Earth is rotating, the altitude and azimuth coordinates of a star or constellation are changing from minute to minute, causing the time datum requirement. The annual motion of the Earth around the Sun causes a seasonal change to the constellations which causes the date datum requirement. An observer in Australia has a completely different horizon from an observer in England.

Because the correct altitude and azimuth depends on these extra data, an observer using the horizon system must record the date, time, and location along with the altitude and azimuth as part of an observation. If the observer fails to record these data, the altitude and azimuth coordinates are meaningless. These problems are ultimately created by using a coordinate system with a starting point and reference lines that are fixed to the Earth. It would be better to use a coordinate system fixed to the moving sky.

2.5 The equatorial coordinate system

The preferred coordinate system among most astronomers is the **equatorial coordinate system**. The equatorial system is made independent of observation time and date, and the observer's location by using an origin and reference lines fixed to the sky. Thus, this system moves with the sky. The equatorial system is very similar to the terrestrial coordinate system except we draw our celestial meridians and parallels on the celestial sphere. The equivalents of terrestrial latitude and longitude are called *declination* and *right ascension*. When specifying the equatorial coordinates of an object or a sky-point such as the vernal equinox, we may use abbreviations for right ascension (RA; or the Greek letter alpha, α) and declination (Dec.; or the Greek letter delta, δ).

Declination

Declination is to the equatorial system what latitude is to the terrestrial system. **Declination** measures the angular distance of an object from the celestial equator, along a meridian (Figure 2.21). The celestial equator is directly above the Earth's equator (refer to Figure 2.15). While latitude uses north (N) or south (S) to designate a position north or south of Earth's equator, declination uses plus (+) or minus (−) to designate a position above or below the celestial equator. (In some older astronomy books, you may still see north and south declinations used.) On Earth, a place is at latitude 25° N. On the celestial sphere, an object is at declination +25°. The celestial equator is 0°, the NCP is +90°, and the SCP is −90° declination. We can draw imaginary declination guidelines on the celestial sphere. In the sky, these guidelines run east–west, parallel to the celestial equator. These are the celestial sphere's "parallels."

Declination measures the angular distance (along a meridian) of an object from the celestial equator, using plus (+) to designate a position above (north celestial hemisphere) and minus (−) to designate a position below (south celestial hemisphere) the celestial equator.

In order to determine if an object is currently visible (above the horizon) the observer must know the sky location of the celestial equator, because the declination of an object is measured from the celestial equator. A description

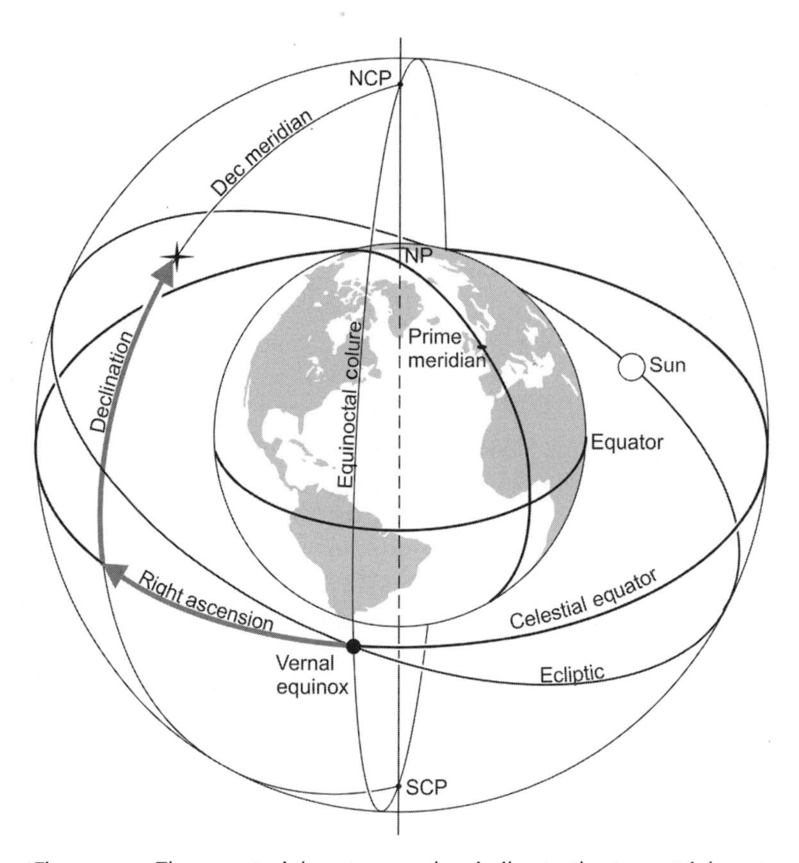

Figure 2.21 The equatorial system works similar to the terrestrial system. Declination measures angles above (+) and below (–) the celestial equator in units of degrees. The north celestial pole is declination $+90°$. The south celestial pole is declination $-90°$. Right ascension measures along the celestial equator in units of time. The origin is the vernal equinox.

of the sky location of the celestial equator depends on the observer's latitude. We start with the northern hemisphere (Figures 2.23 and 2.24), then discuss the southern hemisphere's view (Figure 2.25) and, finally, talk about the view from the Earth's equator and poles. The drawings in this chapter showing the celestial equator and similar drawings in the following chapters dealing with solar motion are based on a teaching tool called a "Tycho globe" – shown in Figure 2.22.

Appearance of the celestial equator in the northern hemisphere The celestial equator is a line running from due east to due west across the southern sky. Because we cannot actually *see* this line, we must calculate and mentally picture its position in the sky. For our purposes, we shall always calculate the **transit altitude** of the celestial equator, which is the point where it crosses the local meridian. In the northern hemisphere, the transit altitude of

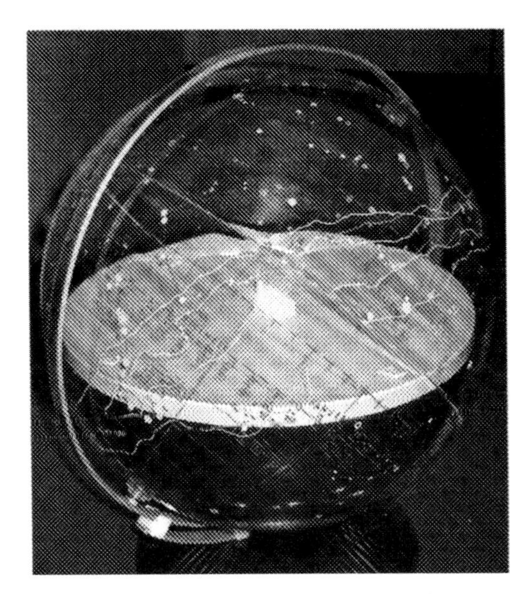

Figure 2.22 A *Tycho globe* has a clear plastic sphere representing the celestial sphere. Inside is a plastic half-sphere, which floats on a small amount of water on the bottom of the celestial sphere. The position of the celestial sphere can be adjusted for observer latitude with the metal ring, which acts like the local meridian. The celestial sphere can rotate, simulating the daily motion of the sky.

the celestial equator is measured from the southern horizon. Here is the general rule:

> *The transit altitude of the celestial equator is equal to 90° minus the observer's latitude.*

This simple rule is created by using two facts: the north celestial pole and the celestial equator are 90° apart and the due north and due south horizon points are 180° apart.

Let's look at how this rule is created. Looking at Figure 2.23, the three angles running along the local meridian must add to the angle between due north and due south. That is,

$$40^\circ + 90^\circ + 50^\circ = 180^\circ.$$

We can use our understanding of Figure 2.23 to rewrite this to see that the transit altitude of the celestial equator, A_{CE} (point Q), comes from

$$A_{\mathrm{CE}} = 50^\circ = 180^\circ - 90^\circ - 40^\circ.$$

Because the celestial equator is *always* 90° from the NCP (Polaris) and the due north and due south horizons are always 180° apart, the 180° and 90° are always in this equation, regardless of the observer's latitude. We can therefore do the subtraction between these two numbers to get,

$$A_{\mathrm{CE}} = 50^\circ = 90^\circ - 40^\circ.$$

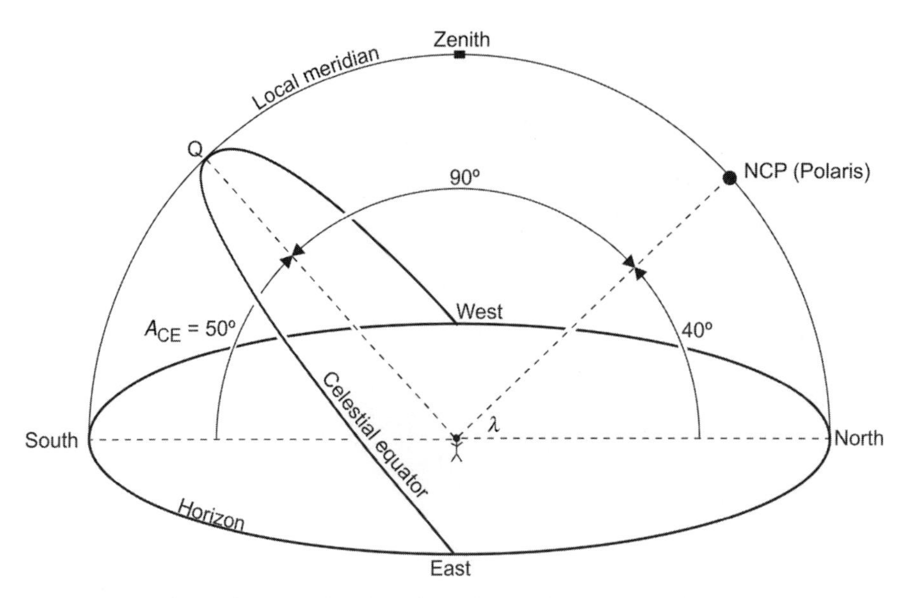

Figure 2.23 At latitude 40° N the altitude of the north celestial pole is 40° and the transit altitude of the celestial equator is $90° - 40° = 50°$ off the southern horizon.

Figure 2.23 shows us that the 40° in this equation is simply the altitude of the NCP (Polaris) as seen by the observer. Now recall the relationship between the altitude of the north celestial pole and the latitude of the observer given on p. 20. When given the latitude of a location, we can find the altitude of the celestial equator with this formula:

$$\text{Altitude of celestial equator} = 90° - (\text{observer's latitude})$$

or, in symbols,

$$A_{\text{CE}} = 90° - \lambda,$$

where the Greek letter λ (lambda) is used to represent the observer's latitude.

Miami, Florida, is at latitude $\lambda = 26°$ N. While moving to Miami from 40° N, the dotted line from the observer to the NCP shown in Figure 2.23 would rotate clockwise, to the position shown in Figure 2.24 (compare the two figures). The 40° angle becomes 26°. In Miami, Polaris is at altitude 26° in the north and the celestial equator (point Q) is at altitude $90° - 26° = 64°$ from the southern horizon.

In the northern hemisphere, the altitude of the NCP is equal to the observer's latitude. The celestial equator rises due east, crossing the local meridian in the southern sky at 90° minus the observer's latitude and sets due west.

On a more practical note, as can be seen on the *Equatorial Chart*, the celestial equator is found near Orion's Belt and also about 10° below the southern tip of the Summer Triangle (the bright star Altair). When seeing these stars in the sky, you are looking at the approximate position of the celestial equator.

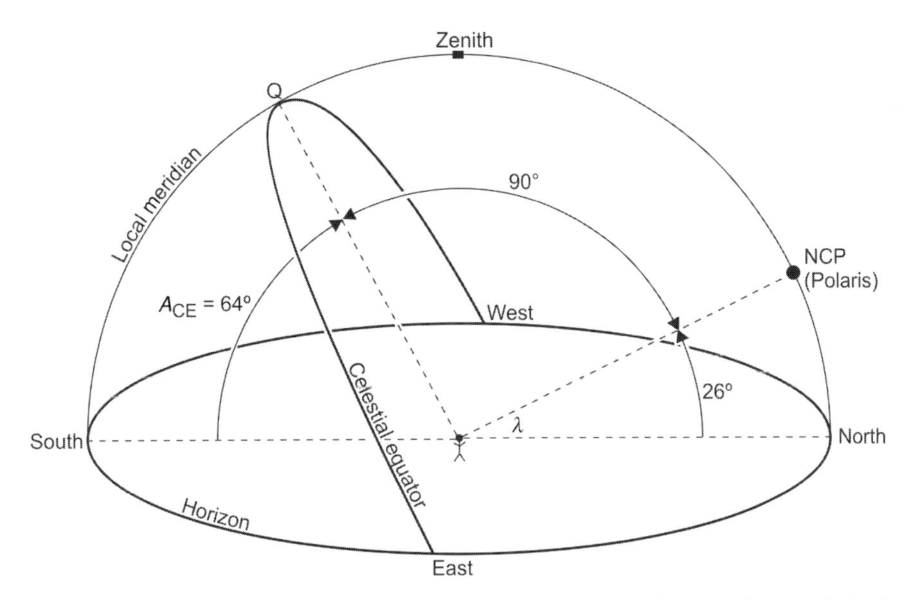

Figure 2.24 At latitude 26° N (Miami, Florida), the altitude of the north celestial pole is 26° and the transit altitude of the celestial equator is $90° - 26° = 64°$ off the southern horizon.

Appearance of the celestial equator in the southern hemisphere The celestial equator runs from due east to due west across the northern sky. Finding the transit altitude of the celestial equator in the southern hemisphere works the same way as in the northern hemisphere, except that the observer's latitude is equal to the altitude of the south celestial pole off the southern horizon, and the transit altitude of the celestial equator is measured from the northern horizon. This is shown in Figure 2.25. Thus, southern hemisphere observers should also use

$$A_{CE} = 90° - \lambda$$

to find the transit altitude of the celestial equator off their northern horizon.

In the southern hemisphere, the altitude of the SCP is equal to the observer's latitude. The celestial equator rises due east, crossing the local meridian in the northern sky at 90° minus the observer's latitude and sets due west.

Compare Figures 2.23, 2.24, and 2.25 with the drawings in Figure 2.26. Notice how the celestial equator moves from the southern sky in the northern hemisphere to directly overhead at the Earth's equator, to the northern sky in the southern hemisphere.

Altitude of the celestial equator at the equator and poles The Earth's equator is at latitude 0° ($\lambda = 0°$). There, the altitude of both the NCP and the SCP is 0° and the altitude of the celestial equator is $90° - 0° = 90°$.

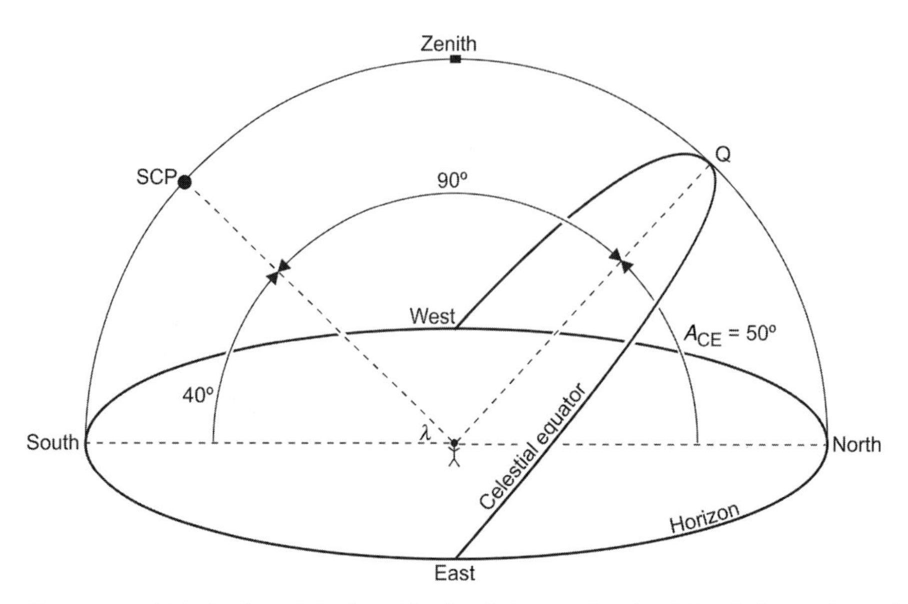

Figure 2.25 At latitude 40° S, the altitude of the south celestial pole is 40° from the southern horizon and the altitude of the celestial equator is $90° - 40° = 50°$ off the northern horizon.

> *The celestial equator is directly overhead when standing on the Earth's equator. It rises straight up off the eastern horizon, passes through the zenith, and sets straight down on the western horizon.*

At the North Pole (latitude 90° N or $\lambda = 90°$), the NCP (and Polaris) is at the zenith (altitude = 90°), the SCP is on the nadir, and the celestial equator is at altitude, $90° - 90° = 0°$. At the South Pole, the SCP (σ Octantis) is at the zenith and the NCP (Polaris) is at the nadir.

> *At either of the Earth's poles the celestial pole is at the zenith and the celestial equator matches the horizon. Because the celestial equator no longer has two intercept points with the horizon line, the east and west compass points are now undefined. The horizon and the celestial equator are the same line on the celestial sphere.*

Declination on the sky charts On the *Equatorial Chart*, declination only reaches +60° or −60°, because the polar regions would be too distorted if they were plotted along the top and bottom (Figure 2.27). It is an effect similar to a rectangular map of the world, where Antarctica is spread out over the bottom of the map. This chart is called an *Equatorial Chart* because the declination coordinate is limited to this range around the celestial equator.

The north-polar region of the sky is plotted on the *North-polar Chart*. The south-polar region is plotted on the *South-polar Chart*. These charts use a polar coordinate layout and declination is measured from the center of the chart to +30° for the *North-polar Chart* and −30° for the *South-polar Chart* (Figure 2.28). The circular grid lines represent positions of equal declination.

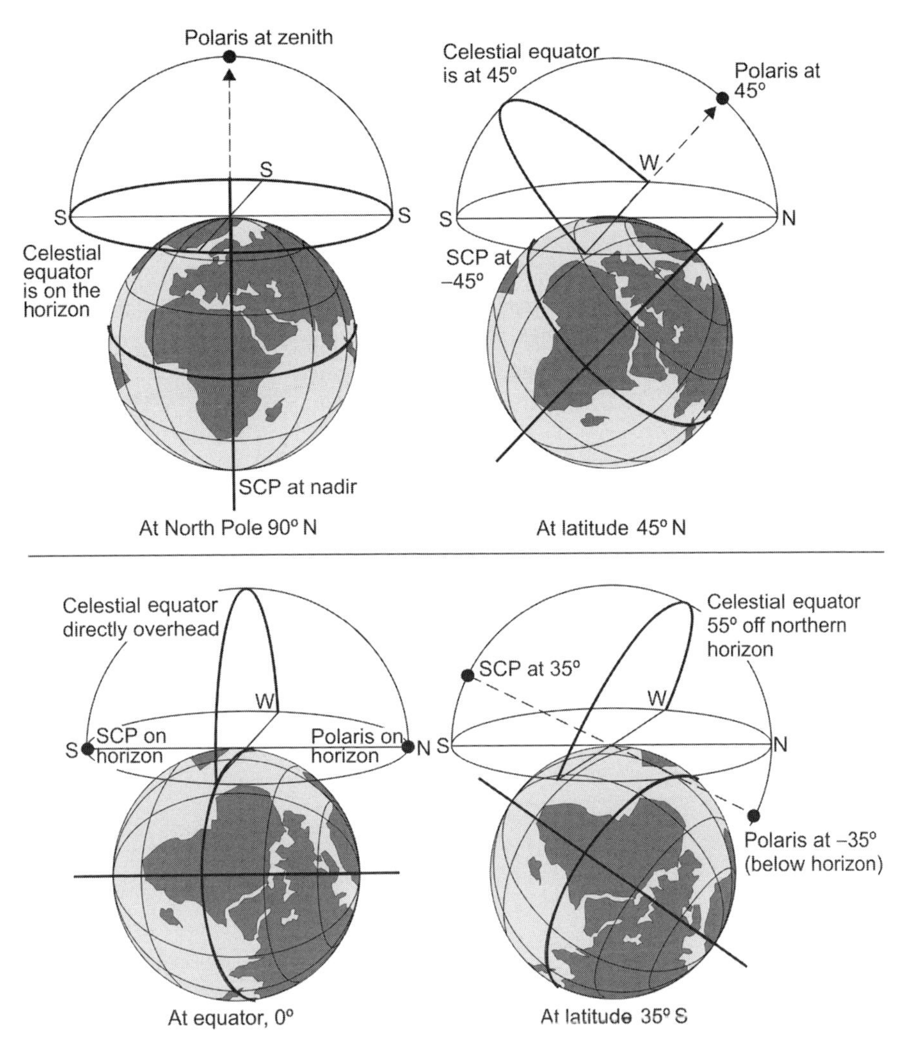

Figure 2.26 The altitude of the celestial poles and the celestial equator depend on the latitude of the observer. At the North or South Pole, the celestial equator is on the horizon. As an observer moves from the North Pole to the South Pole, the NCP gets lower in altitude, the SCP gets higher, and the celestial equator moves from the southern sky to the northern sky.

Right ascension

The celestial equivalent of terrestrial longitude is *right ascension*. The **right ascension** coordinate of an object tells observers when (at what time) an object transits (or upper culmination, p. 69) their local meridian. Right ascension is a set of meridians running from $0^h\,0'\,0''$ to $23^h\,59'\,59''$. The zero hour line starts at the NCP, passes through the celestial equator at the *vernal equinox* and ends at the SCP (Figure 2.21). The vernal equinox is the starting point of the equatorial system and is discussed further on p. 84.

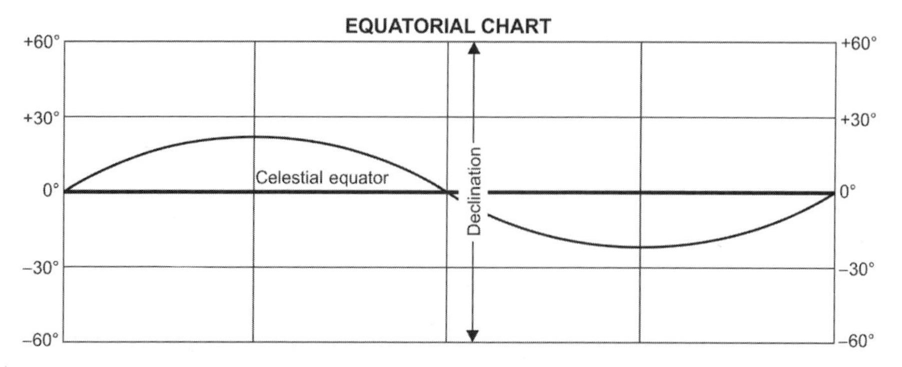

Figure 2.27 The *Equatorial Chart* shows the declination coordinate along each edge of the chart and along the major right ascension hour lines. The heavy horizontal line through the middle is the celestial equator, where declination equals zero.

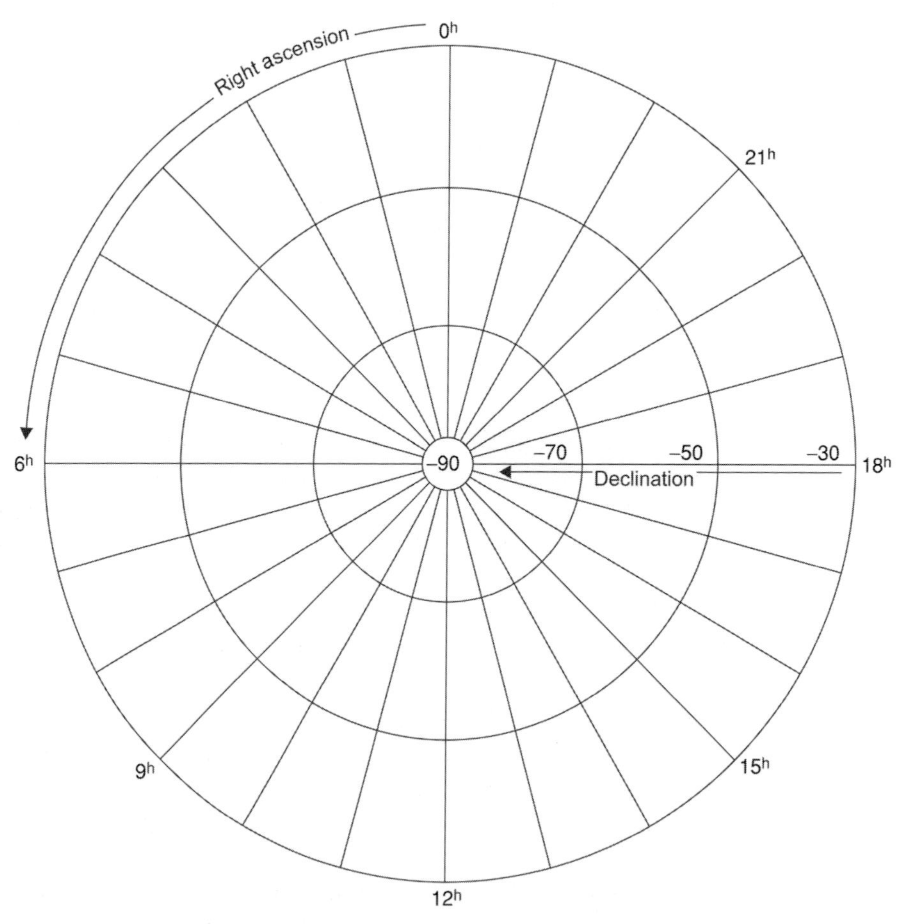

Figure 2.28 On the polar charts, declination is the radial distance from the center and positions of equal declination are shown with circles. Right ascension is measured by angular position with zero hours at the top of the charts. Positions of equal right ascension are shown as radial lines. On the *North-polar Chart* right ascension runs clockwise.

When your right ascension clock reads $0^h\ 0'\ 0''$ ("stellar midnight"), the vernal equinox is on your local meridian. A clock that keeps track of right ascension – a **sidereal clock** (sī-dir´-ē-ell) – is different from the wall clock that runs our daily routine. *Sidereal* is a Latin word meaning, "with reference to the stars." Thus, a sidereal clock keeps time with reference to the stars. A wall (solar) clock keeps **solar time**. *Solar* means "with reference to Sol." "Sol" is the Latin name for the Sun. The difference between these two time measurements is due to the apparent annual motion of the Sun, which is due of course, to the annual revolution of the Earth around the Sun. This is discussed in greater detail on p. 95.

Because sidereal clocks and solar clocks are different, the vernal equinox can pass through your local meridian at any time of day. As examples, on 21 March, the Sun is on the vernal equinox, so the vernal equinox passes through your local meridian at high noon. On the 21 September, the Sun is on the autumnal equinox (as it is called in the northern hemisphere), 12 hours away from the vernal equinox, so the vernal equinox passes through your local meridian at midnight.

An object's right ascension coordinate is the reading of your local sidereal clock when that object is on your local meridian.

Orion's Belt has $\alpha = 5^h\ 30'\ 5''.5$. This means the belt of Orion passes through your local meridian approximately 5 hours and 30 minutes after the vernal equinox passes through. Because every observer has their own zenith and local meridian, every observer must also have their own sidereal clock. Every observer's sidereal clock must be set to their local sidereal time.

The correct local sidereal time can be calculated from the reading of a solar clock, the date, and the observer's longitude. Any planetarium program running on a personal computer can display the local sidereal time. (The bibliography contains some suggested planetarium programs.) Transforming local solar time to local sidereal time is demonstrated in a number of books (Burgess, 1982; Duffett-Smith, 1988, 1997; Kaler, 1996; Meeus, 2000; Montenbruck and Pfleger, 1994). Those readers wishing to learn the details of time conversion should consult these books. Sidereal time can also be estimated by observing the objects currently on your local meridian and looking up their right ascension.

Right ascension on sky charts On the *Equatorial Chart*, the right ascension coordinate is printed in three places – along the top, along the equator, and along the bottom. See Figure 2.29 or the *Equatorial Chart*. It is marked off in 5-minute, then 15-minute, and then one-hour increments, increasing in time toward the left.

Here are example coordinates for some objects. Regulus, the brightest star in Leo: $\alpha = 10^h\ 8'\ 21''.9$, $\delta = +11^\circ\ 58'\ 2''$. Vega, the brightest star in Lyra: $\alpha = 18^h\ 36'\ 56''.0$, $\delta = +38^\circ\ 47'\ 0''$. Again, some older books may use *N* (north) for (+) and *S* (south) for (−) in the declination coordinate. A more compact form of notation is found in some catalogs and star charts. The coordinates of Regulus

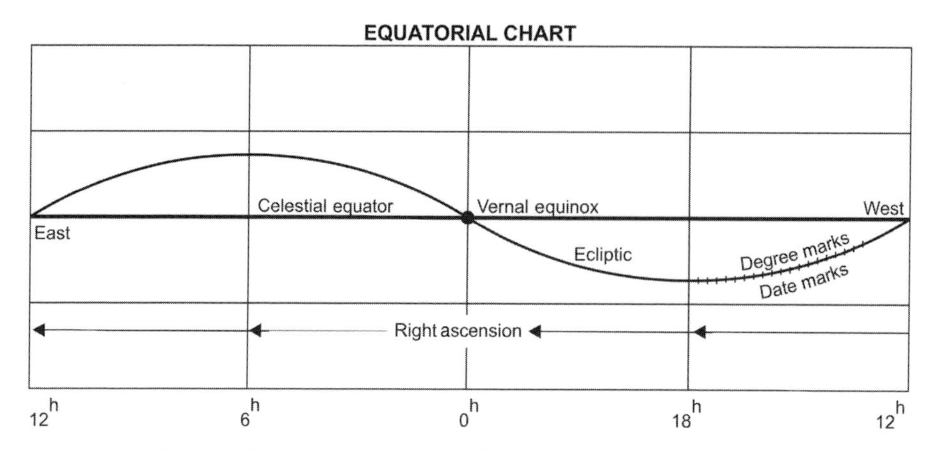

Figure 2.29 The vernal equinox is an intersection of the celestial equator with the ecliptic, the apparent annual path of the Sun. This is the origin point for the equatorial system. Looking at this chart, objects pass through the local meridian from left to right (east to west). Thus, the hours of right ascension increase to the left (to the east).

could be written as (1008219 +115802). The equatorial system allows astronomers to pinpoint objects even when they are not visible by naked eye.

On the polar charts, right ascension is the angular coordinate (Figure 2.28). Zero hours is the radial line toward the top of the chart. Right ascension then increases (in time) clockwise for the *North-polar Chart* and counterclockwise for the *South-polar Chart*. It is marked in five-minute intervals along the outer edge of each chart.

Advantages of the equatorial system

The equatorial system has one major advantage over the horizon system. *Each sky object always has the same declination and right ascension.*

> *The equatorial coordinates of an object are independent of the observer's location, and the date and time of the observation. For any observer at any location on Earth, the right ascension and declination coordinates of any object are the same, at least through the observer's lifetime.*

The equatorial coordinates of an object can be converted to horizon coordinates. This conversion tells observers whether the object is visible at their latitude and, if so, when. The conversion is not trivial, but it can be approximated (see Section 4.2). A detailed discussion of the coordinate conversion process is given in the books previously listed for time conversion details.

Problems with the equatorial system

Although the equatorial system is extremely accurate, it does have one problem. The right ascension and declination of an object are not quite constant. By naked eye, changes in star coordinates from night to night, season to season, or even over a lifetime, cannot be perceived. They do change, however, over decades, centuries, and millennia. It takes 72 years for the right ascension coordinate of an object along the celestial equator to change by four minutes of arc (about one degree). Modern, highly accurate, computer-controlled telescopes can detect changes in right ascension in a matter of months.

At the top of a typical star or constellation chart is a subtitle such as "*Epoch 2000*" or "*Epoch 2000.0*." The **epoch** tells the observer the year for which the right ascension and declination of the objects shown on the chart have been adjusted. This adjustment is necessary because of the precessional motion of the Earth.

Because it takes 26 000 years for the celestial poles to make one circuit of the ecliptic poles, these corrections are small, and affect only telescopic observations. The right ascension of an object near the celestial equator is adjusted by about one minute in time – or 15′ 20″ of arc – every 18 years. At the top of the chart we may find "Epoch 2000.0." This tells us the right ascension and declination coordinates of the objects have been plotted for the year 2000. The ".0" signifies the beginning of the year. Precessional motion and the resulting motion of the vernal equinox is covered in Section 4.6.

Adjustment of star coordinates is also necessary because of the *proper motion* of the stars. Proper motion is the movement of the stars across the celestial sphere due to their spatial motion relative to our sun. This is created by the fact that we are all orbiting the center of the Milky Way galaxy. We based the prime meridian of the terrestrial system on the location of a city (Greenwich, England). Why not base the equatorial system's 0^h line on a celestial city – a star? The spatial motion of the stars is not consistent and, to some extent, not predictable, while the precessional motion of the Earth is consistent, well measured and understood.

> *The equatorial coordinates of objects must be updated to compensate for the precessional motion of the Earth and the proper motion of the stars. These updates are small and cannot be perceived by naked-eye observation.*

2.6 Ecliptic coordinates

The **ecliptic coordinate system** is used for Solar System objects because the majority of Solar System objects move on paths close to this line. The first observation project (Section 8.1) uses one of the ecliptic coordinates for discussion of the Moon's phases. The ecliptic is the apparent path of the Sun on the celestial sphere. The exact definition and cause of this motion of the Sun is discussed in Section 4.4, on p. 83.

The ecliptic coordinate system is very similar to the equatorial and terrestrial systems except that the *north* and *south ecliptic poles* (p. 103) are used to define the equivalent of the equator which in this system is the ecliptic. The two coordinates are called **celestial longitude** (symbol λ) and **celestial latitude** (symbol β). Don't confuse the λ of celestial longitude with the λ of observer's latitude. The context should indicate the difference.

Celestial longitude is measured in degrees along the ecliptic starting from the vernal equinox. The range of value is 0° to 359°. On the *Equatorial Chart*, the celestial longitude degrees are marked off along the top of the ecliptic line (Figure 2.29). Celestial latitude is measured along a meridian from −90° at the south ecliptic pole to 0° at the ecliptic to +90° at the north ecliptic pole.

Questions for review and further thought

1. What do we mean by a coordinate system's *origin*?
2. How many dimensions (coordinates) are required to map the sky as seen from Earth?
3. What are the three basic units of measure for angles and how are they related to each other?
4. Describe the position and orientation of the celestial sphere relative to the Earth.
5. What are the names of the two coordinates used in the horizon coordinate system? Describe the measurement made by each coordinate.
6. What is the meaning of negative altitude?
7. Where (at what altitude) is the zenith? Describe its location relative to the observer.
8. What is the range of possible values for altitude?
9. Describe a quadrant. What coordinate does it measure?
10. What is the possible range of values for azimuth? Along what line is it measured?
11. What are the azimuth angles for the eight major compass points?
12. Why do the altitude and azimuth of a star depend on the date, time, and the location of the observer?
13. Describe an observer's *local meridian*. Why is every observer's local meridian unique?
14. What are the names of the coordinates of the equatorial coordinate system and how are they measured?
15. How is the altitude of a celestial pole (north or south) related to an observer's latitude? (Discuss both hemispheres.)
16. Describe the imagined view of the celestial equator for a naked-eye observer. (Discuss both hemispheres, the equator, and both poles.)
17. How is the altitude of the celestial equator (on the local meridian) related to an observer's latitude? (Discuss both hemispheres.)
18. In what region of the Earth would an observer be if the altitude of Polaris is negative?
19. Why does the *Equatorial Chart*'s declination stop at +/− 60°?
20. What is a sidereal clock?
21. What are the equatorial system's advantages over the horizon system?
22. What is the meaning of "*Epoch 2000.0*" on a star chart?

Review problems

1. If Polaris has an altitude of 32°, what is the observer's latitude?
2. An observer is located at latitude 38° N and longitude 79° W. Calculate the altitude of Polaris (the altitude of the north celestial pole), the altitude of the south celestial pole, and the altitude of the celestial equator for this observer.
3. Rework Problem 2 for an observer located at latitude 38° N and longitude 92° W. How do the results compare?
4. Rework Problem 2 for an observer located at latitude 46° N and longitude 95° E. How do the results compare?
5. Rework Problem 2 for an observer located at latitude 38° S and longitude 30° E. How do the results compare?
6. Calculate the altitude of the celestial equator for an observer at either of the Earth's poles. Which important sky line does the celestial equator match when standing on the poles?

3
Stars and constellations

When we hear the word "astronomy," most of us think first of the stars and constellations. The stars are other suns. Our Sun is a star. We recognized this only in the later 1800s with the spectroscopy work of Father Pietro Secchi (1818–1878), who is (by some historians) considered the founder of modern astronomy. We find stars in all regions of the sky. Some stars have tints of twinkling color caused mainly by Earth's turbulent atmosphere. As we watch the stars, we notice they move across the sky from east to west, and they are not all the same brightness. Some are easy to see and some are difficult. Most stars are visible only with the aid of a telescope, but there are plenty of stars for viewing by naked eye.

As part of describing a star (or any other object) to someone else, we should like to be able to specify its location. One way of doing this is to create pictures in the sky and locate the stars in relation to those pictures. The stars form recognizable patterns we call *constellations*. Eventually, we learn to see the sky in terms of its constellations and to locate the stars within them.

3.1 Constellations

One way of describing the sky is called, the "city-state picture." Here, we think of stars within constellations like we think of cities within states. A **constellation** is a group of stars forming a recognizable pattern. ("Constellation" may come from the Latin words *cum*, for "with," and *stellatum*, for "starry.") There are 88 officially recognized (by the International Astronomical Union) constellations. They are listed in Appendix B, see p. 247. Many constellation names date back thousands of years to the ancient Mediterranean cultures. They are named for important cultural icons, like feared or respected animals, or for important deities. Some cultures believed the constellations were the actual dwelling place of these animal spirits or deities. In some cases, the presence of the creature in the sky was more important than the existence of the associated star pattern.

Figure 3.1 Orion, the Hunter, in the early morning sky as seen from the James C. Veen Observatory at 43° N latitude. The Great Nebula in Orion (M42) can be clearly seen in the position of the sword, hanging from his belt. (Photo by Kevin S. Jung.)

The earliest sky maps from Egypt, show only the creatures or deities and few, if any, stars.

Different cultures created different constellations. Forty-eight of our 88 constellations come from the mythology of ancient Greece. These are called the **ancient constellations**. These 48 constellations are described by Claudius Ptolemy in book seven of the *Almagest*, which describes his (geocentric) system of the world. Greek writings from as early as the third century BC tell us that not even the Greeks were sure of the origin of many of these constellations. Generally, these constellations have more than one mythological story behind them. Probably the most famous of the ancient constellations is Orion, the Hunter (Figure 3.1).

Some constellations, were named as late as the seventeenth century. The constellations named in these later centuries are called the **modern constellations**.

The book *Uranometria* written by the German astronomer Johannes Bayer (1572–1625), and published in 1603, is one of the earliest and most respected constellation atlases. Bayer himself created some of the modern constellations for inclusion with this book. Most of the modern constellations are found in the southern hemisphere, much of which the Greeks and other Mediterranean cultures could not see.

"Microscopium" and "Telescopium" are two modern constellations found in the southern hemisphere. They were named by the European sailors who first crossed the equator on their 'round-the-world voyages. Because these objects were recent and important inventions, their choice of names for these constellations lends credence to the claim that constellations are named for things that are important to the people or cultures naming them.

A few of the modern constellations fill areas between large ancient constellations and so some ancient constellations were reduced in size. An example of this case would be the constellation "Lynx." Its sky region once belonged to Ursa Major. Lynx was created by the German astronomer Johannes Hevelius (1611–1687) as part of his star atlas *Uranographia*, published in 1690. Over the last four centuries many names and representations for new constellations have been suggested and rejected.

A constellation's star pattern usually bears little resemblance to the object it represents. While this is disappointing to some, the response to their complaint is to understand that the original purpose of constellations was to honor or represent, not to depict or portray. When we name a ship after a person, we don't expect the ship to look like the person it honors. The stick figure for Lynx doesn't look anything like a cat – but you need the eyes of a Lynx to see its faint stars.

In the constellation Ursa Major (the Big Bear or the Great Bear), the easiest set of stars to find looks like a large soup dipper or pan. The bowl of the dipper is in the body of the bear. Have you noticed that the Big Bear has a tail (the handle of the dipper) and real bears do not? (H. A. Rey redrew the bear and eliminated the tail problem [Rey, 1988a].) The mythology of ancient Greece has a very imaginative explanation for the long tails of Ursa Major and Ursa Minor.

The tale is of course based on one of the many affairs that occur between Zeus and mortal maidens. After the birth of a son (fathered by Zeus), Zeus' wife, Hera, takes revenge on the former maiden (because she can't take revenge on the king of the gods) by turning her into a bear – doomed to wander the forest for the rest of her life. Her son grows up to become a great hunter. One day, the hunter sees this bear and draws his arrow on it. Zeus, recognizing the bear as the boy's mother, grabs the big bear by its short tail, whirls it around his head – which stretches its tail – and throws it into the sky, where it becomes the Big Bear. Then Zeus decides to make it a family. He turns the boy into a bear. He whirls the little bear around his head, stretching its tail as well, and throws it into the sky. Thus, both the Big Bear and the Little Bear have longer than normal tails, and they are blood relatives.

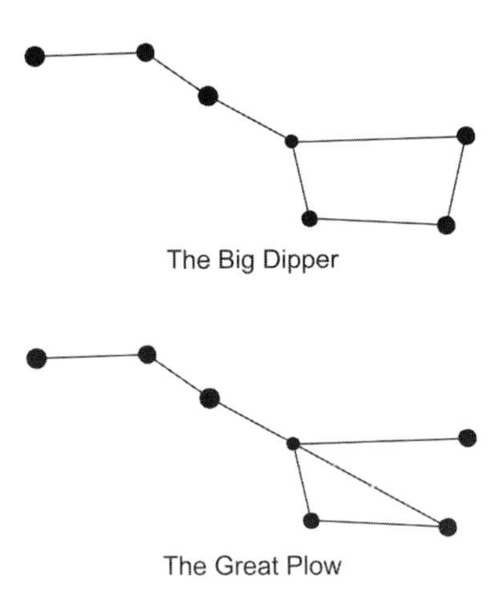

Figure 3.2 Different cultures used the same stars to create different constellations. Although the Big Dipper is not a constellation, it makes an easy example for this claim. Many people recognize the "Big Dipper," but some do not know that it is also known in England as the "Great Plow." In Egypt, it is known as the "Evil Man," and in China, it is known as the "Bureaucrat's Cart" (Sagan, 1980).

William Tyler Olcott describes some of the stories surrounding Ursa Major (Olcott, 2004). On p. 348, Olcott describes an interesting set of stars. The leg of an animal is one of many star patterns depicted on the zodiac of Denderah on the Nile. This group of stars has been identified as "the Thigh." There is no question that this group of stars belongs to the constellation we know as Ursa Major. The Greeks called this group *Άρχτος μεγάλη* (Archos megale), which is the source of the English word *arctic*.

In England, the "Big Dipper" is also known as the "Great Plow" (Figure 3.2). For the ancient Greeks, Gemini represents the twin sons of Zeus, Castor and Pollux. To the Incas of Bolivia and Peru, this set of stars is a pair of pumas. Another well-known Greek constellation, Cassiopeia, the Queen, is shaped like the letter *W*. Today, constellations are just convenient patterns in the sky. We use them for describing the general location of stars and other objects.

> *Constellations are defined by their boundary lines, not by their "stick figures." The stick figures are not standardized and each map maker is free to choose or create a constellation's stick figure drawing.*

The constellation boundary generally lies beyond the visible stars in the stick figure pattern. The boundary shape is irregular, like the political boundary of a state. Figure 3.3 illustrates the irregular shape of the boundary of Ursa Major.

Because the constellations completely cover the sky, all stars lie within the boundary of (and thus, belong to) a constellation. The International Astronomical Union established an international standard for the constellations in 1922. This standard defines the boundaries for the 88 official constellations. The boundary lines for the constellations, just like the boundary lines for states,

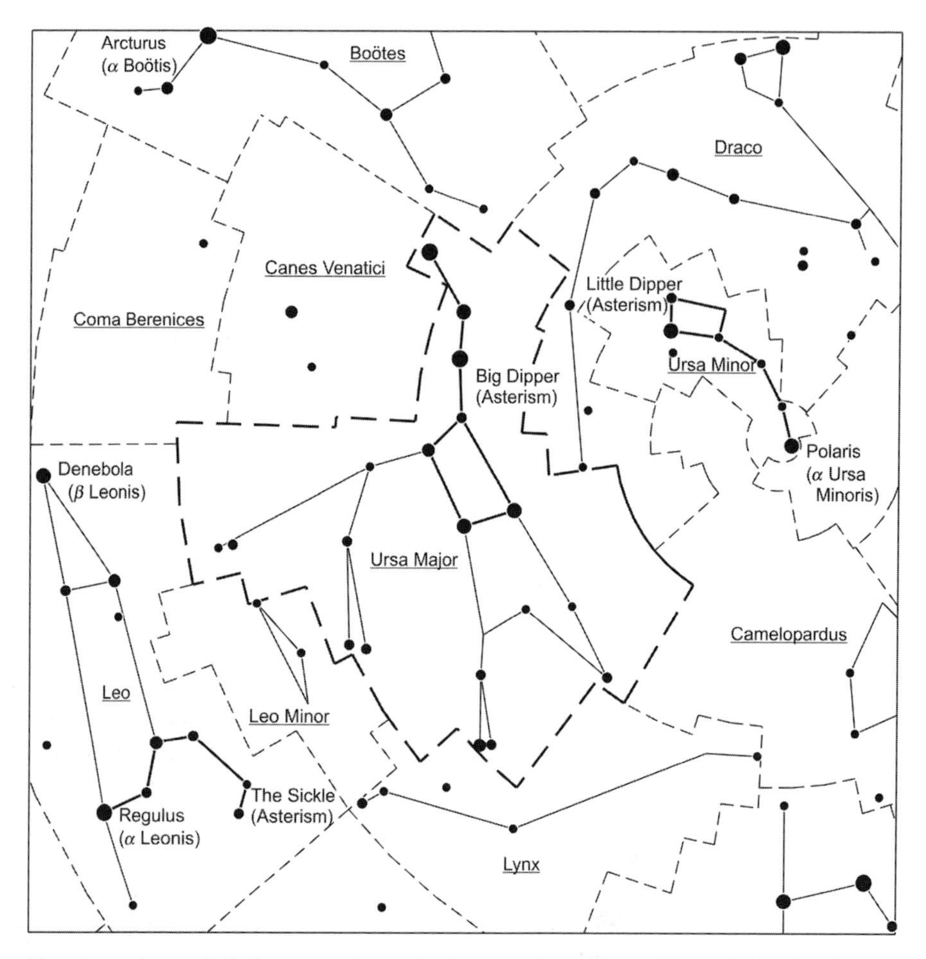

Figure 3.3 A constellation map shows the irregular boundary of Ursa Major, the Big Bear. The 88 constellations and their boundaries were established by international convention in 1922. Because the constellations are defined by their borders, the stick figures can be drawn in any way desired.

must be known precisely. For state boundaries we use the terrestrial coordinate system. For the constellation boundaries, we use the equatorial coordinate system (p. 25). The constellation boundary lines lie along right ascension and declination grid lines.

Asterisms

Another way we group stars is by *asterisms*. An **asterism** is a recognizable pattern of stars, but is not one of the official 88 constellations.

| *An asterism is defined by its stick figure – it has no boundaries.*

For example, the "Big Dipper" is an asterism and is only part of the constellation, Ursa Major. The "Great Square" is the body of Pegasus, the Flying Horse.

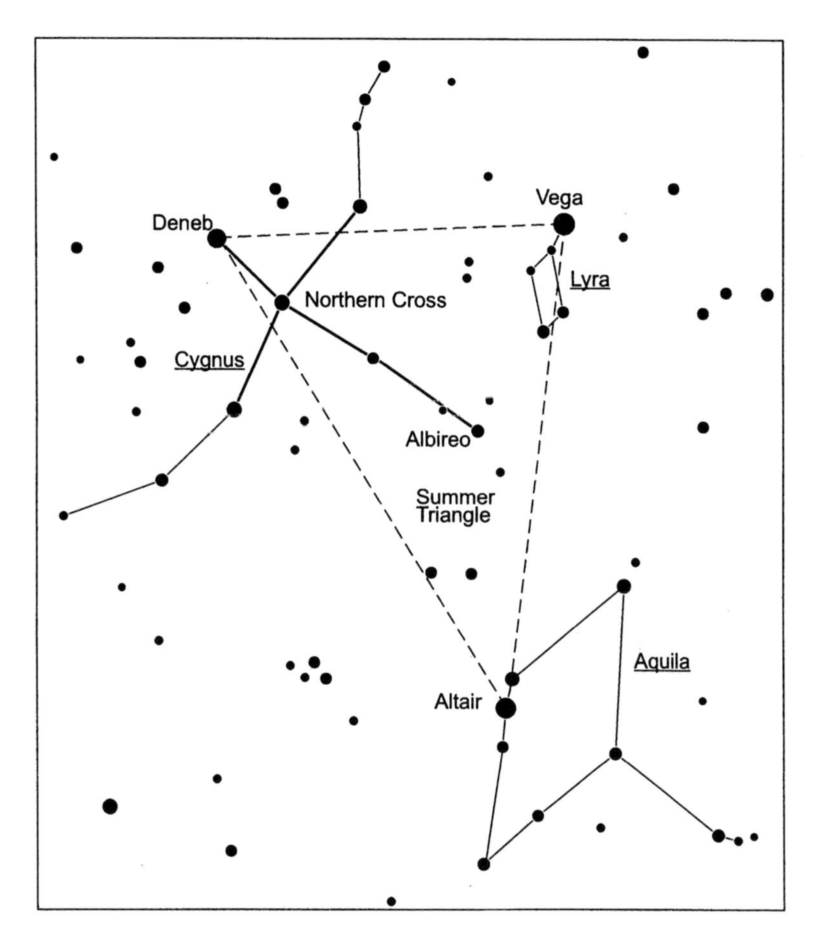

Figure 3.4 The Summer Triangle is an asterism made of three stars from three different constellations. The constellation names are underlined. This may be called the Winter Triangle by observers in the southern hemisphere. Within the triangle (and the constellation Cygnus) is the Northern Cross asterism.

The "Summer Triangle" (Figure 3.4) is an asterism formed by three stars from three separate constellations, which dominate the northern hemisphere's summer sky: Vega, in the constellation Lyra; Deneb, in the constellation Cygnus; and Altair, in Aquila. It is seen high in the northern hemisphere's sky from June through October. To southern hemisphere observers, this is called the "Winter Triangle."

The northern hemisphere's Winter Triangle (Figure 3.5) is made of three stars: Betelgeuse in Orion, Procyon in Canis Minor, and Sirius in Canis Major. For southern hemisphere observers, this asterism is called the Summer Triangle. Just like the constellations, asterisms come in many sizes. One of the smallest asterisms is the Belt of Orion, made from his three "belt stars." Observers use asterisms to orient themselves, or as "guideposts" when searching for constellations

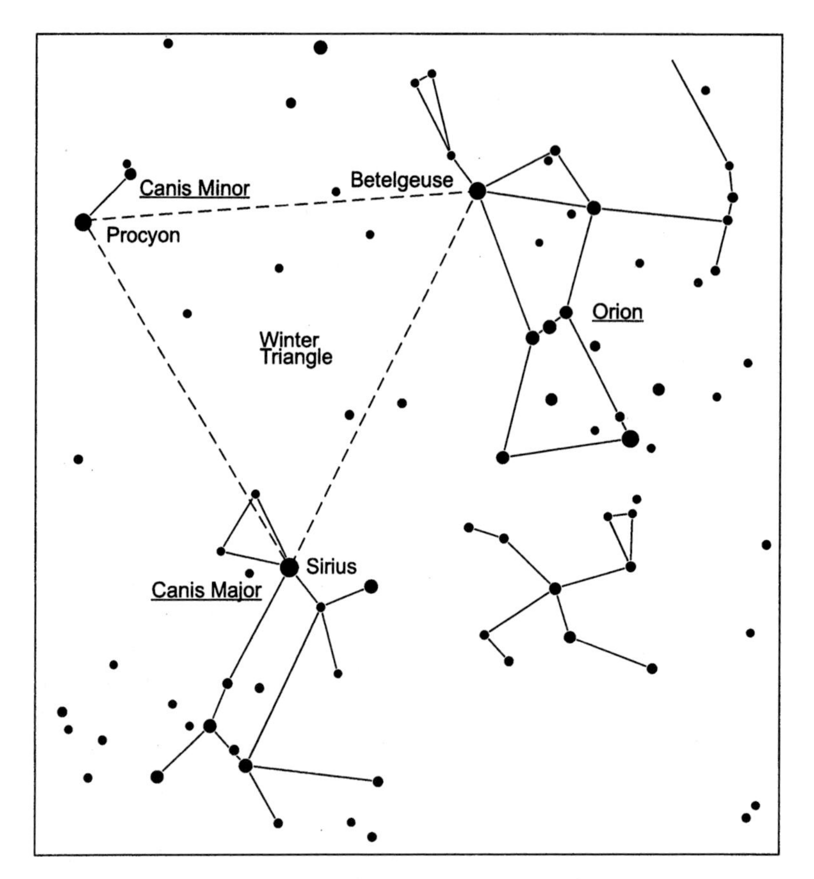

Figure 3.5 The Winter Triangle (northern hemisphere) is an asterism made of the stars Procyon, Betelgeuse, and Sirius. Sirius is the brightest star in the sky and Orion is one of the most prominent constellations, making this asterism easy to find. The Winter Triangle may be called the Summer Triangle in the southern hemisphere.

and stars. Any observer is free to create new asterisms. You might make some new asterisms for yourself, to make finding the stars easier for you.

3.2 Stars

Stars are like cities within the "celestial states" we call constellations. The brighter stars have individual (proper) names that are mostly Arabic in origin. During Europe's Dark Ages when very little science was being done, the people of the Middle East (mainly Arabic) were the world's leading mathematicians and astronomers. Examples of proper names, such as Polaris (in Ursa Minor) and Regulus (in Leo), are shown in Figures 3.3 through 3.7. Generally, star names beginning with *al* are Arabic in origin because the prefix *al-* is an article in the Arabic language.

Regulus was known as *Malikiyy* (Kingly) in Arabic. It is one of the four royal stars of the heavens and one reason that lions are associated with royalty

Table 3.1 *The Greek alphabet*

Name	Uppercase	Lowercase	Name	Uppercase	Lowercase
Alpha	A	α	Nu	N	ν
Beta	B	β	Xi	Ξ	ξ
Gamma	Γ	γ	Omicron	O	o
Delta	Δ	δ	Pi	Π	π
Epsilon	E	ϵ, ε	Rho	R	ρ, ϱ
Zeta	Z	ζ	Sigma	Σ	σ, ς
Eta	H	η	Tau	T	τ
Theta	Θ	θ, ϑ	Upsilon	Υ	υ
Iota	I	ι	Phi	Φ	ϕ, φ
Kappa	K	κ	Chi	X	χ
Lambda	Λ	λ	Psi	Ψ	ψ
Mu	M	μ	Omega	Ω	ω

throughout the world. Ptolemy labeled this star as $\beta\alpha\sigma\iota\lambda\acute{\iota}\sigma\kappa o\varsigma$ and Copernicus labeled it in Latin form as *Rex*. This mutated into English as *Regulus*. Richard Allen outlines the sketchy evidence for the stars Fomalhaut, Regulus, Aldebaran, and Antares as being the four rulers of the heavens – the four royal stars (Allen, 1963, p. 256). Not all star names are Arabic. Some Greek and Roman names are still used. *Sirius* is from the Greek word for "scorching."

These star names are known as the **proper names** for the stars. The problem with proper names – as they are handed down from generation to generation and from culture to culture – is that the same names were being used for the bright stars of different constellations. Slight variations of a name also appeared. For example, Deneb is in the constellation Cygnus, while Denebola is in Leo. The use of the same names and similar names led to a great deal of confusion. It also made the star names hard to remember. Adding to the difficulty were some unusual names such as "Zubenel Genube," (also spelled "Zubenelgenube") in the constellation Libra. Astronomers of the sixteenth and seventeenth centuries began looking for solutions to this confusion of names.

Bayer designations

When Johannes Bayer published the *Uranometria*, he assigned lowercase Greek letters to the stars of each constellation, alphabetically, in order of the stars' apparent brightness. Table 3.1 lists the Greek alphabet. The letter alpha (α) denotes the brightest star in a constellation, beta (β) the next brightest, gamma (γ) the next, and so on (Figure 3.6).

> *A star's* **Bayer designation** *is formed by combining the Greek letter with the Latin genitive (possessive) form of the constellation's name (see Appendix B, p. 247). The Greek letters are assigned in order of apparent brightness, or apparent magnitude.*

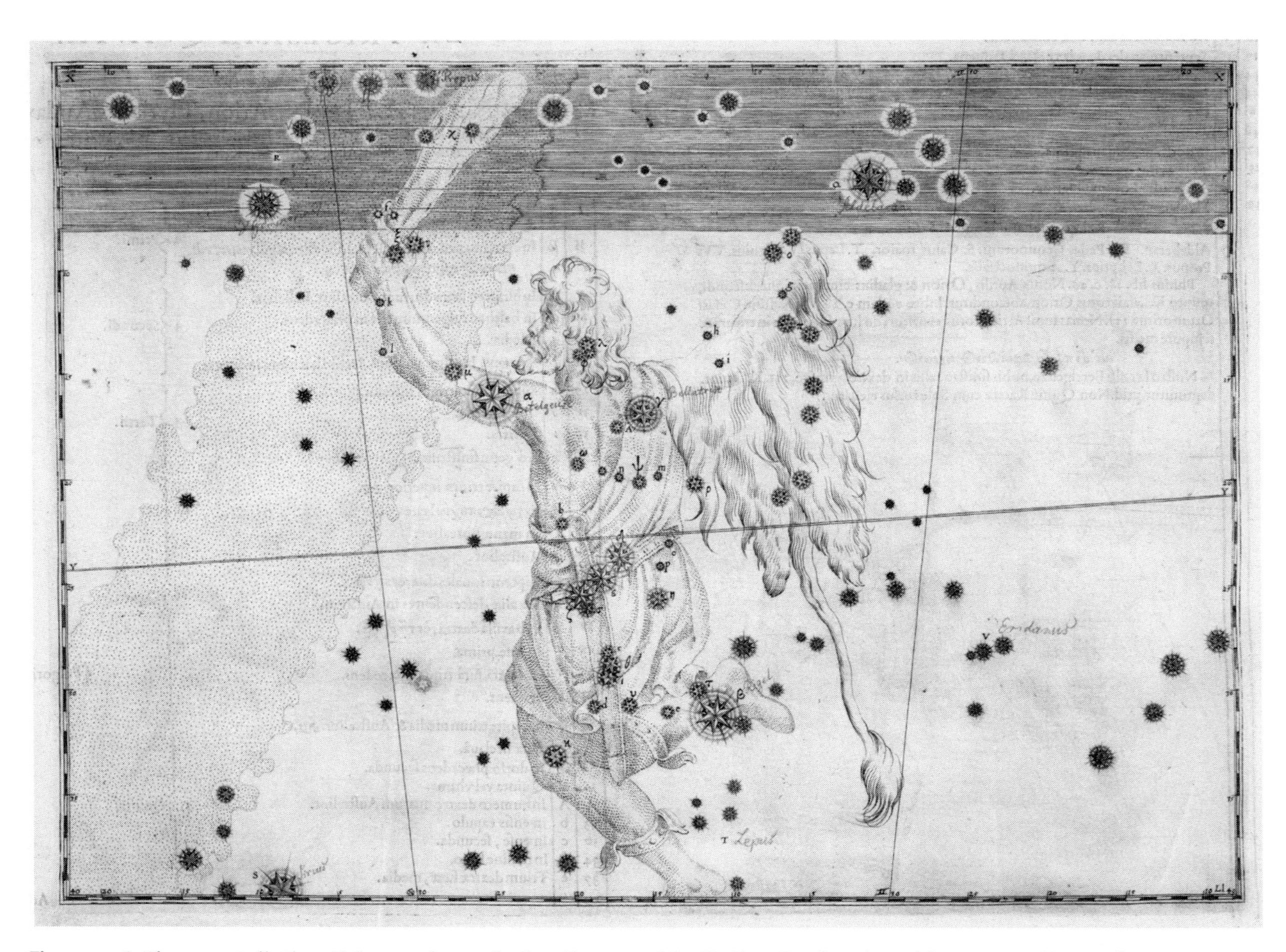

Figure 3.6 The constellation Orion as drawn in the *Uranometria*. Notice the Greek and lowercase Roman letters next to the brighter stars and the celestial equator passing through his waist. (Reproduced with permission of the Linda Hall Library of Science, Engineering & Technology.)

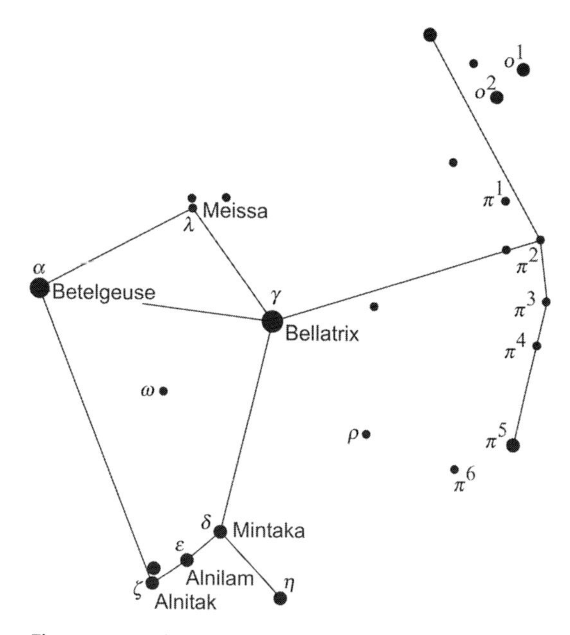

Figure 3.7 Johannes Bayer introduced a system of labeling stars with lowercase Greek letters and the possessive form of the constellation name in the *Uranometria*. The system was extended by adding superscripts to some letters. Notice the brighter stars also have proper names.

Allen describes the publication of *Uranometria*, where Bayer uses this designation system now named for him (Allen, 1963, p. 13). He notes that Bayer was not the first person to use this system. It was previously used by Piccolomini of Siena, Ptolemy, and possibly the Persians and Hebrews. The Bayer designation system didn't really catch on for common use until the next century after its publication. "Alpha Ursa Minoris" (α Ursa Minoris) is the brightest star in Ursa Minor, better known by its proper name Polaris. Regulus is also called "Alpha Leonis" (α Leonis) and Rigel, in the constellation Orion, is also "Beta Orionis" (β Orionis).

There are only 24 Greek letters and there may be hundreds of naked-eye stars within a constellation. When the Greek letters ran out, Bayer extended the scheme by continuing with lowercase and then uppercase Roman letters, progressing to the letter Q. You can find examples of this near the stinger of the Scorpion (G Scorpii) or in a few other constellations on the *Equatorial Chart* and the polar charts. This particular method of extending the star designation system fell into disuse. Another method, which is now the preferred method, is to use numerical superscripts on Greek letters that label stars of nearly the same apparent brightness. Thus, many stars are labeled by a Greek letter with a superscript, such as the string of stars on the west end of Orion, labeled π^1 Orionis to π^6 Orionis (Figure 3.7).

Flamsteed numbers

John Flamsteed (1642–1719), a celestial cartographer and first Astronomer Royal of England, created a companion star catalog for his great star atlas first published (in complete form) in 1729. In 1712, the Royal Society, under the direction of Sir Isaac Newton (1642–1728), published an unauthorized version of his catalog (Newton considered the work of the Astronomer Royal to be public property), edited by Sir Edmund Halley. The resulting dispute between Flamsteed and Newton was one of many between them. Flamsteed was a meticulous observer but very slow at publishing results. This characteristic of Flamsteed's was frustrating to Newton and other astronomers. Flamsteed succeeded in obtaining, and publicly burning, all the unsold copies (Willmoth, 1997).

Flamsteed was the first astronomer to use a clock for the measure of right ascension. His data was eventually completed, supplemented, and reissued by his assistants in 1729. In this catalog (*Historia coelestis Britannica*, or *British History of the Heavens*, where *history* is being used in the sense of the word *data*), Flamsteed listed the stars of each constellation in order of their right ascension. Later, the French astronomer Joseph La Lande (1732–1807) assigned numbers to this ordering. Although Flamsteed never used these numbers himself, they became known as the **Flamsteed numbers**. See the books by Arthur Berry (Berry, 1961, Section 198) or John North (North, 1995) for more about Flamsteed.

Flamsteed numbers are assigned to the stars within the boundaries of a constellation in order of right ascension. Each star is then designated by its Flamsteed number and the Latin genitive of its constellation.

The brighter stars are known by all three designations (Table 3.2). For instance, Vega, in the constellation Lyra, is also known as α Lyrae and as 3 Lyrae. The most famous stars that have only Flamsteed numbers are probably 61 Cygni, the first star for which stellar parallax was measured, and 51 Pegasi, the first star similar to our own, around which we have found planets. Stellar parallax was first measured by Friedrich Wilhelm Bessel (1784–1846) in 1838. He chose 61 Cygni because its large proper motion (motion relative to the vernal equinox) suggested it was a nearby star. The planets around 51 Pegasi were discovered in 1995 by Swiss astronomers Michel Mayor and Didier Queloz (Mayor and Queloz, 1995).

3.3 Double stars

Telescopes have shown us that many of the bright, single stars we see by naked eye are actually multiple star systems. We use the terms **double star** and **binary star** to describe those systems where there are only two stars. Binary star systems are the most important to astronomers because they allow us to use Kepler's laws to make direct measurements of stellar mass.

A multiple star system is defined as such only when it is demonstrated that the stars are gravitationally bound to each other. This means the stars orbit each

Table 3.2 *Names and designations of some bright stars*

Proper name	Bayer designation	Flamsteed number	Constellation	Meaning[a]
Albireo	β Cygni	6 Cygni	Cygnus	The hen's beak (not an Arabic name)
Aldebaran	α Tauri	87 Tauri	Taurus	The follower of the Pleiades
Algol	β Persei	26 Persei	Perseus	The demon star
Altair	α Aquilae	53 Aquilae	Aquila	The bird
Antares	α Scorpii	21 Scorpii	Scorpius	Like Mars
Arcturus	α Boötis	16 Boötis	Boötes	The bear
Betelgeuse	α Orionis	58 Orionis	Orion	Arm pit of the giant
Capella	α Aurigae	13 Aurigae	Auriga	Little she goat (Latin)
Deneb	α Cygni	50 Cygni	Cygnus	The hen's tail
Denebola	β Leonis	94 Leonis	Leo	The lion's tail
Mizar	ζ Ursa Majoris	79 Ursa Majoris	Ursa Major	The waist-cloth
Polaris	α Ursa Minoris	1 Ursa Minoris	Ursa Minor	The pole star
Regulus	α Leonis	32 Leonis	Leo	The lion's heart
Rigel	β Orionis	19 Orionis	Orion	The foot
Sirius	α Canis Majoris	9 Canis Majoris	Canis Major	Scorching (Greek)
Spica	α Virginis	67 Virginis	Virgo	The ear of wheat
Vega	α Lyrae	3 Lyrae	Lyra	The harp star

[a] Meanings are derived mostly from *Star Names, Their Lore, and Meaning* (Allen, 1963).

other (actually they orbit their center of mass, or barycenter) and they are not simply in the same line of sight. When two stars are in the same line of sight, they are called an *apparent binary*. Other examples of double star systems are Rigel, Sirius, and Procyon. The companion star to Sirius is a white dwarf – the first one discovered. More than 20% of all bright, naked-eye stars resolve into binary stars in telescopes. When spectra of stars are analyzed we find that more than 50% of all stars are included in some kind of multiple star combination – binary, ternary, etc. Castor appears as a single star by naked eye but the telescope shows us it is actually a system of six stars!

3.4 Apparent magnitude

When we look at the stars, it is obvious they are not all the same brightness. The perceived brightness of a star is described by its **apparent magnitude**. We use the symbol m, to represent apparent magnitude. A star's apparent magnitude corresponds to its apparent brightness, which in turn corresponds to the light energy we receive from the star. On the original scale, a bright star was magnitude one, or $m = 1$. The dimmest star (visible without a telescope) was (and still is) $m = 6$. The *dimmer* the star, the *greater* its magnitude number. This seems backward, but it has its advantages and you get used to it.

The original magnitude scale was created by the Greek astronomer Hipparchus as part of a new star catalog. He may have created this catalog as a result of the appearance of a new star in the Scorpion in 134 BC. The catalog contained 1080 stars divided into six classes, according to their brightness. This catalog remained the standard for nearly 16 centuries (Berry, 1961, Section 42). With the publication of Ptolemy's *Almagest*, Hipparchus' work became of lesser importance and was unfortunately lost. What we know of Hipparchus is from references made in the *Almagest* and other works.

As our ability to measure the brightness (energy output) of stars and our ability to see fainter stars (with telescopes) improved, we needed to expand the range of the magnitude system. We also wished to include objects not originally included by Hipparchus – namely, the planets (Figure 3.8). For objects brighter than $m = 1$, we use zero and the negative numbers. The full Moon has, $m = -12.6$. For Venus at its brightest, $m = -4.7$. The Sun has apparent magnitude, $m = -26.73$. Polaris, the stars in Orion's belt, and the Big Dipper are all about magnitude +1. For objects dimmer than $m = +6$, we use greater positive numbers.

The dimmest stars we can observe with ground-based telescopes are about magnitude +26 (the 10-m Keck telescopes). The Hubble Space Telescope can see objects down to about magnitude +29. This is where there is an advantage to the apparent backward scale. Because the number of faint objects is much larger than the number of bright ones, finding fainter objects means we extend the scale into the positive numbers rather than the negative numbers. Thus, we need not worry about accidentally dropping minus signs.

When we talk about the apparent brightness of an object, we mean the intensity of the light energy we receive from the object. At this point we must be more careful because astronomers have a very specific definition for the word *brightness*. Apparent magnitude is actually a numerical description of a star's *energy flux* as measured from Earth. **Energy flux** is a measure of the flow of energy through a surface of known area. It is measured in units of watts per square meter (W/m^2) or, equivalently, as joules per second per square meter ($J/sec/m^2$). For this discussion, the word *brightness* is used only as a descriptive term in place of the more correct term *energy flux*.

The actual energy output of a star is not the only thing affecting its apparent brightness in our sky. Its distance from us, and other factors (such as interstellar dust), has an effect. When we use the term *apparent visual magnitude*, the measurement of energy flux is restricted to the visible light portion of the spectrum.

Systems of measurement

A numerical relationship between magnitude and brightness was first discovered by Sir William Herschel (1738–1822) around 1817. Herschel used two telescopes to observe various pairs of stars. He covered a portion of the aperture on one telescope until the brightness of its star was equal to the brightness of the dimmer star in the other telescope. The ratio of the telescopes' open aperture is then equal to the ratio of the stars' apparent brightness. His original

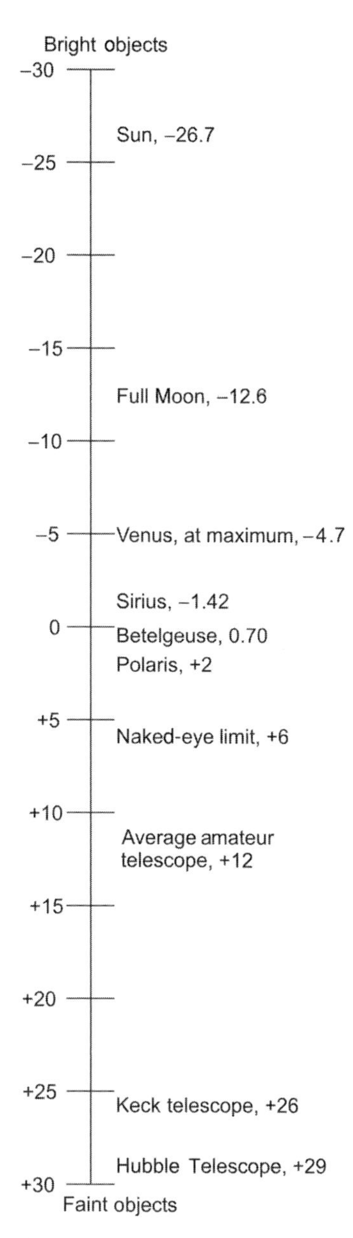

Figure 3.8 The apparent magnitude scale uses positive numbers for faint objects and negative numbers for bright objects. This figure shows the relative magnitude of some well-known objects.

intention for this observation was to determine the validity of his hypothesis concerning stellar distance. However, he discovered this magnitude–brightness relationship:

> *When the difference in magnitude between two stars is five, the ratio of their brightness (ratio in aperture opening) is* 100.

Table 3.3 *The apparent magnitude of some stars*

Star name	Apparent magnitude
Aldebaran	0.86
Arcturus	−0.06
Deneb	1.26
Polaris	+1.97
Sirius	−1.42
Sol	−26.73

Herschel, and others, thought each star produced the same amount of energy and, therefore, the apparent brightness was an indicator of a star's distance from Earth. This turns out to be wrong, as his discovery of binary star systems proves. There are binary star systems (two stars, gravitationally bound to each other and therefore in orbit about each other) where the two stars are clearly not the same apparent brightness. Because the stars must obviously be the same distance from Earth, their difference in apparent brightness can only be attributed to a difference in their energy output. Therefore, Herschel's hypothesis that each star produces the same energy and so its apparent brightness indicates its distance, was clearly wrong.

The Pogson method A method for measuring magnitude was suggested by the English astronomer Norman Pogson (1829–1891) in 1856 (Pogson, 1856). His measurements of the apparent magnitude of some of the minor planets (asteroids) supported Herschel's earlier discovery of the relation between magnitude and brightness. Pogson suggested the relationship be fixed as a difference in one magnitude corresponding to a ratio in brightness of the fifth root of 100 ($\sqrt[5]{100}$), or 2.512. This creates a logarithmic relationship between apparent magnitude and apparent brightness. Altair and Aldebaran were assigned the magnitude of $m = +1.0$. The photometric (photographic) measurement of any other star was then compared to these two stars for assignment of a magnitude number (Table 3.3).

This definition gives the result that a difference of five magnitudes, $m_2 - m_1 = \Delta m = 5$, represents a ratio in brightness $f_1/f_2 = 100$. That is, if two stars (or objects) differ by five magnitudes, then the brighter object provides 100 times greater light-energy flux than the dimmer one (Table 3.4). Appendix A gives the detailed mathematical relations between apparent magnitude and brightness.

If for star A, $m = 1$, and for star B, $m = 2$, and for star C, $m = 3$; we can make the following observations:

(a) Star A is 2.5 times brighter than star B.
(b) Star B is 2.5 times brighter than star C.
(c) Star A is $2.5 \times 2.5 = 6.25$ times brighter than star C.

Table 3.4 *The difference in magnitude related to the ratio of brightness*[a]

$\Delta m = m_2 - m_1$	Approximation: $2.5^{\Delta m}$	f_1/f_2 (Actual)
0	1	1.000
1	2.5	2.512
2	$2.5^2 = 2.5 \times 2.5$	6.310
3	$2.5^3 = 2.5 \times 2.5 \times 2.5$	15.85
4	$2.5^4 = 2.5 \times 2.5 \times 2.5 \times 2.5$	39.81
5	100 (by definition)	100
6	$2.5^6 = 2.5^{5+1} = 100 \times 2.5$	251.2
7	$2.5^7 = 2.5^{5+2} = 100 \times 6.31$	631.0
10	$2.5^{10} = (2.5^5)^2 = 100^2$	10 000
15	$2.5^{15} = (2.5^5)^3 = 100^3$	1 000 000

[a] See Appendix A, p. 230, if you do not follow all the arithmetic here.

A simplistic calibration The apparent magnitude scale can be calibrated in a simple way by using the apparent brightness of the Sun. The Earth receives 1350 W/m^2 of energy from the Sun. This measurement is called the *solar constant*. The solar constant is the average amount of energy we receive from the Sun – measured just above the atmosphere. The *constant* part of the name is a slight misnomer. This value takes into account the varying distance between the Earth and the Sun and the slight variations in solar output due mainly to the solar cycle.

The Sun's apparent magnitude is $m = -26.73 \pm 0.03$. With this as a calibration standard, a magnitude zero star produces about 2.7×10^{-8} W/m^2 at the Earth. This is not the method used for modern measurements of apparent magnitude. Currently, the ultraviolet blue (UBV) system, created by Harold Johnson and William Morgan (1953) is used. However, understanding the standards for this system requires knowledge of stellar spectra. Therefore any further discussion of stellar magnitude and brightness must wait.

3.5 Variable stars

Many stars vary in magnitude with a regular or semi-regular period. These are called **variable stars**. The variation in apparent magnitude is typically 1 magnitude, but it may be as small as 0.1 magnitude or as great as 4 magnitudes. The variations are usually caused by an instability established in the structure of the star, related to the star's ability to transport energy from its core to its surface. Most variable stars are associated with stages of stellar birth or death. There are many variable stars, but professional astronomers have time to study only the very interesting ones. There is a worldwide network of amateur astronomers, i.e. the Association of Amateur Variable Star Observers (AAVSO), a very active group of amateurs, studying the characteristics of variable stars.

This area of observation and research is one of the most important for amateur astronomy.

One of the more important of these characteristics is the star's light curve. The light curve is a plot of the star's changes in magnitude with time. The shape of the light curve can tell us to which class of variable stars a particular star belongs. Classes of variable stars are named after the first star discovered in a class. For instance, two of the most important classes of variables (because they are used to measure distance) are the *Cepheid variables*, named after δ Cephei, and the *RR Lyrae variables*. Because these stars change their physical characteristics (their size and temperature), they are called *intrinsic variables*. These types of variable stars have short periods of hours or days.

The majority of variable stars have longer periods. Usually the length of time of each variation is slightly different and the change in magnitude is not consistent between cycles. These types of long period variables are called *Mira variables* after their prototype, Mira (o Ceti) in the constellation Cetus. Betelgeuse is a long period, Mira-type variable. It regularly changes in diameter, from about the orbit of Mars to the orbit of Jupiter, with a period just short of one year.

One of the most famous variable stars is Algol, the "Demon Star," or the "Eye of Medusa." This star is an eclipsing binary. Its magnitude varies not because of a change in the characteristics of a particular star, but because it is a system of two stars in orbit around each other. From our point of view, they are alternately passing in front of (eclipsing) each other. When one star is eclipsed by the other, the total amount of light we receive from the system decreases slightly, until the eclipse is over. For the Algol system, the variation in magnitude is about one, with a 2.9-day period. Eclipsing binaries are identified by their light curve.

Naming variable stars

Many variable stars already have proper names. For instance, Betelgeuse is a variable star. Some dimmer stars are variable and already have Bayer designations or Flamsteed numbers. These stars keep their names or designations. However, there are so many variable stars that a standard for designating those variables that did not already have designations was needed. The German astronomer, Friedrich Argeländer (1799–1875) devised a system in the 1800s.

The practice of extending the Bayer designations with lowercase and uppercase Roman letters never exceeded the letter Q. Argeländer started labeling the new found variable stars with the uppercase Roman letter R and continued to the letter Z. Argeländer apparently never imagined there could be more than ten variable stars within a constellation. It turns out there are generally hundreds of variable stars within a constellation. After Z, the letters are doubled, starting with RR. The sequence goes as RR, . . . , RZ, SS, . . . , SZ, and so on, to ZZ. As more variables were discovered, the sequence was continued with AA, . . . , AZ, BB, . . . , BZ, CC, . . . , CZ, on to QQ, . . . , QZ, excluding the letter J. (In many type faces, I and J are hard to distinguish.) This list contains 334 letters or letter combinations. At this point astronomers gave up and started using numbers, starting with V335. The "V" denotes a variable star. Thus, variable stars have names like T Tauri, RR Lyrae, V404 Cygni, and BL Lacertae. Although the extended system of Bayer was not completely adopted it still has an influence on our star names.

3.6 Observing the stars

The "morning star" is the last star visible in the morning and the "evening star" is the first star visible in the evening. Neither star is a star at all – both of them are the planet Venus. Whether Venus shows as the morning star or as the evening star depends on its current position in its orbit around the Sun. In general, observation of the stars depends on the magnitude of the star we are trying to find, the position of the Sun near the horizon, and current atmospheric conditions.

Twilight

There are three periods of twilight: civil, nautical, and astronomical. They occur just after sunset or just before sunrise. The beginning at sunrise or the ending at sunset of each period of twilight is defined by the distance between the Sun and the observer's zenith. Twilight is caused by sunlight being scattered to the ground by the upper layers of the atmosphere. Once the Sun is far enough below the horizon there is insufficient light to scatter and twilight ends.

Civil twilight begins in the morning or ends in the evening when the center of the Sun is 96° away from the zenith – 6° below the horizon. Under civil twilight, the brightest stars can be seen and the sea horizon is clearly defined. Nautical twilight begins or ends when the center of the Sun is 12° below the horizon. At this point the sea horizon is no longer visible and altitude measurements with reference to the horizon for navigation can no longer be made.

Astronomical twilight begins or ends when the center of the Sun is 18° below the horizon. At this point the scattered sunlight is less than the average level of starlight. It is about the same brightness as the aurora or the zodiacal light. In all these periods the actual brightness of the sky depends on meteorological conditions and azimuthal direction of observation. The length of twilight also depends on the observer's latitude, as discussed on p. 100.

Atmospheric conditions

Atmospheric conditions control our ability to see the stars. These conditions are described with the three terms, *sky brightness*, *transparency*, and *seeing*. Sky brightness is caused by aurora, moonlight, and light pollution. While this generally does not interfere with the observation of bright objects such as the Moon, the naked-eye planets, or the bright stars, it does reduce our ability to see the dimmer stars or those extent objects such as galaxies or nebulae that we view through a telescope.

Transparency is the optical clarity of the atmosphere. It is affected by clouds, air moisture content, and particulate matter created by industrial pollutants or by natural causes such as volcanos. The quality of the transparency can be judged by looking at the Mizar/Alcor star pair. Mizar is the bright, middle star in the handle of the Big Dipper. Alcor is a much dimmer star next to Mizar. With good transparency Alcor is clearly visible. Good transparency usually occurs after a rain storm or rain shower cleans the air of particulate matter or when the air pressure is high. Dry air is heavier than moist air.

Seeing is actually a judgment of conditions while observing through a telescope. "Good seeing" is the condition of being able to discern details in the telescopic image of the Moon, the planets, or extended objects. Poor seeing occurs with greater atmospheric turbulence or turbulent conditions of the air within the telescope's tube.

The *limiting apparent magnitude* is the lowest (most positive) magnitude visible under the current sky conditions. It is usually judged by counting the number of faint stars visible within 20° of the zenith.

3.7 Star names, sky maps, and catalogs

In the last 300 years we have created many star catalogs. Most stars are known only by their catalog number. Catalogs are used by astronomers to keep track of the position and properties of individual stars and other objects. The most commonly used catalogs are the *Harvard Revised Catalog*, the *Messier Catalog*, the *New General Catalog*, the *Index Catalog*, the *Henry Drapier Catalogue*, and the *Smithsonian Astrophysical Observatory Star Catalog*. Example designations of objects found in these catalogs are M42, NGC6566, IC324, HD3956, or SAO3562. The *Messier Catalog* (objects designated by "M...") is probably the best known by amateur astronomers. The French astronomer Charles Messier (1730–1817) was a comet hunter. His famous list of fuzzy objects was intended as a list of objects to be avoided – that is, objects not to be confused with comets.

Star names and sky maps

On modern star maps, the proper name of the star is used if the star is of first magnitude ($m = +1$) or lower (brighter), or if the star is well known by its proper name. For the remaining naked-eye stars, the Bayer designation is used for those stars that have one and otherwise the Flamsteed number is generally used. The stars below the naked-eye limit ($m > +6$) are only known by their catalog numbers. The catalog used is the choice of the map maker.

The *Yale Bright Star Catalog* (*Bright Star Catalog*) has over 800 proper star names. Most of these are the Latinized versions of the names given to the stars by Arabic astronomers. There are far too many proper names to remember. The Bayer designations and Flamsteed numbers have a few problems, but they are better suited for describing where a star may be found than simply using a proper name.

There are some exceptions to the brightness rule for Bayer designations. For instance, β Orionis (Rigel) is actually slightly brighter than α Orionis (Betelgeuse). This exception is known only because of our modern, more accurate, magnitude measuring techniques. A few constellations, such as Puppis and Vela, do not have an alpha star because these constellations were once part of the larger, Ptolemaic constellation, Argo Navis. This is one result of the formalization of the constellation boundaries.

Another problem caused by the formalization of the constellations occurs with the Flamsteed numbers. A few stars are now found on the wrong side of the border. For example, because of its large proper motion, 30 Monocerotis is now found in the constellation Hydra. The Flamsteed numbers were not used in the southern constellations, so many southern hemisphere stars only have Bayer designations with many of them still using the lower and upper case Roman letters. For examples, see the *South-polar Chart*.

Star catalogs

As bigger telescopes were built, fainter stars could be seen. All these stars needed to be labeled, so star catalogs, each with their unique star designation system, started proliferating in the mid 1800s. In 1859, Friedrich Argeländer, working at the Bonn Observatory in Germany, started compiling the *Bonner Durchmusterung* (the Bonn Survey). This survey split the sky into 1° strips in declination and numbered the stars within each strip in order of right ascension. Constellation borders were ignored. The first version of the survey covered the sky from the north celestial pole to −2° declination.

Stars in the *Bonner Durchmusterung* are designated as, for example, BD +35° 2471, which is the 2471st star in the strip between +35° and +36°declination. The survey was extended southward to −23° as the *Southern Bonner Durchmusterung* (SBD). Later still, it was extended to the south celestial pole as the *Cordoba Durchmusterung* (CD or CoD). The complete survey, the visual *Durchmusterung* (DM), covers the entire sky and contains 1 071 800 stars, down to about magnitude +9.5. The *Durchmusterung* (DM) became the standard catalog for astronomers for nearly 100 years. However, while the positions of the stars were pretty accurate, the magnitude measurements were not. Many of the magnitude measurements were simply judgment calls on the part of the observers.

The *Harvard Revised Photometry* catalog was published in 1908. This catalog listed the magnitudes of the brightest 9110 stars, reaching down to about $m = +6.5$. The stars from this catalog are designated by "HR" and listed in order of magnitude. This catalog forms the basis of the *Yale Bright Star Catalog*.

The next important catalog produced is the *Henry Drapier Catalogue* (HD), created by Annie Jump Cannon (and others) at Harvard College. It was published from 1918 to 1924, containing 225 300 stars. The extended version (HDE) raised the number to almost 350 000 stars. Any star in these catalogs has had its spectrum taken and classified.

The *Smithsonian Astrophysical Observatory Star Catalog* (SAO) was also produced at Harvard, and published in 1966. This is probably the most commonly used catalog. The stars in the SAO are listed in order of right ascension in 10° declination strips from the north celestial pole to the south celestial pole. It gives very accurate positions for 258 997 stars down to magnitude +9.

The largest catalog produced so far is the *Hubble Guide Star Catalog* (GSC). This catalog was created by computer analysis of photographs taken by many observatories (mainly Palomar). It is used for guiding the Hubble Space Telescope during observations. This list contains a total of 18 819 291 objects with positions measured to within one arc second and magnitude measured to a few tenths down to +14. Of the total, 15 169 873 are listed as stars, with the remainder listed as faint galaxies or other objects. An object in the GSC is listed by a serial number within one of 9537 regions of the sky. This list is so large it is available only on a CD-ROM.

There are other more specialized catalogs for visual binary systems (the *Aitken Double Star*, catalog ADS, or the *Index Catalogue of Visual Double Stars*, IDS, from Lick Observatory) or variable stars (*Star Catalog 2000.0*, Vol. 2, from Sky Publishing). There are currently over 1000 star catalogs available.

Questions for review and further thought

1. What are stars and how do they relate to our Sun?
2. What are constellations? How are they defined?
3. How many official constellations are there?
4. How are constellations named?
5. What are the ancient constellations? Where did they come from?
6. Where are most of the modern constellations found?
7. What is the original intent of constellations?
8. How does modern astronomy use constellations?
9. How many stars do not belong to a constellation?
10. What is an asterism? How does it differ from a constellation?
11. What do we mean by a star's *proper* name?
12. What problems are created by using proper names for stars?
13. How are Greek letters assigned to the stars within the Bayer designation system of star naming?
14. Describe a binary or double star system.
15. What is *apparent magnitude*?
16. Who invented the apparent magnitude scale? When did he invent it?
17. How is the relationship between apparent magnitude and apparent brightness defined?
18. List the three kinds of twilight, describe the observation conditions they create, and describe the position of the Sun that defines them.

19. List and describe the terms we use for the description of observation conditions.
20. What is a star catalog? What general information does it provide?

Review problems

1. Star A has $m = +4$ and star B has $m = +3$. Which is brighter? What is the ratio of their brightness?
2. Star A has $m = +5$ and Star B has $m = +1$. Which is brighter? What is the ratio of their brightness?
3. Star A has $m = +2$ and Star B has $m = -1$. Which is brighter? What is the ratio of the brightness?
4. Star A has $m = -1$ and star B has $m = +6$. Which is brighter? What is the ratio of their brightness?

Use the formula provided in Appendix A, p. 230, for the next three problems.

5. Star A has $m = +2.3$ and star B has $m = +3.6$. What is the ratio of their brightness?
6. Star A has $m = -0.7$ and star B has $m = +2.62$. What is the ratio of their brightness?
7. Verify the calculation of energy flux for an $m = 0$ magnitude star given in the simplistic calibration section on p. 53.

4
Motions of the Earth

It is said that Galileo mumbled the words "and yet it moves," while getting up from his knees after making a confession of his errors. Some historians don't believe he actually said this in such a dangerous situation. Galileo and those who followed him changed our view of the world. However, the ancient (not quite correct) point of view does have its uses.

As we watch the night sky, the stars move. This can easily be verified by watching a star rising on the eastern horizon for just ten minutes. When I speak of the motion of the stars in our sky, I do not mean to say that star *A* is passing star *B*. Although the stars *are* in motion relative to each other (and to us, resulting in what we call *proper motion*), this motion appears too slow for us to notice. Because true stellar motion appears so slow, the same constellations have adorned our skies for more than 10 000 years. The Big Dipper has been in our night sky for many millennia, and will be there for many more. We can use a computer and the measured proper motion (changes in right ascension and declination with time) of the stars to understand how the constellations change shape. In some cases we find it may take more than 25 000 years to see any major change.

This chapter is about the daily (or nightly), seasonal, and long term apparent motions of the stars, the constellations, and the Sun, which are actually caused by the motions of the Earth. The apparent motion of the sky is a combination of Earth motions: *diurnal*, *annual*, *precessional*, and *nutational*. Over 100 different motions of the Earth have been identified, mainly by making precise observations of position changes of celestial radio sources. The position changes of these sources are caused by the various motions of the Earth as it spins on its axis. Most of these Earth motions are caused by the gravitational influences of other members of the Solar System. The most important of those influences is from the Moon and the Sun.

The history of astronomy is filled with arguments over the motion of the Earth. It was not until Galileo's remarkable observations during the early 1600s that we began to accept the idea of a moving Earth, and only in the eighteenth and nineteenth centuries did we finally find observational and experimental proof for its motion (stellar aberration and the Foucault pendulum).

We can ignore all of this for the purposes of this chapter because, for the topics covered here, it ultimately makes no difference whether the Earth is rotating or the celestial sphere is rotating. The relative motion of the observer and the celestial sphere is all that really matters. Therefore an observer might speak of the rotation of the Earth and the rotation of the sky (the celestial sphere) interchangeably. When two observers are in relative motion to each other, either observer can completely understand the motion by considering themselves at rest while the other observer is moving. This is a fundamental concept of Albert Einstein's relativity theories.

> *For observation and understanding of the motion of the celestial sphere, it does not matter whether one speaks of the motion of the Earth or of the motion of the sphere. All that matters is their relative motion.*

Because a motionless Earth was the original philosophical view of the heavens, our language is built from this point of view. It is therefore easier to speak of the motion of the sky rather than the motion of the Earth.

4.1 Diurnal motion of the sky

Diurnal motion is the daily motion of the sky caused by the rotation of the Earth on its axis. "Diurnal" (Latin) is the opposite word of "nocturnal." The Earth has diurnal motion and astronomers have nocturnal lives. To *rotate* means to spin like a top. The Earth rotates once every 24 hours, making all celestial objects move across the sky (or move "around the Earth") once every 24 hours. The appearance of the sky's diurnal motion depends on the observer's latitude, and the horizon being viewed.

Look at Figures 4.2 through 4.9. In each of these drawings, the solid lines with an arrow represent a portion of the diurnal path of the stars. If we point a camera at the stars and leave the shutter open for a few minutes, the moving stars create lines (or star streaks) in the photo (Figure 4.1). The pattern of the star streaks for the various horizons would be like those drawn in these figures. The arrows indicate the direction of the stars' motion as the Earth rotates. Stars rise on the eastern horizon and set on the western horizon. When facing north, the stars ascend on your right and descend on your left.

> *All sky (celestial sphere) motion (diurnal and annual) occurs around the celestial poles and all celestial objects trace paths (small circles) that are parallel to the celestial equator.*

The northern hemisphere

Look at Figures 4.2, 4.3, and 4.4. When observing from any place other than the Earth's equator, objects close to the celestial poles never go below the horizon

Figure 4.1 This is a time-exposure photograph of the northern sky from 43° N latitude. The NCP is toward the lower right and the motion is counterclockwise. This is about a 10-minute exposure.

(Figures 4.2a, 4.4, 4.7b, and 4.8). In other words, the stars and constellations (or any other objects) that are close to the celestial pole, never set. They are called **circumpolar stars** or **circumpolar constellations**. When facing north, the stars to the right of the NCP are rising higher in altitude, the stars to the left of the NCP are moving to lower altitudes, moving counterclockwise around the NCP. The constellations Ursa Major, Ursa Minor, Draco, Cassiopeia, and Cepheus are all circumpolar at the higher northern latitudes.

Which stars and how many stars and constellations are circumpolar depends on the observer's latitude. All objects whose declination is greater than +50° (all objects within 40° of the NCP) are circumpolar objects at 40° N latitude (Figures 4.2a and 4.4). At 26° N, all stars closer than 26° to the NCP (declination greater than +64°) are circumpolar (Figure 4.3a). With this smaller angle, fewer objects are circumpolar at this lower latitude. At the equator, there are *no* circumpolar stars. However, at the North Pole and South Pole, *all* stars and constellations are circumpolar.

An object is circumpolar if the following condition is met:

$$\text{Object declination} > 90^\circ - \text{Altitude of NCP}$$

or, in symbols,

$$\delta_{\text{Object}} > 90^\circ - A_{\text{NCP}},$$

where δ is the object's declination (p. 25). This conditional test works only in the northern hemisphere.

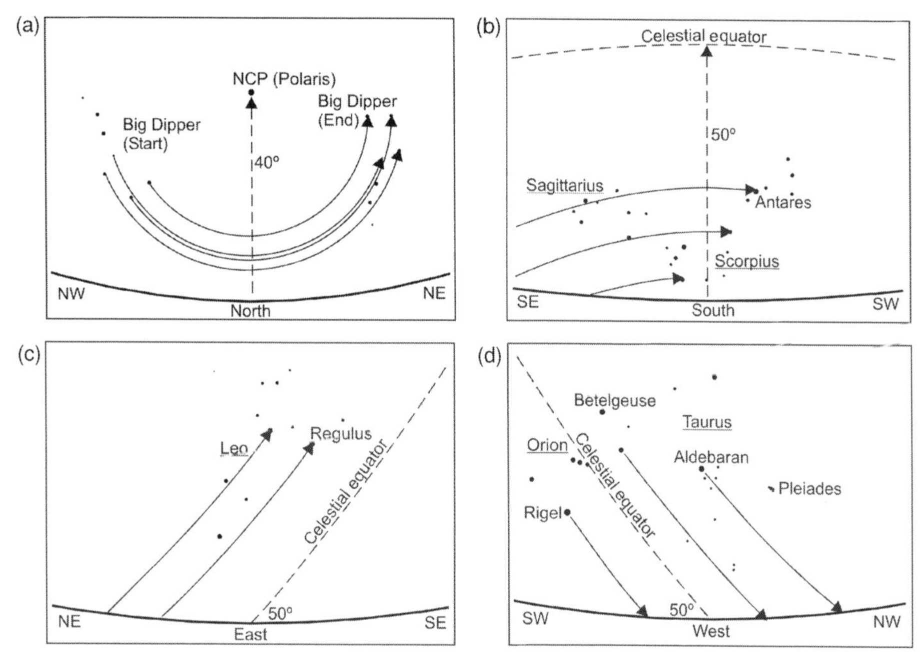

Figure 4.2 The solid lines indicate star streaks created by a long term photographic exposure, where the camera is held still. For clarity, not all of the possible star streaks are shown. Note that the angle shown between the celestial equator and the horizons is true only for locations at 40° N. (a) In the northern sky, stars like those in the Big Dipper revolve around the north celestial pole (Polaris). If these stars are close enough to Polaris that they never go below the horizon, they are called circumpolar stars. Note that the altitude of the NCP is equal to the observer's latitude. (b) In the southern sky, all stars rise in the southeast and set in the southwest. Their arched paths are parallel to the celestial equator, which is 50° above the southern horizon. Stars whose declination is lower than the altitude of the celestial equator ($\delta < -50°$) never come above the horizon and, thus, are never seen. (c) The stars of Leo rise in the east on paths parallel to the celestial equator. At 40° N latitude, the equator rises at an angle of 50° to the horizon. (d) The stars of Orion and Taurus set in the west along paths that parallel the celestial equator. At 40° N latitude, the equator meets the horizon at an angle of 50°.

In the northern hemisphere, an object is circumpolar if its declination is greater than 90° minus the altitude of the north celestial pole at the observer's latitude.

Recall that the altitude of the north celestial pole is equal to the observer's latitude.

The declination of the stars passing through the zenith is also equal to the observer's latitude. At 40° N, the declination of the stars that pass directly overhead is +40°. Some objects that pass very nearly overhead are Deneb (+45°), Vega (+40°), Capella (+46°), and the Andromeda galaxy (+41°). At 26° N, objects with declination +26° pass through the zenith. As the observer gets closer to the Earth's equator, the declinations of zenith-transit objects comes closer

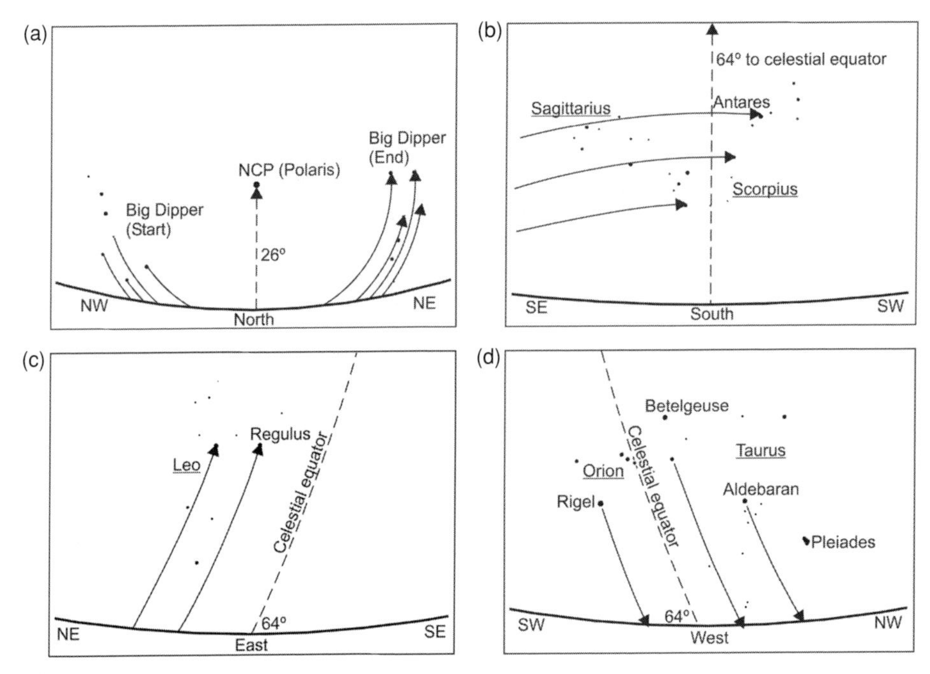

Figure 4.3 The diurnal motion of the stars as seen at 26° N latitude. Note that the angles shown between the celestial equator and the eastern and western horizons are true only for locations at 26° N, such as Miami, Florida. (a) In the northern sky, stars in the Big Dipper are no longer circumpolar, as they are at greater latitudes. There are still some circumpolar stars, like those in the Little Dipper. However, because the NCP is closer to the northern horizon, there are fewer circumpolar stars at the lower latitudes. Note that the altitude of the NCP is equal to the observer's latitude. (b) In the southern sky, all stars rise in the southeast and set in the southwest. Their arched paths are parallel to the celestial equator, which is 64° above the southern horizon (higher than this field of view). Stars whose declination is lower than the altitude of the celestial equator never come above the horizon and, thus, are never seen. Because the celestial equator is higher at lower latitudes, more stars are visible, compared to higher latitudes. (c) The stars of Leo rise in the east on paths parallel to the celestial equator. At 26° N, the equator rises at an angle of 64° to the horizon. (d) The stars of Orion and Taurus set in the west along paths that parallel the celestial equator. At 26° N, the equator meets the horizon at an angle of 64°.

to the celestial equator. This is expected, because the celestial equator moves closer to the zenith as the observer moves closer to the Earth's equator (see Section 2.5).

At 40° N the celestial equator's transit altitude is 50° above the southern horizon. Therefore, any object with declination below −50° cannot be seen anytime of the year (Figure 4.2b). At 26° N the celestial equator's transit altitude is 64° above the southern horizon. Therefore, any object with declination below −64° cannot be seen anytime of the year (Figure 4.3b).

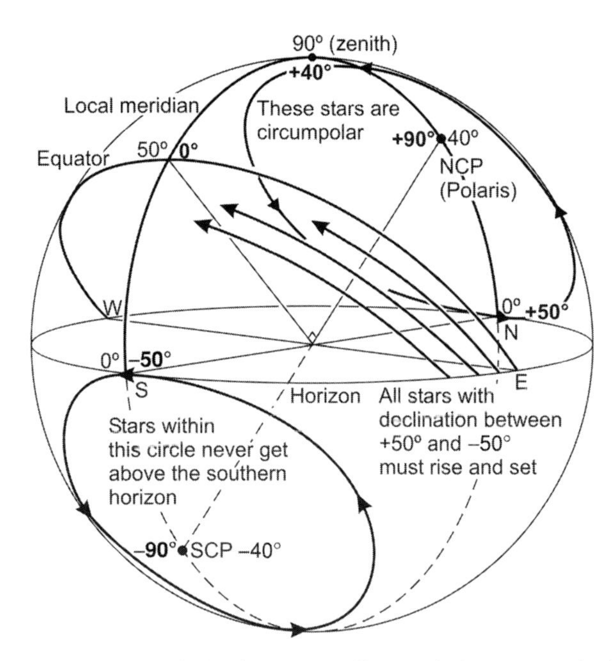

Figure 4.4 At latitude 40° N, all stars below −50° declination never come above the southern horizon. All stars above +50° declination are circumpolar. All stars between these declinations must rise in the east and set in the west. These declination values are given by the altitude of the celestial equator at the given latitude (Section 2.5). The altitudes of important points are shown in normal print, while their equivalent declination is shown in bold.

In the northern hemisphere, any object whose declination is lower than the (negative) altitude of the celestial equator is never visible to the observer (Figure 4.4).

Because Polaris is very near to the north celestial pole, we see Polaris next to the NCP throughout the night, throughout the year. This is the most important characteristic of Polaris. Contrary to popular beliefs, Polaris is *not* the brightest star in the sky, nor is it straight up in the sky. Only at the North Pole will an observer see Polaris at the zenith. What makes Polaris important is that it is the only star that does not move (at least, not very much) with the rotation of the Earth.

Because Polaris is so close to the NCP it is usually referred to as the "North Star." Use Polaris to find the north celestial pole and to find due north along the horizon for your informal observations.

Also, it is important to realize that though they are moving from horizon to horizon, the stars (and other objects) seen in the night sky at any given latitude are the same stars seen around the World by observers at that same latitude. If the Moon is visible at midnight (local time) in New York City, it is also visible at midnight (local time) in Hong Kong (ignoring weather).

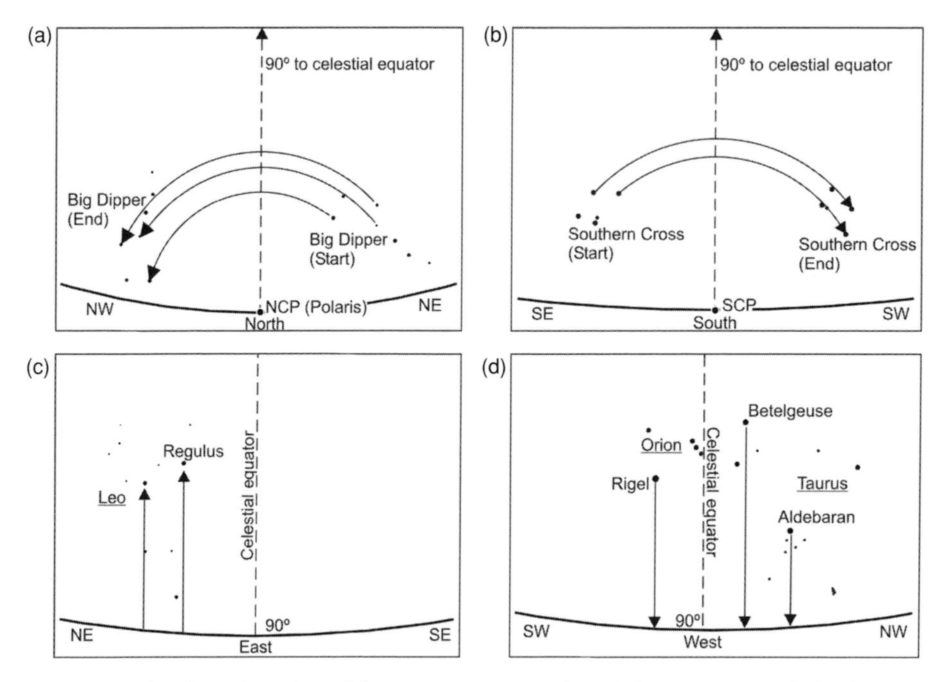

Figure 4.5 The diurnal motion of the stars as seen at the Earth's equator, 0° latitude. The angles shown between the celestial equator and the eastern and western horizons are true only for locations on the Earth's equator. (a) In the northern sky, there are no circumpolar stars, because the NCP is located directly on the northern horizon. The celestial equator is 90° from the northern horizon because at the Earth's equator the celestial equator passes through the zenith. Notice the Big Dipper revolves around the NCP counterclockwise. (b) In the southern sky, there are no circumpolar stars, because the SCP is located directly on the southern horizon. The celestial equator is 90° from the southern horizon. Notice the Southern Cross revolves around the SCP clockwise. Because both celestial poles are on the horizon, all sky objects are observable at the Earth's equator. (c) The stars of Leo rise in the east on paths parallel to the celestial equator. At 0° latitude, the equator rises at an angle of 90° to the horizon. (d) The stars of Orion and Taurus set in the west along paths that parallel the celestial equator. At 0° latitude, the equator sets on the horizon at an angle of 90°.

The Earth's equator (latitude 0°)

At the Earth's equator the celestial poles are located on the horizon so there are no circumpolar stars (Figures 4.5 and 4.6). The altitude of both the NCP and SCP is zero degrees. The altitude of the celestial equator is 90° off either the northern or southern horizon. It rises straight up in the east, passes through the zenith, and sets straight down in the west. Because all sky motion is parallel to the celestial equator, the stars also rise straight up in the east and set straight down in the west. Objects near the celestial equator pass through the zenith. Orion's belt passes right next to the zenith. Stars on the northern horizon make a semi-circle moving counterclockwise and stars on the southern horizon make

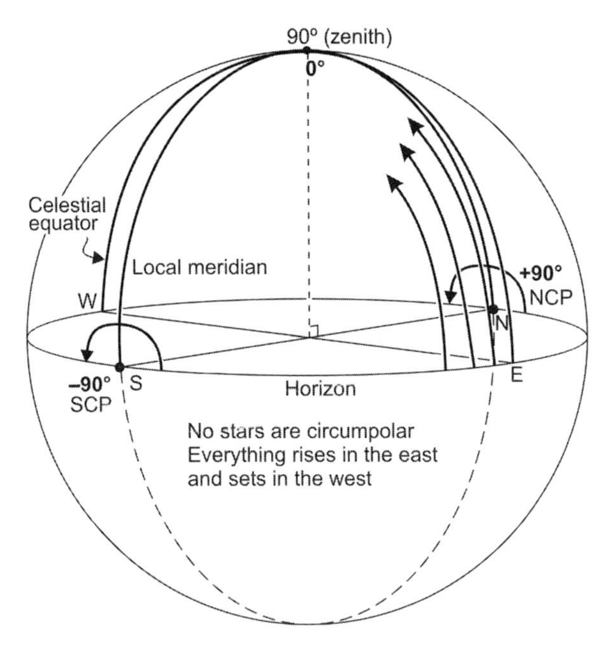

Figure 4.6 When observing from the Earth's equator, nothing is circumpolar. Everything must rise straight up in the east and set straight down in the west. The celestial equator passes through the zenith (see Section 2.5). The altitudes of important points are shown in normal print, while their equivalent declination is shown in bold.

a semi-circle moving clockwise. Given a clear horizon, every celestial object is visible for the constant 12 hours of night.

The southern hemisphere

For observers in the southern hemisphere the celestial equator passes through the northern sky (Figures 4.7 and 4.8). Thus the altitude of the celestial equator is measured off the northern horizon. To a person who lives in the northern hemisphere, the constellations are upside-down. Indeed, a person living in the southern hemisphere usually has to get used to maps printed in the northern hemisphere and thus has to look at the printed maps upside-down.

When facing south (Figure 4.7b), the stars to the left (east) of the SCP are rising higher in altitude, the stars to the right (west) of the SCP are moving to lower altitudes. Circumpolar stars (and constellations) move clockwise around the SCP. For southern latitudes of South America, Africa, or Australia, the circumpolar constellations include Octans, Mensa, Hydrus, and Crux. The large and small Magellanic Clouds can easily be seen as circumpolar objects.

An object is circumpolar if the following condition is met:

$$\text{Object declination} < \text{Altitude of SCP} - 90^\circ$$

or, in symbols,

$$\delta_{\text{Object}} < A_{\text{SCP}} - 90^\circ,$$

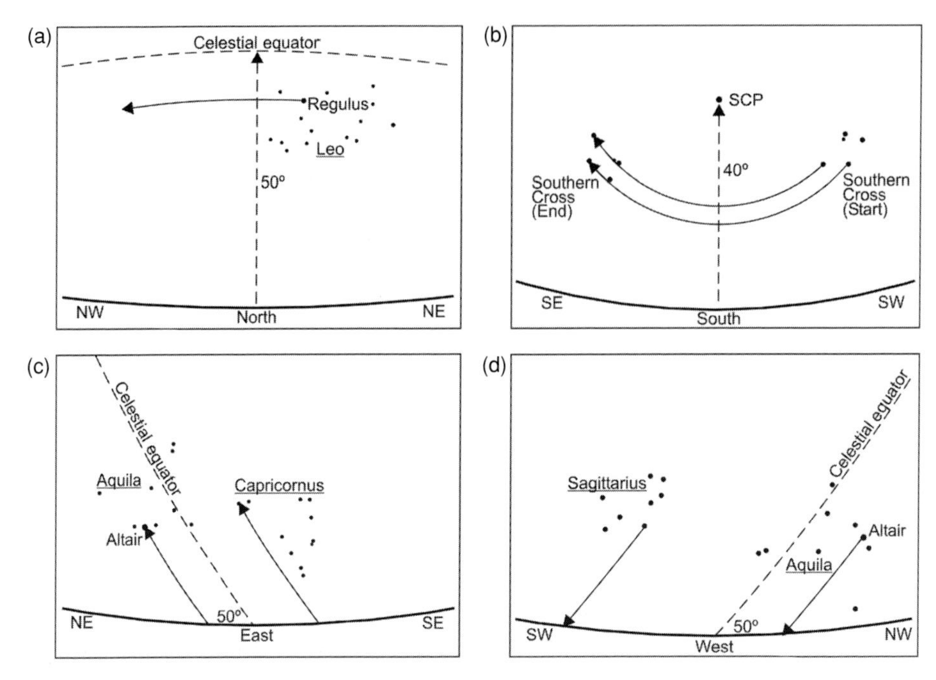

Figure 4.7 The diurnal motion of the stars as seen at 40° S latitude. The angles shown between the celestial equator and the eastern and western horizons are true only for locations with latitude of 40° S. (a) In the northern sky, the NCP is 40° below the horizon. The celestial equator is 50° above the horizon. Any object with declination above 50° ($\delta > +50°$) is never visible. Polaris and the Little Dipper are never seen. (b) In the southern sky, the south celestial pole is 40° from the horizon. The stars of the Southern Cross are circumpolar. Notice the Southern Cross revolves around the SCP clockwise. (c) The stars of Aquila and Capricornus rise in the east on paths parallel to the celestial equator. The left arrow is pointing to the bright star Altair. (d) The stars of Aquila and Sagittarius set in the west along paths that parallel the celestial equator. The arrow on the right is the path of the bright star Altair.

where δ is the object's declination (p. 25). Note that A_{SCP} is a positive number. This conditional test works only in the southern hemisphere.

> *In the southern hemisphere, an object is circumpolar at the observer's latitude if its declination is lower than the altitude of the south celestial pole minus 90°.*

Recall that the altitude of the south celestial pole is equal to the observer's latitude. The nearest star to the SCP is a very dim star labeled σ (Sigma) Octantis.

The declination of the stars passing through the zenith is equal to the observer's latitude. At 40° S, the declination of the stars that pass directly overhead is −40°. Some objects that pass very nearly overhead at this latitude are Lupus, Corona Australis, Ankaa, and Centaurus. At 40° S, the celestial equator's altitude is 50° above the northern horizon. Any object with declination above +50° cannot be seen at anytime of the year.

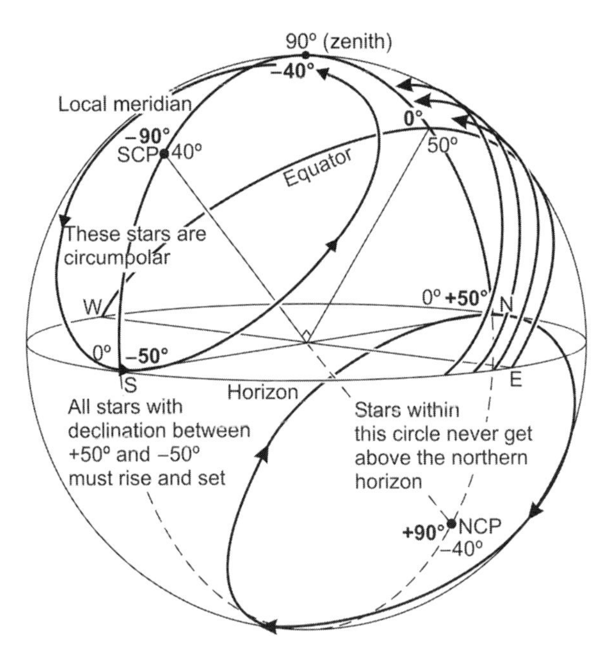

Figure 4.8 Apparent motion of the celestial sphere at 40° S latitude. At latitude 40° S, all stars above +50° declination never come above the northern horizon and all stars below −50° declination are circumpolar. All stars between these declinations must rise in the east and set in the west. These declination values are given by the altitude of the celestial equator at the given latitude.

In the southern hemisphere, any object whose declination is greater than the altitude of the celestial equator is never visible to the observer.

At the poles

When observing from the North or South Pole the respective celestial pole is at the zenith (Figure 4.9). This means the celestial equator matches the horizon. Because all diurnal motion is parallel to the celestial equator, the stars move along almucantars. Everything is circumpolar, no stars rise or set. The Sun, Moon, and planets rise and set because they move along or near to the inclined ecliptic line – not because of diurnal motion. At the North Pole, the stars move from left to right. At the South Pole, they move right to left.

4.2 The culminations of an object

Converting from equatorial coordinates to horizon coordinates requires the use of spherical trigonometry – not a trivial process. But there is one case where finding the altitude of an object is simple – when the object is crossing the local meridian – the object's *culminations*. This section discusses this case.

Recall that we use the word *meridian* to indicate the combination of the local meridian and the antimeridian – the entire great circle. The local meridian and

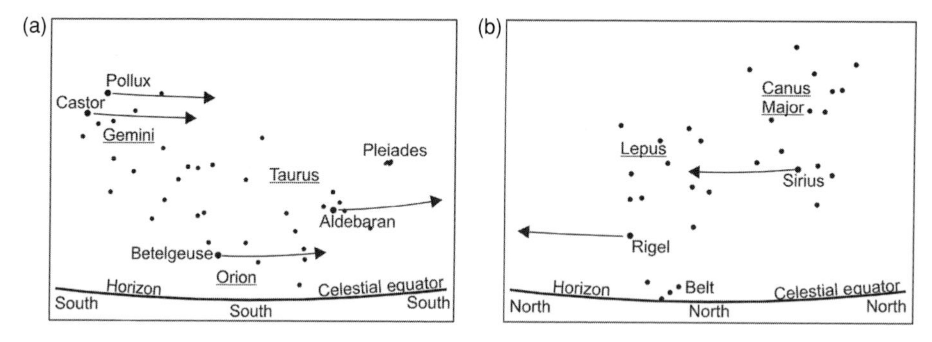

Figure 4.9 The diurnal motion of the stars as seen at the Earth's poles. At either pole, the altitude of the celestial equator is 0°, i.e. it matches the horizon and all motion is parallel to the horizon. Everything is circumpolar – nothing rises or sets (except the Sun, Moon, and planets). (a) At the North Pole only the upper half of Orion is visible, moving along the horizon from left to right. (b) At the South Pole only the lower half of Orion is visible, moving along the horizon from right to left. Notice the rotation is the opposite of the North Pole.

the antimeridian are created by dividing the meridian in half with the horizon line. There is another way to divide the meridian in half, using the celestial poles or the polar axis. With this division, the **upper meridian** is the half containing the zenith and the **lower meridian** the half containing the nadir (Figure 4.10).

All objects must transit the meridian twice each day – 12 hours apart. These transits are called the *upper culmination* (above the pole) and the *lower culmination* (below the pole). The **upper culmination** is the object's altitude when it crosses the upper meridian. The **lower culmination** is the object's altitude when it crosses the lower meridian. Here, we will concern ourselves with only the visible transits – those above the horizon.

For non-circumpolar objects only the upper culmination is above the horizon and we simply refer to their *transit altitude*. For the circumpolar objects both culminations are visible. We discuss both culminations, which are sometimes referred to as the *upper transit* and the *lower transit* altitude. Neither culmination is visible for those objects which are always below the horizon. Appendix A discusses upper and lower culminations in more detail, including the culminations which are not visible.

Given the declination of an object (Figure 4.11), calculating its culminations is easy. We have already discussed how to figure out the altitude of the celestial equator on the local meridian (see the formula on p. 28). Once this is known, finding an object's transit altitude is simply an addition or subtraction problem.

It is important to note that these procedures are used for the culminations of any celestial object, including those of the Solar System.

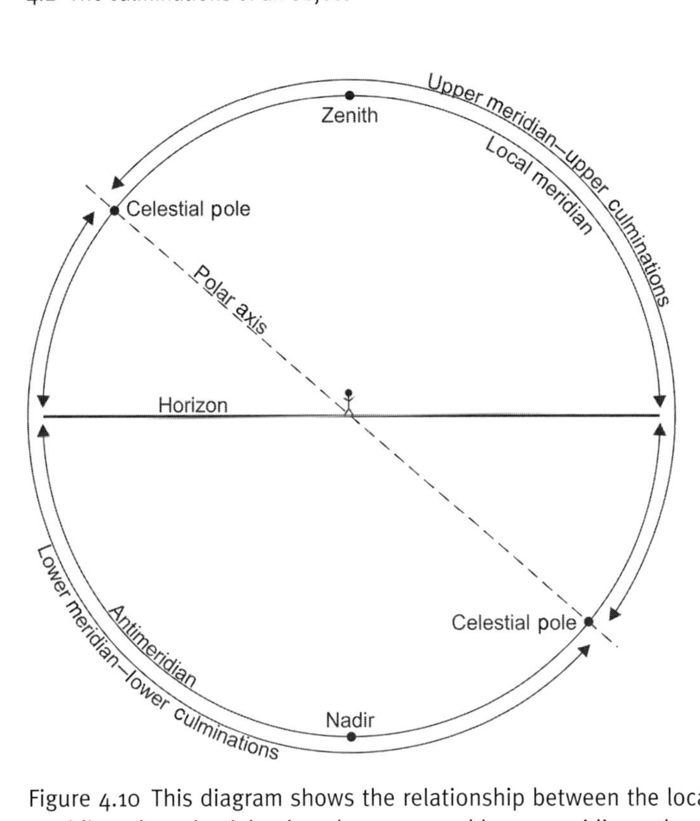

Figure 4.10 This diagram shows the relationship between the local meridian, the antimeridian, the celestial poles, the upper and lower meridians, the zenith, the nadir, and the upper and lower culminations. The meridian is the entire great circle. The celestial poles are not labeled with north or south so the diagram works for either hemisphere.

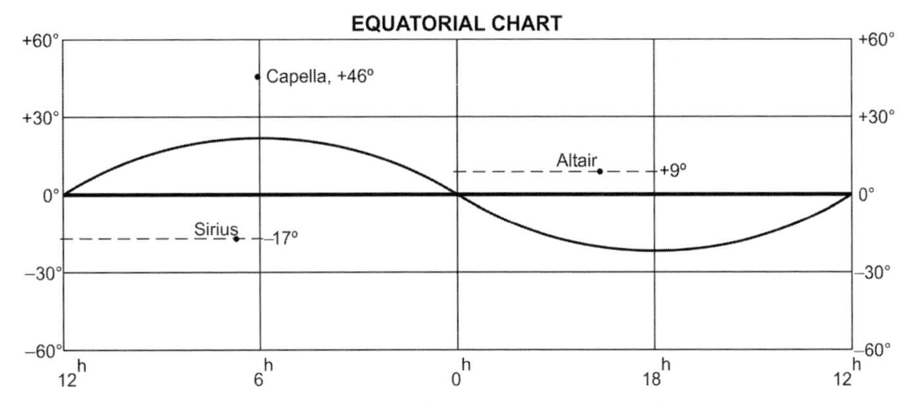

Figure 4.11 This drawing shows the declination of a few example objects from the *Equatorial Chart*. The culminations of an object is determined from its declination. Add the object's declination to the transit altitude of the celestial equator. The transit altitude of the celestial equator depends on the observer's latitude (Figures 2.23 through 2.26).

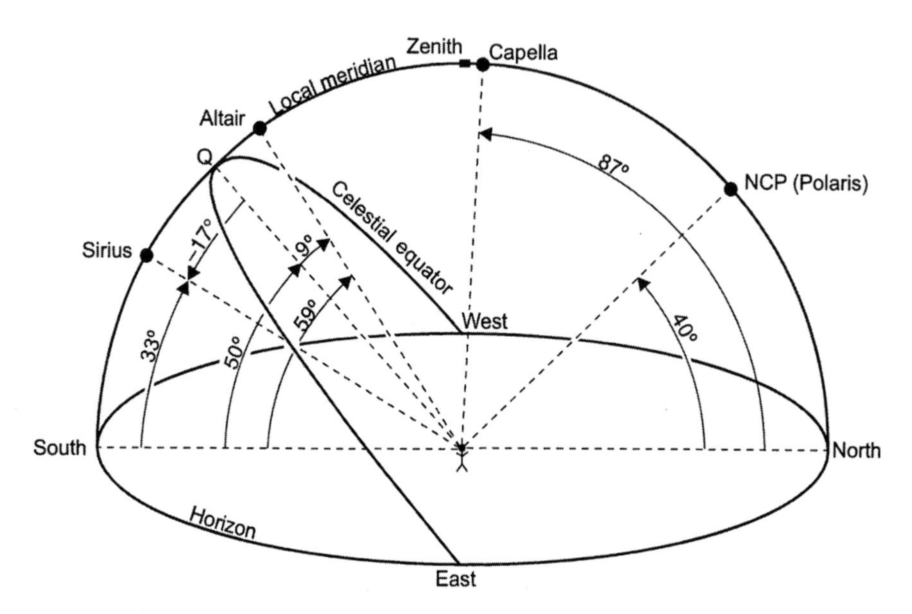

Figure 4.12 This figure shows the transit altitudes of Altair and Sirius as seen at 40° N. Note that it is not intended to imply that Altair and Sirius transit the local meridian at the same time. They transit according to their right ascension coordinate. They are shown together simply for example and comparison.

Northern hemisphere

At 40° N, the altitude of the celestial equator on the local meridian is 50° from the southern horizon (Figure 4.12).

Example: Altair has a declination coordinate of +9° (Figure 4.11). To find the transit altitude of Altair, simply add its declination to the altitude of the celestial equator. Thus, Altair crosses the local meridian at $50° + 9° = 59°$ from the southern horizon.

Example: Sirius has a declination coordinate of −17°. Therefore, it transits the local meridian at altitude $50° - 17° = 33°$ from the southern horizon.

Example: Capella is at declination +46°. This gives a transit altitude of $50° + 46° = 96°$. However, because altitudes are never greater than 90°, this result tells us that Capella transits the local meridian at $180° - 96° = 84°$ off the *northern* horizon.

In general then, for the northern hemisphere,

Transit altitude = celestial equator altitude + object declination

or

$$A_{\mathrm{T}} = A_{\mathrm{CE}} + \delta_{\mathrm{Object}}.$$

Be careful not to drop the declination sign. The addition and subtraction operations in this section are algebraic. That is, if the declination is negative, this

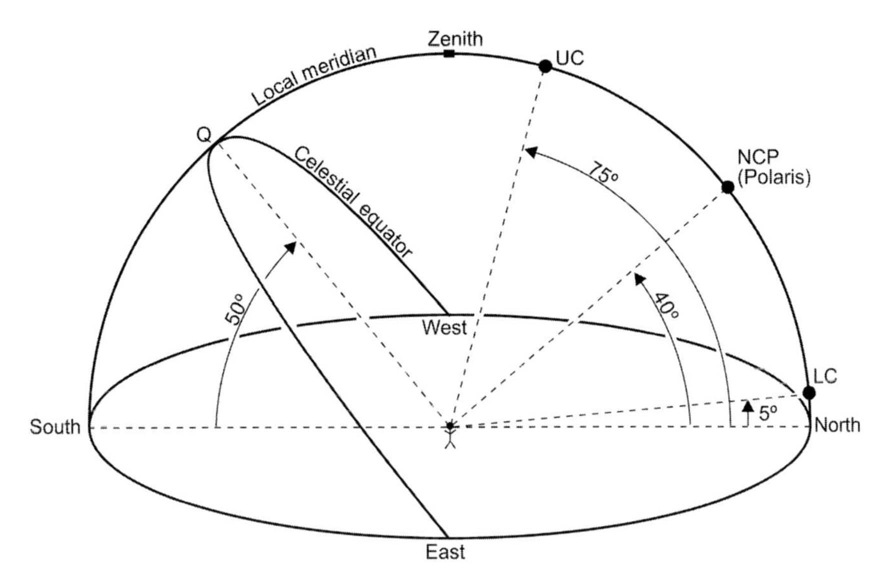

Figure 4.13 The upper (UC) and lower culmination (LC) of Mizar is shown for 40° N latitude. Below 35° N latitude, Mizar is not circumpolar and would not have a visible lower culmination. At latitudes higher than 40° N, the culminations move toward the zenith with the NCP.

becomes a subtraction problem. The formula for the altitude of the celestial equator on the local meridian, A_{CE}, is given on p. 28.

To find the transit altitude of an object, add its declination to the transit altitude of the celestial equator. The resulting altitude is measured from the southern horizon. If the result from this formula exceeds 90° then subtract the result from 180° and use this new value as the transit altitude off the northern horizon.

Circumpolar objects Because circumpolar objects never go below the horizon, they transit the local meridian twice a day. An object's lower culmination occurs when it transits the local meridian below the pole, or between the pole and the horizon. The upper culmination is when the object transits the local meridian at an altitude greater than the pole's altitude (Figures 4.10 and 4.13). The altitude of the upper culmination is given by

$$\text{Upper culmination} = \text{pole altitude} + (90^\circ - \text{object declination})$$

or

$$A_{UC} = A_{NCP} + (90^\circ - \delta_{Object}).$$

The lower culmination is given by

$$\text{Lower culmination} = \text{pole altitude} - (90^\circ - \text{object declination})$$

or

$$A_{LC} = A_{NCP} - (90^\circ - \delta_{Object}).$$

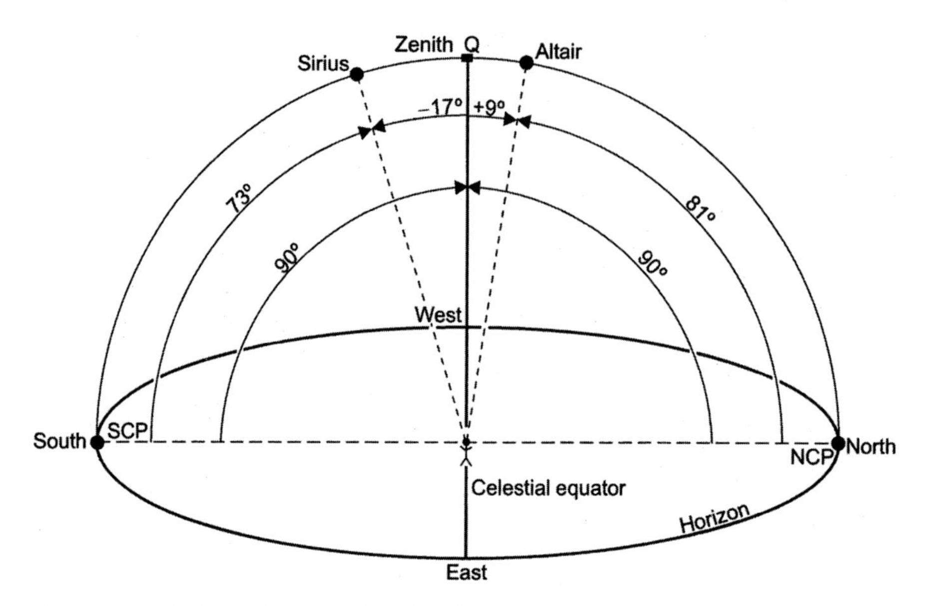

Figure 4.14 This figure shows the transit altitudes of Altair and Sirius as seen at 0° latitude. Because Sirius has a negative declination, its altitude is measured off the southern horizon. Altair has a positive declination, so its altitude is measured off the northern horizon.

For both culminations, the altitude is measured off the northern horizon. If the upper culmination is greater than 90°, subtract it from 180° and measure the upper culmination from the southern horizon.

Example: The declination of Mizar, the middle star in the handle of the Big Dipper, is +55°. At 40° N latitude, the altitude of the NCP is 40°. Using the circumpolar test from p. 62, we see that $+55^\circ > 90^\circ - 40^\circ$ and, thus, Mizar is a circumpolar star. Its upper culmination is

$$40^\circ + (90^\circ - 55^\circ) = 75^\circ$$

and its lower culmination is

$$40^\circ - (90^\circ - 55^\circ) = 5^\circ,$$

where both altitudes are measured off the northern horizon.

At the Earth's equator

The situation at the Earth's equator becomes a strange looking special case because the celestial equator passes through the zenith so the altitude of the celestial equator is 90° (Figure 4.14). The calculation becomes

$$\text{Transit altitude} = 90^\circ - |\,\text{object declination}\,|$$

or

$$A_\text{T} = 90^\circ - |\,\delta_\text{Object}\,|,$$

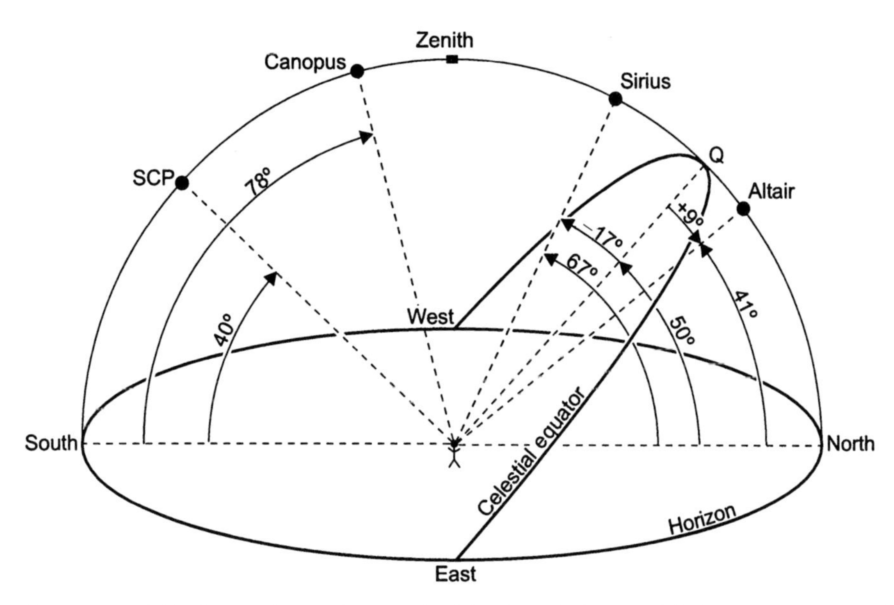

Figure 4.15 This figure shows the transit altitudes of Altair and Sirius as seen at 40° S.

where the vertical bars mean absolute value. Ignore the positive or negative sign on the declination while calculating and always subtract the value from the altitude of the celestial equator. The sign of the declination is then used to decide from which horizon the transit altitude is measured. For positive declinations, the transit altitude is measured from the northern horizon. For negative declinations it is measured from the southern horizon.

Example: For Altair, the transit altitude is $90^\circ - |+9^\circ| = 90^\circ - 9^\circ = 81^\circ$. Because Altair's declination is positive, the transit altitude is measured off the northern horizon.

Example: For Sirius we have, $90^\circ - |-17^\circ| = 90^\circ - 17^\circ = 73^\circ$. Because Sirius has negative declination, its transit altitude is measured off the southern horizon.

At the Earth's equator, the transit altitude of an object is equal to 90° minus the absolute value of the object's declination. If the declination is positive the transit altitude is measured from the northern horizon. If it is negative the altitude is measured from the southern horizon.

Southern hemisphere

Finding the transit altitude of an object in the southern hemisphere is a similar process to that of northern latitudes, but the declination is now subtracted from the altitude of the celestial equator rather than added (Figure 4.15). The equation becomes,

$$\text{Transit altitude} = \text{celestial equator altitude} - \text{object declination}$$

or

$$A_T = A_{CE} - \delta_{Object},$$

and the result is the altitude of the object off the northern horizon. If the result is greater than 90°, subtract it from 180° and measure the altitude from the southern horizon. At 40° S latitude the altitude of the celestial equator is 50° off the northern horizon.

Example: Altair, with $\delta = +9°$, transits at $50° - (+9°) = 41°$ from the northern horizon.

Example: Sirius, with $\delta = -17°$, transits at $50° - (-17°) = 67°$ from the northern horizon.

Example: Canopus has $\delta = -52°$. Its transit altitude is $50° - (-52°) = 102°$, which is greater than 90°. Thus, Canopus transits at $180° - 102° = 78°$ off the southern horizon.

To find the transit altitude of any object, subtract its declination from the transit altitude of the celestial equator. The resulting altitude is measured from the northern horizon. If the result from this formula exceeds 90° then subtract the result from 180° and use this new value as the transit altitude off the southern horizon.

Circumpolar objects For southern hemisphere observers, the upper and lower culminations of the circumpolar objects are found in a similar way as for the northern hemisphere observers. To use these formulas, it is important to remember that the declination probably has a negative value. Thus, when it says to add the declination, it is actually a subtraction problem. The altitude of the upper culmination is given by

$$\text{Upper culmination} = \text{pole altitude} + (90° + \text{object declination})$$

or

$$A_{UC} = A_{SCP} + (90° + \delta_{Object}).$$

The lower culmination is given by

$$\text{Lower culmination} = \text{pole altitude} - (90° + \text{object declination})$$

or

$$A_{LC} = A_{SCP} - (90° + \delta_{Object}).$$

For both culminations, the altitude is measured off the southern horizon. If the upper culmination exceeds 90° then subtract it from 180° and measure this upper culmination from the northern horizon.

Example: The declination of α Centauri is $-62°$. At 40° S latitude, α Centauri is a circumpolar star because $-62° < 90° - 40°$. Alpha Centauri's upper culmination

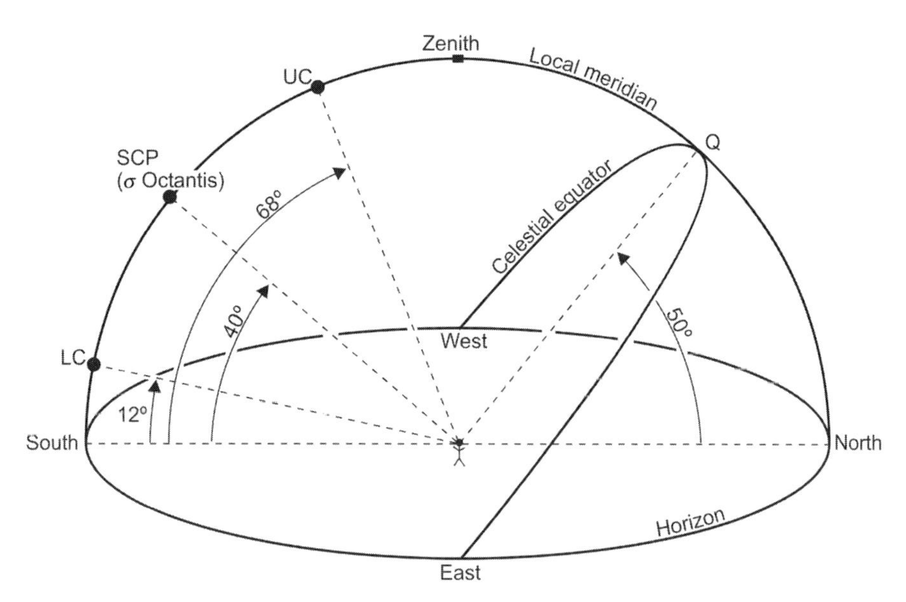

Figure 4.16 The upper (UC) and lower culmination (LC) of α Centauri at 40° S latitude. At latitudes higher than 38° S, α Centauri is no longer circumpolar and the lower culmination is not visible, leaving only the transit altitude (upper culmination) of α Centauri to think about. At greater southern latitudes, the culmination points move toward the zenith.

is then (Figure 4.16)

$$40^\circ + (90^\circ - 62^\circ) = 68^\circ.$$

Its lower culmination is

$$40^\circ - (90^\circ - 62^\circ) = 12^\circ,$$

with both altitudes measured from the southern horizon.

When further south in latitude it is possible for the upper culmination to exceed 90°. In this case we subtract the result from 180° and take the altitude measurement off the northern horizon.

Transit time

The right ascension of the stars Altair and Sirius are $19^h\,50'$ and $6^h\,45'$ respectively. Thus, when Altair transits the local meridian, the sidereal clock reads 19 : 50 hours. When Sirius transits the local meridian ten hours and 55 minutes later, the sidereal clock reads 06 : 45 hours. The reading on the solar clock during these events depends on the time zone and date.

Because an observer can make a reasonable estimate of the current sidereal time (p. 33) and a sidereal hour is nearly the same as a solar hour (p. 97), an observer can also make a reasonable estimate of the transit time of an object.

Example: The right ascension of Sirius is $6^h\ 45'$. If Orion's Belt is currently near the local meridian, then the current sidereal time is roughly $5^h\ 30'$. This tells us that Sirius will transit in about 1 hour and 15 minutes.

Note however, that because we can see about six hours in right ascension on either side of the local meridian (at least along the celestial equator), we can see Sirius well before it actually reaches the local meridian. The exception to this statement would be for the far northern latitudes where Sirius' culmination is only a few degrees above the horizon.

4.3 The annual motion of the sky

The **annual motion** of the sky is caused by the yearly revolution of the Earth around the Sun. To **revolve** means to orbit or circle a central point or object. The Earth revolves around the Sun once every 365 days, while it rotates on its axis once per day. The number of days has been rounded off for the purposes of the current discussion. Generally it is quoted as 365.25 days and, actually, even this number is not quite correct. A discussion of the number of days in a year starts on p. 164, where of course we must be careful of the definition of both *year* and *day*. The Earth's annual motion causes the Sun to appear to move against the starry background and results in different constellations being seen during each season.

The seasonal migration of the constellations is the same east to west motion as the diurnal motion, but it takes a year instead of a day. As we view the stars at the same (solar) time each night, say 22:00 hours (10:00 p.m. in some countries), we see the stars gradually move from east to west. Thus, different constellations adorn the skies during each season. For observers in the northern hemisphere, Leo (the Lion) roars across the sky during spring. Scorpius (the Scorpion) stings the southern horizon during summer. Pegasus (the Winged Horse) flies high into the sky during autumn and Orion (the Hunter) seeks his prey during the winter. For observers in the southern hemisphere, these constellations appear in the opposite seasons. Leo in the autumn, Scorpius in the winter (along the northern horizon), Pegasus in the spring, and Orion in the summer.

The circumpolar stars, like those of the Big Dipper in the northern hemisphere, are visible every night, all year. The Greek poet Homer wrote

> ...the Big Bear never bathes...

passing along the seafarer's knowledge that Ursa Major never dips below the watery Mediterranean horizon. Because the Big Dipper is a familiar circumpolar asterism for most observers, let's use it to study the yearly motion of the stars. For observers in the southern hemisphere this discussion can be applied to Crux, the Southern Cross, a very familiar constellation.

Each day, the Earth rotates on its axis, bringing the Big Dipper back to the same position in the sky night after night. However, while the Earth rotates

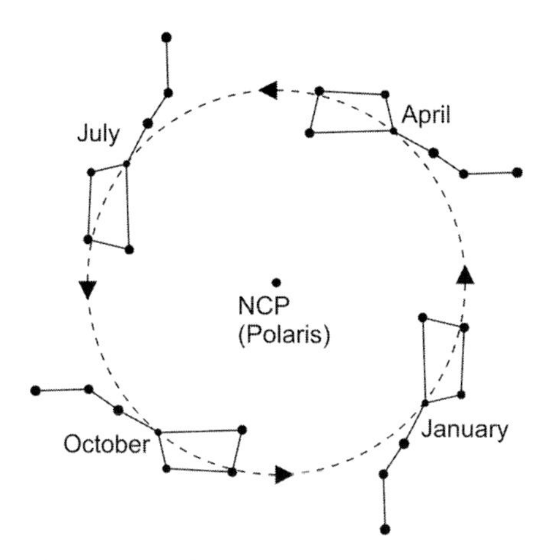

Figure 4.17 The annual motion of the Big Dipper. This is the approximate position of the Big Dipper on the first day of the months shown at 22 : 00. The annual motion of the Big Dipper is caused by the Earth revolving around (orbiting) the Sun.

on its axis it also revolves around the Sun. After one Earth rotation any given star returns to its original sky position about four minutes sooner than the Sun returns to the local meridian. There are 365 days in a year and 360° in the Earth's orbit around the Sun. The result is a change in the position of the Big Dipper (from night to night) of about 1° per night. From 22:00 one night to 22:00 the next night (solar time), the Big Dipper revolves 361° around Polaris.

The difference in length of sidereal and solar days is about four minutes. Four minutes is $\frac{1}{360}$ of 1440 minutes (24 hours). Each night the stars rise about four minutes earlier than the previous night. However, the Earth is in a slightly elliptical orbit around the Sun so its orbital speed is not constant, while its rotational speed is very nearly constant. This means the difference in the length of sidereal and solar days is not constant, contributing to an effect called the "equation of time." (In ancient astronomy the word *equation* meant "difference.") The equation of time is discussed on p. 94.

Let us review what has just been said. We observe the position of the Big Dipper each night at 22:00 (Figure 4.17). Remember, the clock is running according to the Sun's daily motion. When we look at the Big Dipper tonight, it has moved 1° more than a complete 360° circle from last night, because it returned to last night's position four minutes before 22:00. On each successive night, the Big Dipper moves one degree farther than a complete circle along its counterclockwise path around the NCP (Polaris). The one degree per day is an average over the 365 days of the year.

If we wait for longer periods the change in position is more dramatic. Look at the Big Dipper on 1 January and note its position. Then look again on 15 January. The Big Dipper is about 15° farther along its circumpolar trail. Look again at the end of January and it is about 30° farther along. After 90 days – about one season – it has moved (about) 90° around its circumpolar path from its original

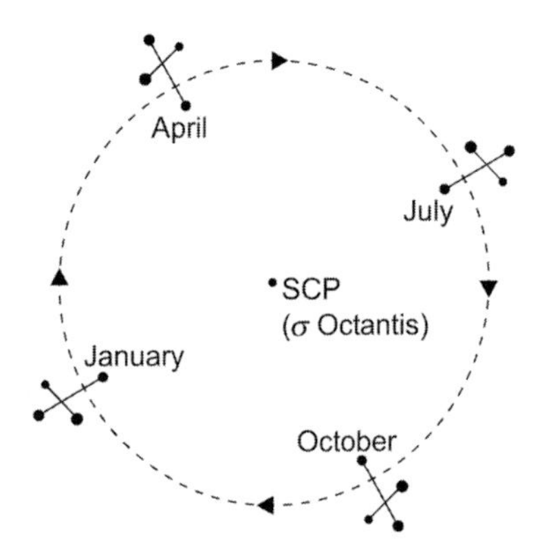

Figure 4.18 The annual motion of the Southern Cross is clockwise around the south celestial pole. This is the approximate position of the Southern Cross on the first day of the months shown at 22 : 00 hours.

position on 1 January. Remember, the motion is counterclockwise and we are viewing the Big Dipper at exactly 22 : 00 solar time each night.

We have just seen that the Big Dipper changes its position around Polaris about 90° per season. Figure 4.17 is a sketch of a multiple exposure photograph of the Big Dipper at the beginning of each season. The dashed circle shows the Big Dipper's counterclockwise circumpolar path around the NCP. Compare Figure 4.17 to Figure 4.2a. Notice the daily and annual motions of the circumpolar stars appear the same. The extra 1° per day builds up during the year to an extra revolution around Polaris. The circumpolar stars revolve around Polaris once each day and once each year. That is, they revolve around Polaris 366 times (367 times in a leap year) each year.

In the southern hemisphere, these same arguments can be applied to the Southern Cross as it circles the south celestial pole. Each day the Southern Cross revolves clockwise around the pole 361°, creating the same seasonal changes to its position as is observed for the Big Dipper in its motion. Compare Figure 4.18 to Figure 4.7b. Also compare the motions depicted in Figures 4.2a, 4.4, and 4.17 for the northern hemisphere with Figures 4.7b, 4.8, and 4.18 for the southern hemisphere.

Because of the orbital motion of the Earth, the sky revolves around the celestial poles approximately 361° per 24 solar hours, giving the sky a slow annual motion of one degree per day from east to west.

4.4 The apparent motion of the Sun

The Earth is 12 756 km (7926 miles) in diameter. The Sun is 1 391 980 km (864 936 miles) in diameter. Given these numbers, the Sun has about 109 times the diameter and about 1 300 000 times the volume of the Earth. If the Sun were the size of a basketball, the Earth would be the size of a pinhead. Although

Figure 4.19 An American penny shown actual size.

the Sun is much larger than the Earth, the Sun is so far away (149 597 870 km or 92 955 807 miles), it appears to us to be about 0°.5 in diameter. This is the apparent size of an American penny (Figure 4.19) at a distance of 204 cm (80.4 inches). It is also about the size of the full Moon which, as we see in Section 7.4, causes some spectacular sights. From Earth, the Sun appears small enough to be blotted out with the tip of your finger.

As far as we can determine, the Sun has been burning its fuel for about five billion years, and will continue to burn for another five billion. During its lifetime, it has slowly increased in size, and energy output, while remaining fairly constant in surface temperature. The Sun is yellow in color and circular (spherical) in shape. It is rotating once per month and therefore it is slightly flattened at the poles, just like the Earth. In this section, we shall look only at the apparent motion of the Sun (caused by the motions of the Earth) as we observe it in our skies.

The Earth rotates on its axis while it orbits the Sun. Making this statement might have cost you your life 400 years ago. (Partly for his support of Copernican ideas, Giordano Bruno was burned at the stake in the Vatican Square in 1600. There is now a statue of him at the place where he was burned.) Today, we take it for granted. These two Earth motions combine to create the sky motion described in the previous sections of this chapter. They also create the apparent motion of the Sun – its diurnal motion and its annual motion. The annual motion has both a west–east component and a north–south component. Only the west–east component is discussed in this chapter. The north–south component is discussed in Chapter 5, "The seasons."

The diurnal motion of the Sun

The diurnal motion of the Sun is the same as that of the stars or any other celestial object. The Sun rises in the east and sets in the west every day. (However, the Sun may become circumpolar for latitudes near the Earth's poles during some part of the year.) The time of day we may call **high noon** occurs when the Sun is on the local meridian. However, because of time zones and politics, high noon probably does not occur at 12 : 00 hours. Depending on your location (both longitude and latitude) within the time zone, and whether *Daylight Savings Time* (DST) is in effect, the wall (solar) clock may read anywhere from 11 : 20 to 13 : 40 hours (these are estimates) when it is your high noon. Daylight Savings Time is simply a social preference and implemented only on a local basis.

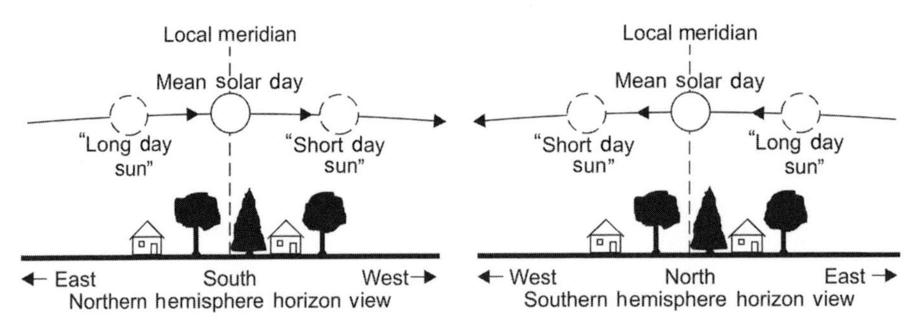

Figure 4.20 The Sun rises in the east and sets in the west once per day – its diurnal motion. The amount of time for successive local meridian transits is one solar day. The Earth does not have a constant speed as it orbits the Sun while its rotational speed is nearly constant. Thus, the amount of time between transits is slightly different from day to day.

Some of the wide variation in the time of high noon is due to the political boundaries of the time zones. In theory, the time zones are each 15° wide in longitude. In practice, there is quite a variation in time zone width and this contributes to the variation in the apparent time of high noon. There is further discussion of this topic on p. 93.

The time it takes for the Sun to move from the local (or time zone) meridian, "around the Earth" and back to the local meridian is one **solar day**.

Due to variations in the speed of the Earth's motion around the Sun (described by Kepler's laws of planetary motion), there is about a 30 minute variation in the time of the Sun's transit of the local meridian during the year. This variation is in addition to the constant deviation caused by location within a time zone and occurs even if one ignores time zones. It also brings us back to the equation of time, mentioned previously and discussed on p. 94. Averaging the length of the day through the year allows us to create a time interval called the **mean solar day**.

When the Earth is closer to the Sun than average (perihelion) it moves faster in its orbit, passing through a larger than average orbital angle. Therefore, the Earth must rotate more than average (established by the mean solar day) to bring the Sun back to the meridian. That is, the Sun returns to the local meridian later than average. The Sun appears to be behind its average position. This is the "long day sun" shown in Figure 4.20. When the Earth is farther from the Sun (aphelion) and thus moving slower than average, the Earth passes through a smaller than average orbital angle. The Earth does not need to rotate as far to bring the Sun back to the meridian and the Sun is ahead of its average position. This is the "short day sun." The wall clocks by which we run our lives are based on the mean solar day. A deeper discussion about the relationship between the orbital and rotational motion of the Earth starts on p. 95.

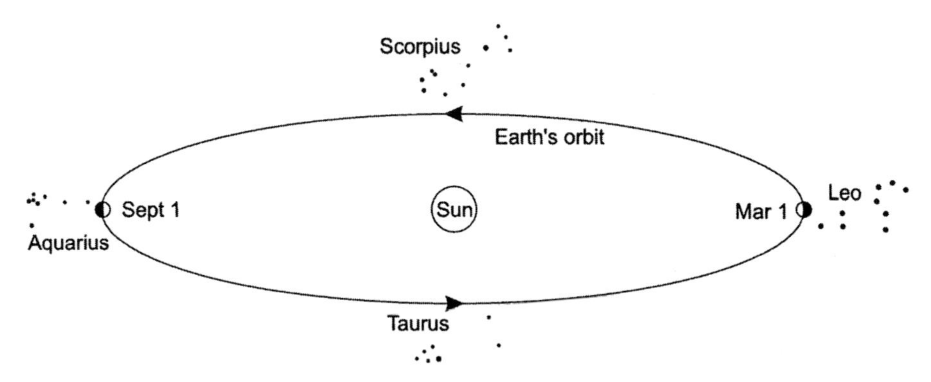

Figure 4.21 The annual motion of the Earth around the Sun causes the "seasonal migration" of the constellations. On 1 March, we can see Leo but not Aquarius. On 1 September, the Sun blocks our view of Leo but we can see Aquarius.

The wall clock's reading is based on the average time it takes for the Sun to return to the time zone's (not the local) meridian each day – the mean solar day for the time zone's meridian. Some countries use a true 24-hour clock with hours running from 0 to 23. In other countries the clock hours run from 1 to 12 and the designations, a.m. and p.m. describe the position of the Sun relative to the local (or time zone) meridian: a.m. is the abbreviation for the Latin words *ante meridiem*, meaning "before midday," p.m. is the abbreviation for *post meridiem*, "after midday."

Mean local solar time is a clock keeping time with the Sun, but based on the local meridian, ignoring time zones. Such a clock will still keep mean solar time, but the Sun will be on the *local* meridian at 12 : 00 hours at least four times per year (the zero crossings of the equation of time). **True local solar time** is kept relative to the true motion of the Sun, where the Sun is on the local meridian at high noon every day. This time is most easily kept with a sundial.

Because of the Earth's diurnal rotation, the Sun rises in the east and sets in the west each day. The length of the day is defined by the apparent motion of the Sun across the local meridian. However, because of the Earth's varying orbital speed compared with its comparatively constant rotational speed, there is a small variation in the true length of the solar day.

The annual motion of the Sun

When the Sun is within the boundaries of a constellation, that constellation (and those surrounding it) is not visible. The Sun's brilliance blocks out the dimmer stars. However, the constellation *is* seen during *some* months of the year (Figure 4.21). Due to the motion of the Earth around the Sun, the Sun appears to move through the constellations with a regular period. This is what we mean by the Sun's **annual motion** or *yearly* motion.

On 1 September, the Sun blocks our view of Leo; but at night, we can see Aquarius. We say, "The Sun is in Leo." On 1 March, the Sun blocks our view

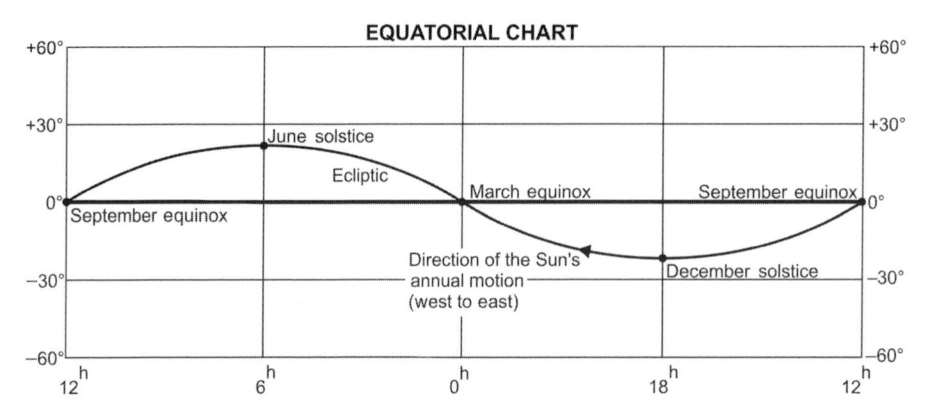

Figure 4.22 The ecliptic is the path of the Sun through the zodiac constellations. The positions of the equinoxes and solstices are shown along the ecliptic line.

of Aquarius, but we can see Leo. We say, "The Sun is in Aquarius." So, when the Sun – or any other Solar System object – is "in a constellation," it simply means that as seen from the Earth, the Sun (or other object) is in front of the stars forming that constellation. Although we know it is the Earth that is moving, it *appears to us* that the Sun moves through the constellations.

The path the Sun follows on its yearly journey through the constellations is called the **ecliptic line** or, simply, the *ecliptic*. This line is so named, because all solar and lunar eclipses occur at points on this line. The Greek word, *έκλείπο* (ekleipo) means "to leave out." Eclipses are covered in Chapter 7.

The ecliptic represents the apparent path of the Sun against the background stars. The Sun's position along the ecliptic changes by about 1° per day from west to east. This is the opposite of its east to west diurnal motion and also opposite to the annual motion of the stars.

The opposite motion of the Sun and stars is explained on p. 100.

Look at your *Equatorial Chart* and Figure 4.22. At the center of the chart is one of two intersections of the ecliptic and the celestial equator. This intersection *defines* the **vernal equinox**. (In order to make the charts equally usable in both the northern and southern hemispheres, the vernal equinox is also labeled the March equinox.) The vernal equinox is the point in the sky where the Sun crosses the celestial equator, from south to north, during its annual motion. Notice the date on the ecliptic line next to the vernal equinox is 21 March. The date of the vernal equinox changes slightly due to adjustments to the calendar, such as leap year. "Vernal" is from the Latin word *ver* for "spring." Equinox comes from the Latin words *aequabilis* for "equal" and *nox* for "night." The Latin word for equinox is *aequinoctium*. In the northern hemisphere, the vernal equinox is also called the *spring equinox*. In the southern hemisphere, it is considered the *fall equinox*.

The other intersection, at the left (or right) edge of the chart, is the September equinox. It is called the *autumnal equinox*, or *fall equinox* in the northern hemisphere and the *spring equinox* in the southern hemisphere. The date here

Figure 4.23 The symbols for the medical sciences usually contain one or two serpents. The zodiac constellation Ophiuchus represents "The Serpent Handler" or "The Doctor."

is 21 September. Between the equinoxes are the (summer and winter) solstices. The solstices are the points where the Sun's north–south annual motion reverses direction. These important solar positions are fully discussed in Section 5.1.

There is a great circle on the celestial sphere, called the **equinoctial colure**. This circle contains both equinox points. The half of the equinoctial colure containing the March equinox is the zero-hour line of right ascension (Figure 2.21). The half containing the September equinox is the 12^h line. There is a second great circle containing the solstice points, called the **solstitial colure**.

> *The vernal equinox (a name that we shall consider independent of the observer's hemisphere) is the sky point where the ecliptic passes through the celestial equator with the Sun moving from negative declination to positive. It is the origin of the equatorial coordinate system.*

The ecliptic passes through a set of constellations called the **zodiac constellations**. In order from the vernal equinox they are: Pisces, the Two Fish; Aries, the Ram; Taurus, the Bull; Gemini, the Twins; Cancer, the Crab; Leo, the Lion; Virgo, the Virgin (or the Goddess of Justice); Libra, the Scales; Scorpius, the Scorpion; Ophiuchus, the Serpent Handler; Sagittarius, the Archer; Capricornus, the Sea-Goat (or the Goat); and Aquarius, the Water Bearer. The Sun moves through roughly one zodiac constellation per month. Seven of the constellations represent an animal of some form. Because of this, the Greeks called these constellations *Zōιδιακοσ κυκλοσ* (zōidiakos kyklos) or "circle of the animals," from which we get *zodiac*.

Twelve of these 13 constellations are named for the zodiac signs of astrology. Notice the ecliptic passes through the lower part of the constellation Ophiuchus (O-fee-euw′-kus). Ophiuchus represents "The Serpent Handler" or "The Doctor." This is one possible reason why most of the symbols related to medical science contain serpents. Another possible reason is the Greek/Roman god of medicine, Asclepius (a son of Apollo) who was usually depicted holding serpents. It's also possible these reasons may themselves be related.

Table 4.1 shows the dates of the Sun's entry into both the zodiac constellations and the zodiac signs. They do not line up very well. When the Babylonians created the (western) zodiac, the astrological signs and the constellations lined

Table 4.1 *Dates for the Sun's entry into the zodiac constellations*

Zodiac sign	Astrological dates[a]	Constellation	Representation	Sun's entry date (approximate)	Days in constellation
Aries	21 Mar–19 Apr	Aries	The Ram	19 Apr	26
Taurus	20 Apr–20 May	Taurus	The Bull	15 May	37
Gemini	21 May–20 Jun	Gemini	The Twins	21 Jun	30
Cancer	21 Jun–22 Jul	Cancer	The Crab	21 Jul	21
Leo	23 Jul–22 Aug	Leo	The Lion	11 Aug	37
Virgo	23 Aug–22 Sep	Virgo	The Virgin	17 Sep	45
Libra	23 Sep–22 Oct	Libra	The Scales	01 Nov	24
Scorpio	23 Oct–21 Nov	Scorpius	The Scorpion	25 Nov	06
*	*	Ophiuchus	The Serpent Handler	30 Nov	18
Sagittarius	22 Nov–21 Dec	Sagittarius	The Archer	18 Dec	33
Capricorn	22 Dec–19 Jan	Capricornus	The Goat	20 Jan	28
Aquarius	20 Jan–18 Feb	Aquarius	The Water Bearer	17 Feb	23
Pisces	19 Feb–20 Mar	Pisces	The Two Fish	12 Mar	37

[a] There is some variation in all these dates due to adjustments in the calendar, such as leap days.

up fairly well. Since then, the Earth's precessional motion (Section 4.6) has created this misalignment. This misalignment has no affect on astrology because the zodiac signs (or houses) are defined as 30° intervals along the ecliptic, starting from the vernal equinox. The constellations were simply used as guides to the position of the ecliptic intervals. Because the vernal equinox starts the ecliptic interval known as Aries, the vernal equinox is also called the **first point of Aries**.

> *The zodiac constellations are those constellations through which the ecliptic line passes. There are* 13 *zodiac constellations. The zodiac signs are twelve,* 30° *intervals along the ecliptic line as defined by astrology. The constellations and signs are misaligned because of the precessional motion of the Earth.*

4.5 Keeping time

Astronomical measurements such as right ascension coordinates are based on time, so time keeping is a major concern for astronomy. Time is ultimately the measure of all things. Atomic clocks allow us to measure time to precision on the order of 10^{-15} seconds. Because time can be measured so precisely, physicists have redefined many units of measure so they are based in time.

Measuring time is fundamental to astronomy and cosmology. When we look into deep space we are looking back in time – back into the past. This happens because it takes time for the light we are seeing to get to us from the object. This is called *look back time*. For instance, when we look at Alpha Centauri, we see that star as it appeared 4.3 years ago because it takes 4.3 years for the light from that star to get to us. (This is, of course, simply using the definition of a light year.) When observing the Andromeda galaxy (the nearest spiral type galaxy, at 2.2 million light years), we see it as it was 2.2 million years ago.

Our basic units of time (days and years) are defined in terms of the Earth's rotation and orbital motion. In the last half of the twentieth century, particularly with the development of radio astronomy, we have learned that the Earth's rotational motion is inconsistent. It has seasonal and yearly variations as well as the decay caused by the Moon's tidal forces. Today, time is defined by a number of atomic vibrations that we can count using the technology of atomic clocks. However, we still measure days and years in terms of the motion of the Earth.

Clocks

Clocks are the devices we use to keep time. Mechanical clocks were used as time standards until the mid-twentieth century when electronic devices and systems increased the precision of our measurements. By the term *precision*, we mean to say the *rate* at which a clock's error increases.

All clocks are built on the idea of counting. Some mechanical or electronic system is set into vibration with a certain, known number of vibrations per second (its *frequency*). The source of the vibrations is irrelevant, except for the consistency of the vibrational period. The clock mechanism simply counts the

Figure 4.24 These are pendulum-based transmitting clocks and chronographs operating at the time service of the US Naval Observatory near Washington, DC, about 1915. (Courtesy of the US Naval Observatory.)

number of vibrations and increments the seconds count when appropriate. Electronic clock systems with very high frequencies have become known as *frequency standards* by the clockmakers.

Pendulum clocks Through the middle of the twentieth century, standard clocks used pendulums to measure intervals of time. As discovered by Galileo, the period of a pendulum is independent of the pendulum's mass and swing amplitude (as long as the swing amplitude is small). A pendulum's period depends only on the strength of gravity and the length of the pendulum arm. However, the strength of gravity depends on the distance of the clock to the center of mass of the Earth, which (because of polar flattening) varies with latitude and also with the topology of the Earth's surface. A clock in New York runs slightly faster than a clock in Denver and slightly slower than one at the North Pole. If the clock stays at one location, this is not a concern because the pendulum swing period of a clock can be adjusted for proper speed according to its location.

The clock's pendulum arm is generally made of a metal. Metal expands and contracts with temperature, changing the length of the pendulum and thus its swing period. Rooms with elaborate temperature controls were built to house the standard pendulum clocks (Figure 4.24). The best of the pendulum clocks were precise to about one second per year.

Figure 4.25 This is the B-12 quartz crystal clock used by the USNO time service in the early 1960s. (Courtesy of the US Naval Observatory.)

Electronic clocks By the mid-twentieth century, the vibrations of quartz crystals were being used as time references. Quartz clocks were precise to about one millisecond per year (Figure 4.25). The original atomic clocks were based on the vibrations of cesium-133 (^{133}Cs) atoms (Figure 4.26). The second generation of atomic clocks is based on the vibrations of hydrogen atoms and, in either case, they are precise to about one second per million years.

Figure 4.26 The "Atomichron," one of the earlier atomic clocks, used by the time services at the USNO in 1961. (Courtesy of the US Naval Observatory.)

A technology known as *trapped ions* was introduced in the 1980s and improvements have made these clocks very stable and reliable. These clocks use the interactions of ionized mercury (or ytterbium) atoms trapped in a magnetic field with microwave radiation (at a frequency of about 40 GHz), which is compared against a master clock to create a countable time base.

Because of the new hi-tech clock devices, time is the most precisely measured physical quantity. Some standards of measure have been converted to a time basis.

Earth's rotation

Neither clocks nor the Earth are perfect mechanisms for periodic motion. The rotation rate of the Earth has many regular and irregular variations. The major regular variations have periods of one year, six months, one (lunar) month, and one-half (lunar) month. They are caused by gravitational interactions between the Earth, Moon, and Sun. Some irregular variations are caused by changes in the oceans, the atmosphere, and interactions between the Earth's core and mantle. An irregular and unpredictable variation in rotation is caused by the Earth's molten core and nearly solid mantle rotating at different rates. A regular variation in rotation is seasonal, linked to the melting and freezing of the ice caps, which in turn causes changes in atmospheric pressure and density. The Earth rotates faster in the (northern hemisphere's) fall and slower in the spring.

There is a periodic variation of 14 months caused by a small wobble (a nutation, see Section 4.7) in the Earth's rotational axis that is caused by a small displacement of the Earth's center of mass from the axis of rotation. Another small nutation is caused by the precession of the lunar nodes (p. 185). Finally, because of tidal interactions between the Moon and the Earth, the Earth's rotation is slowing down.

> *Human measurement of time has always been based in the motions of the Earth. Although we have devices that can keep time far more accurately than the Earth, we adjust our devices to keep pace with the Earth's motions.*

Measuring time Because of these rotational variations, there are many ways (more than 20) of calculating the passage of time and calculating the current time at any location on Earth. While some of these methods differ by only a few milliseconds, others may differ by hours. Most of these time scales are calculated relative to the standard metric second (the SI second), which is kept by atomic clocks. Time kept by a set of almost 200 statistically coordinated frequency standards (located at various sites around the world) all of which are counting SI seconds is called *International Atomic Time* (TAI, from French: Temps Atomique International). TAI is the basis for all general time scales.

> *One SI second is equal to* 9 192 631 770 *periods of the radiation corresponding to the transition between the two hyperfine energy levels of the outer electron in the ground state of cesium-*133 *(^{133}Cs) atoms in zero magnetic field.*

Astronomical time In 1991 the International Astronomical Union introduced new time scales with measurement units consistent with the SI second. Terrestial Time (TT) is the time reference for geocentric ephemerides (positioning Solar System objects with respect to the center of the Earth). This is equivalent to the older Terrestial Dynamical Time (TDT), which was defined in such a way as to maintain continuity with the older ephemeris time scale. Geocentric Coordinate Time (TCG) is the time scale used for a coordinate system with its origin at the center of mass of the Earth. Barycentric Coordinate Time (TCB) is the time scale used for a coordinate system with its origin at the center of mass of the Solar System. Relativistic effects cause these two time scales to differ from Terrestrial Time. They also differ from each other by about 49 seconds per century.

Planetary ephemerides are listed according to Terrestial Dynamical Time (TDT) or Barycentric Dynamical Time (TDB). In the first case the position of objects is stated with reference to the center of the Earth and in the second with reference to the center of the Solar System. TDT differs from TAI by a constant offset chosen to keep consistency with older ephemeris time.

Universal Time (UT) is time kept according to the average time of the Sun's transit of the local meridian at Greenwich, England – the Greenwich mean solar day. Universal time can be calculated from the local solar time by adding the time zone number to the local time.

Example: For Eastern Standard Time, Universal Time is equal to the local solar time plus five hours. If the local time is 14 : 35, UT is 19 : 35 hours.

Because the Sun's transit is ultimately caused by the motion of the Earth, UT includes both the erratic rotation and the effects of the orbital motion of the Earth. These effects are averaged over to make UT as uniform as possible.

UT0 is a time system based on the longitude of the observer and measured according to the diurnal motion of the stars or celestial radio sources. UT1 is calculated from UT0 and includes corrections for the Earth's erratic motions (nutations) around its pole which affect longitude. Astronomers use UT1 as the time reference for observations.

Civil time Civil time (used for recording world affairs) is also known as **Coordinated Universal Time** (UTC) – which is related to Greenwich Mean Time (GMT). GMT has been defined in multiple ways and has become ambiguous in meaning. Therefore GMT is no longer recommended for use in astronomy. In UTC the seconds are all the same length because they are kept by atomic clocks. UTC time differs from TAI time by an integer number of seconds. Humans live their lives according to the motions of the sky – the rotation of the Earth – not by the march of atomic clocks. Because the Earth's rotation is erratic and slowing with time, the two time systems UT1 and UTC must be periodically reconciled.

In 1956, after years of work, two American astronomers and two British astronomers determined the exact length of the year 1900. The number of cycles of radiation quoted above was determined from their work.

$$9\,192\,631\,770\ {}^{133}\mathrm{Cs\ cycles} = \frac{1}{31\,556\,925.9747}\ \text{of the year 1900.}$$

In terms of the rotation of the Earth, atomic time was synchronized to the year 1900. Due to tidal interactions between the Earth and Moon, the Earth's rotation is slowing at the rate of 1.5 milliseconds (0.0015 s) per day per century.

The Earth's rotation is now about 2 milliseconds per day out of sync with the atomic clocks keeping UTC time. Thus it takes 500 days, or about a year and a half for the atomic clocks to become one second out of sync with the rotation of the Earth. Because we humans cannot change the Earth's rotation, we choose to change the readings of our atomic clocks. This is done at the new year, or on 30 June, as required. Generally, at the new year, the news media announces that a **leap second** is being added to the atomic clocks. Because the Earth's rotation is erratic, there is a possibility that a second would need to be subtracted from atomic time. However, leap seconds have only been added (none subtracted) since the UTC time standard was established in 1972.

Time zones National observatories are the principal keepers of the time standards in their respective countries. Until relatively recently, local time was kept by the position of the Sun relative to the local meridian. High noon actually occurred when the clock read 12:00 hours. This was only natural in the lives of ordinary people and astronomers thought nothing of converting star tables between the local meridians of the major observatories.

The change came with our increasing ability to communicate and travel over long distances in short times. It was mainly the telegraph and railroad companies that brought the change to our way of keeping common time. In 1847, most British railroads agreed to keep Greenwich time at all of its stations. The problem was more severe in countries with longer east–west distance. Stanford Fleming, of Toronto, Canada, publicized the idea of using standard time zones, 15° wide ($\frac{1}{24}$ of 360°), starting at some prime meridian. In 1888, an international conference was held in Washington, DC, to discuss the problem. However, the English and many American railroads had already adopted a system with the Royal Observatory at Greenwich as the prime meridian or first time zone (Barnett, 1999; Jesperson and Fitz-Randolph, 1999).

Other countries had strong arguments for using Paris or Berlin as the prime meridian. However, Greenwich had the added advantage that almost all the nautical almanacs in use were based on Greenwich observations. By 1893, most of the world adopted Greenwich as the prime meridian and time zones as the standard of time keeping. A few places, like Detroit, Michigan, held out until 1900, and France held out until 1911. Time zones, like many things, became (and still are) subject to politics and political boundaries. Astronomers, however, still keep time according to the Royal Observatory at Greenwich, England. Local sidereal time is calculated from Greenwich sidereal time by adjusting for the observer's longitude.

Local (zone) time can be calculated from universal time by adding or subtracting a number of hours, according to your time zone, to the current universal time. In the USA, Eastern Standard Time is normally five hours behind

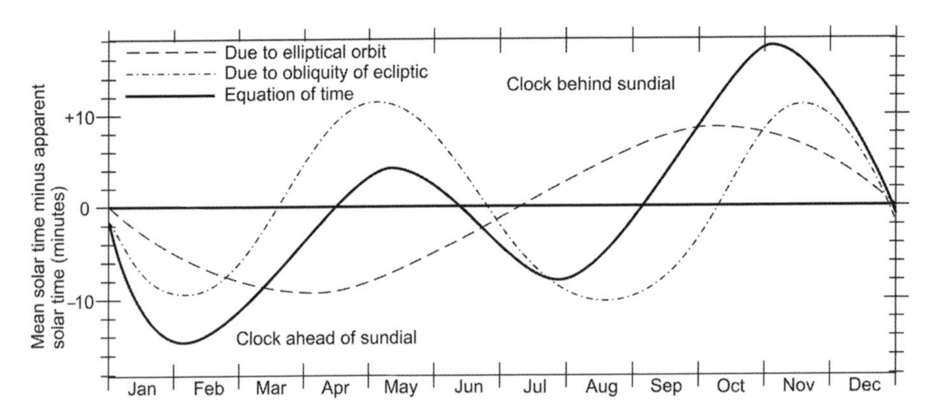

Figure 4.27 The equation of time is the difference between apparent solar time and mean solar time. The two principal causes of the difference are the varying speed of the Earth as it orbits the Sun and the obliquity of the ecliptic. When the variations caused by these two effects are added together, the result is called the *equation of time*.

universal time. During Eastern Daylight Savings Time, it is four hours behind universal time. The minutes are the same.

The equation of time Because of the Earth's elliptical orbit, the Sun moves along the ecliptic at a variable rate. Therefore it cannot be used as an accurate clock. The Earth moves fastest in January and slowest in July. Using the diurnal solar transits to measure the length of an 86 400 second day would then require the length of one second to be longer in January than in July. This is not acceptable.

If the Earth's orbit were a perfect circle so that the Sun moved at a constant rate along the ecliptic, the Sun would still not work well for measuring the length of a day. The obliquity of the ecliptic causes the Sun to move on different small circles around the celestial sphere each day. When the Sun is at the solstices its rate of change in horizon coordinates is greater than when it is at the equinoxes.

To resolve these problems, astronomers invented the *mean sun*. The **mean sun** (or average sun) is an imaginary sun which travels along the celestial equator at a constant rate, representing the average position of the apparent Sun during the year. This imaginary sun keeps *local mean solar time. Local apparent solar* time is measured by the position of the Sun on the ecliptic. The difference between local mean solar time and local apparent solar time is called the **equation of time** (Figure 4.27).

One consequence of the variation in the motion of the true Sun is the dates of latest and earliest sunrise and sunset. The longest and shortest days of the year must be on the solstice days. However, because of the equation of time the earliest sunrise and latest sunset do not occur on the solstice days. Around the December solstice, earliest sunrise occurs about 7 December and latest sunset about 7 January. Around the June solstice, earliest sunrise occurs about 18 June and latest sunset about 27 June.

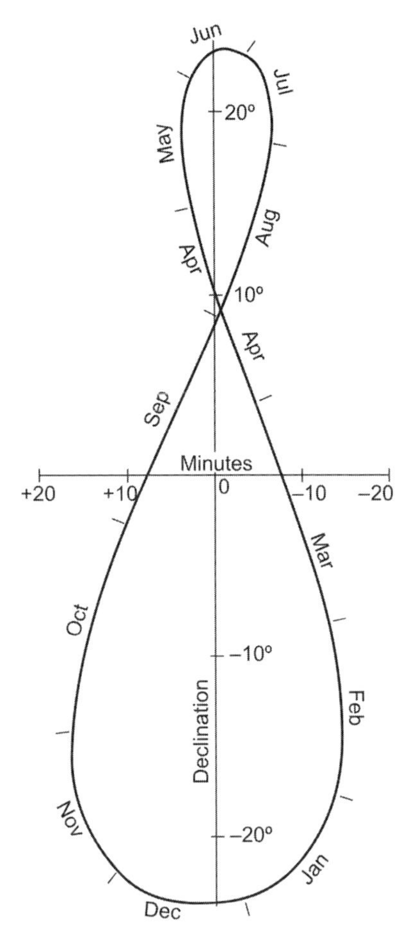

Figure 4.28 The analemma is a plot of the equation of time against the declination of the Sun.

On many of the larger Earth globes, you might find a slightly mis-shaped figure of 8 graphic. This figure is called the **analemma**. (From the Greek word *ἀνάλημμα*, for "sundial.") It is a plot of the equation of time against the declination of the Sun (Figure 4.28). The misshape occurs because the equation of time is not zero at the equinoxes or solstices and this also causes the misalignment with the vertical axis. The figure shows the position of the apparent Sun relative to the mean Sun.

Sidereal and solar time

In our previous discussions we have left the definition of a day somewhat vague. Now we are ready to give it a strict definition. The general definition of one day is one rotation of the Earth. While this is clearly one of the motions of the Earth, any object's motion must be defined relative to some reference – another object. For the rotational motion of the Earth, we have more than one possible reference

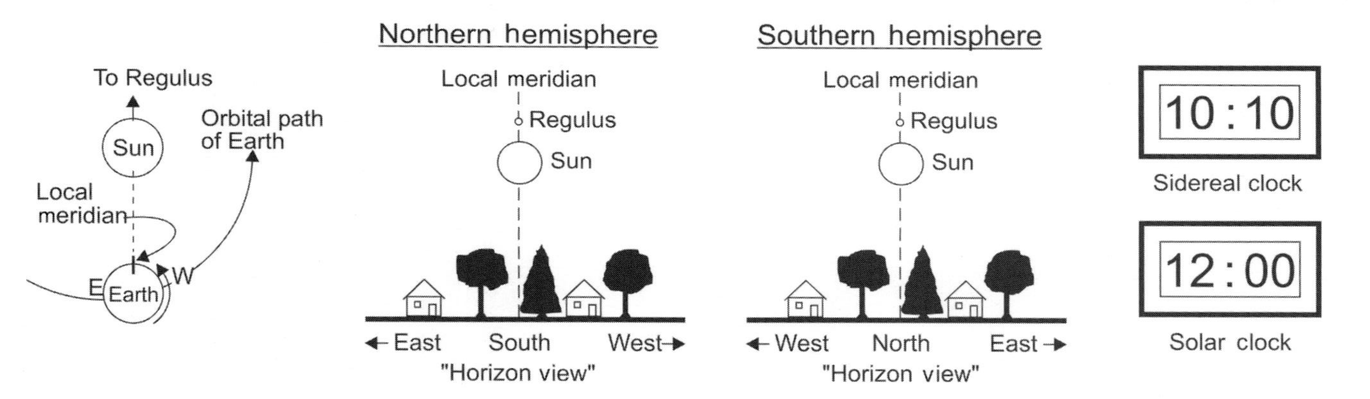

Figure 4.29 We start with the Sun and the star Regulus (in the constellation Leo) on the local meridian. Notice the relationship of the positions of Regulus, the Sun, and the local meridian with the clock readings. This is the first of three drawings showing the relationship of solar and sidereal time.

object. We may use the Sun to define one rotation, or we may choose a star to define one rotation. Choosing different reference objects may cause a day to be defined with different lengths. This is essentially the difference between solar and sidereal time, or days.

Due to the Sun's apparent annual motion, which is of course, caused by the Earth's orbital motion, it takes a little longer for the Sun to return to the local meridian than for a star to return to the local meridian. In fact, it takes about four minutes longer. Figure 4.29 shows both the Sun and Regulus on the local meridian. One day later, in Figure 4.30, the Earth has rotated once on its axis, traveled about 1° along its path around the Sun (the 1° of motion has been exaggerated in this figure to make this concept clearer), and Regulus has returned to the local meridian. This is one *sidereal day*. One **sidereal day** is the length of time required for a particular star to "move around the Earth" and return to the local meridian. The Earth must rotate about 1° farther to get the Sun to return to the local meridian, as seen in Figure 4.31. One **solar day** is the time taken for the Sun to return to the local meridian.

The difference (of about four minutes) in length between solar and sidereal days is caused by the Earth's orbital motion.

The year is defined (in one way) as the time between successive solar transits of the vernal equinox. The number of solar days in a year is 365.2422. The number of degrees in a circle, or one orbit is 360°. On average, the Earth moves

$$\frac{360^\circ}{365.2422} = 0.985\ 647\ 3^\circ \text{ per day.}$$

This is an average because the Earth's orbit is slightly elliptical (eccentric), as described by Kepler's laws of planetary motion. The slight eccentricity causes the Earth to move with varying speed during its orbit. The slight orbital speed variation along with rotational speed variations created by nutations and other causes, create a slight variance in the number of degrees per day in the Earth's motion around the Sun.

On average, one solar day is about four minutes longer than one sidereal day. We can make this relation more exact. Twenty-four hours is equal to 1440 minutes, one degree is equivalent to four minutes. Thus, an hour angle of 0.985 6473° is equivalent to 3′ 56″.56. One solar day is equal to 1.002 738 sidereal days or one sidereal day is 0.997 27 solar days. Converting this into hours (Seidelmann, 1992),

$$24^{h}_{UT} = 24^{h}\ 03'\ 56''.555\ 3678 \text{ mean sidereal day}$$

or

$$\text{Mean sidereal day} = 23^{h}\ 56'\ 04''.090\ 524 \text{ UT.}$$

This also tells us that sidereal hours are only a few seconds different from solar hours in length.

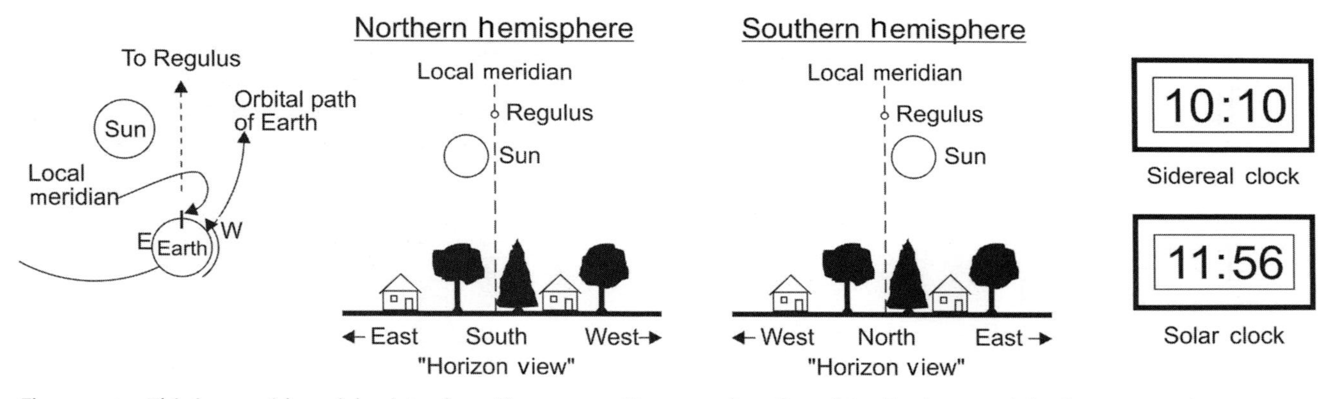

Figure 4.30 This is one sidereal day later from Figure 4.29. The annual motion of the Earth around the Sun causes the apparent motion of the Sun against the background stars. One sidereal day later, the Earth has rotated 360° and revolved around the Sun by $\approx 1^\circ$ (the angle shown here for the change in the orbital position of the Earth is exaggerated for clarity). However, the orbital motion of the Earth causes the Sun to no longer line up with Regulus. The Sun appears to move eastward in our sky relative to the stars. This was labeled the *annual motion of the Sun* in Section 4.4.

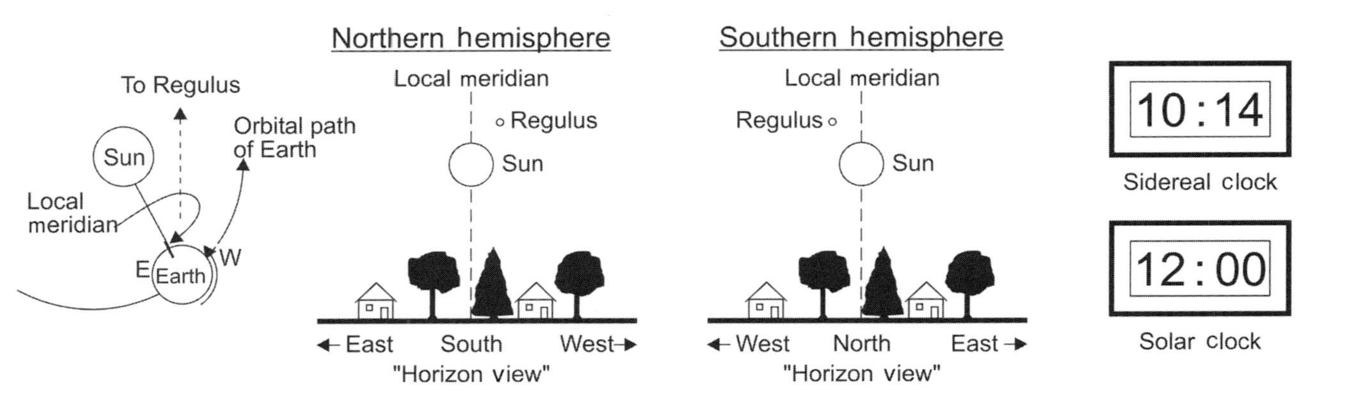

Figure 4.31 About four minutes after Figure 4.30, the Earth has rotated the extra degree so the Sun is now on the local meridian, but Regulus has moved to the west. The stars appear to move westward in our sky relative to the Sun. The Earth's orbital motion is also responsible for the seasonal motion of the Big Dipper (and the other constellations) around Polaris. Refer to the discussions surrounding Figures 4.2 through 4.18, the *annual motion of the sky* in Section 4.3.

There are two times each year that solar time and sidereal time have a special relation. When the Sun is on the vernal (or March) equinox and on the local meridian at the same time, local solar time is $12^h\ 00'\ 00''$ (high noon, ignoring time zones and politics) and local sidereal time is $0^h\ 00'\ 00''$. When the Sun is on the September equinox and the vernal equinox is on the local meridian, then local solar time is $0^h\ 00'\ 00''$ (midnight) and local sidereal time is $0^h\ 00'\ 00''$.

Annual motion of the Sun compared to the annual motion of the stars

Look back to p. 78 where we discussed the motion of the Big Dipper around the NCP, or the motion of the Southern Cross around the SCP. Recall that these stars revolve around their respective celestial poles 361° per solar day. The extra one degree of rotation necessary to bring the Sun back to the local meridian is the source of the extra one degree of revolution per solar day. These stars revolve around the celestial pole once more than the number of solar days in the year. On the other hand, these same stars revolve around their celestial poles exactly 360° each sidereal day. The orbital motion of the Earth is responsible for both the Sun's west to east annual motion and the stars' east to west annual motion. Look at Figures 4.29 and 4.30 again and compare them to each other. Then compare Figure 4.29 with Figure 4.31.

The annual east to west motion of the stars (p. 78) appears only if we measure time with respect to the Sun. The annual west to east motion of the Sun (p. 83) appears only if we measure time with respect to the stars. These two seemingly different motions are actually just two different points of view of the same motion – the orbital motion of the Earth.

Sunrise and sunset

Because we set our clocks by the Sun, we can observe a small shift in the rise and set time of the stars and constellations. The difference in solar time and sidereal time (p. 95) causes the stars to rise and set four minutes earlier each day. Variations in sunrise and sunset times are caused by the changing declination of the Sun as discussed in Section 5.3.

The variation in the length of daylight is related to observer latitude. There is also a difference in the length of twilight (p. 55) for observers at different latitudes (Figure 4.32). At the Earth's equator the diurnal path of the Sun is always perpendicular to the horizon and thus twilight is the shortest possible duration. With greater northern or southern latitude the celestial equator becomes more parallel with the horizon. Because the diurnal motion of the Sun must be parallel to the celestial equator, greater latitude increases the length of the diurnal path of the Sun within the defining intervals of altitude for twilight.

Predicting the exact time of sunrise and sunset is not an easy task. Part of the difficulty comes from the fact that the Sun is a disk and not a point source of light. Sunrise and sunset are defined as those times when the upper point of

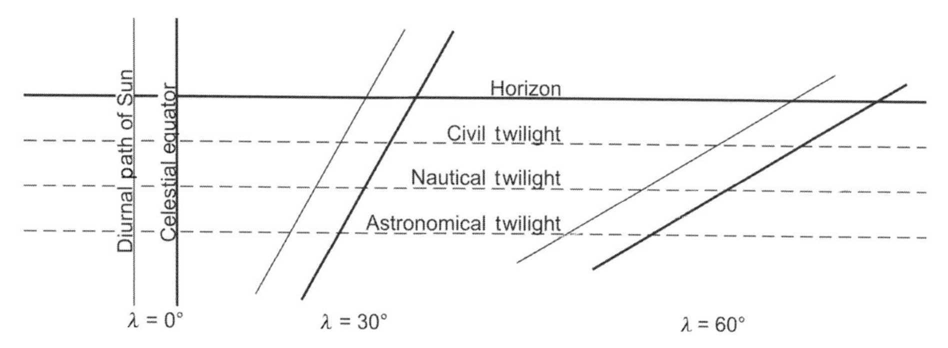

Figure 4.32 The length of twilight depends on the observer's latitude because the angle of the diurnal path of the Sun through the twilight altitudes (the twilight zone?) varies with latitude. At the Earth's equator ($\lambda = 0^\circ$), twilight is the minimum length of time because the path of the Sun through the twilight altitudes is as short as possible. At latitude $\lambda = 30^\circ$, the angle of the Sun's diurnal path with the horizon causes an increase in the amount of time the Sun spends in the twilight altitudes. It is even greater at latitude 60°. North or south latitude was not specified because the effect is the same north or south of the Earth's equator. This diagram shows sunrise in the northern hemisphere or sunset in the southern hemisphere.

the disk becomes visible over the horizon, not when the center becomes visible. The difference in the time between the top of the disk and the center of the disk reaching visibility varies with observer latitude.

Another consideration is atmospheric refraction, which allows us to see the Sun even though it is actually below the horizon. Typically, the shift in the apparent position of the Sun at rise or set (due to refraction) is 36 minutes of arc. However, refraction has some weather dependence (mainly pressure and temperature) and can create unpredictable variations in rise and set time. Kaler gives a rather complete description of the process of predicting the time and azimuth of sunrise and sunset (Kaler, 1994, Chapter 7).

4.6 Precessional motion

Watch a toy gyroscope spinning on its axis while standing on its pedestal. Notice the slow motion of the gyroscope around its pedestal. The gyroscope has at least two motions: the spinning motion of the gyroscope wheel and the rotation motion of the entire gyroscope around the pedestal. The second motion is called **precession**. Precession is caused by a torque (see Appendix A) acting on the spinning wheel. Gyroscopic motion is the most complex mechanical motion known. Entire volumes have been written on this topic and our understanding of it is still limited.

The primary motion of the gyroscope is the rotation of its wheel. The second motion of the gyroscope is called precession (Figure 4.33). Precession is caused by a torque acting on the spinning wheel and is sometimes described as a "wobble" motion.

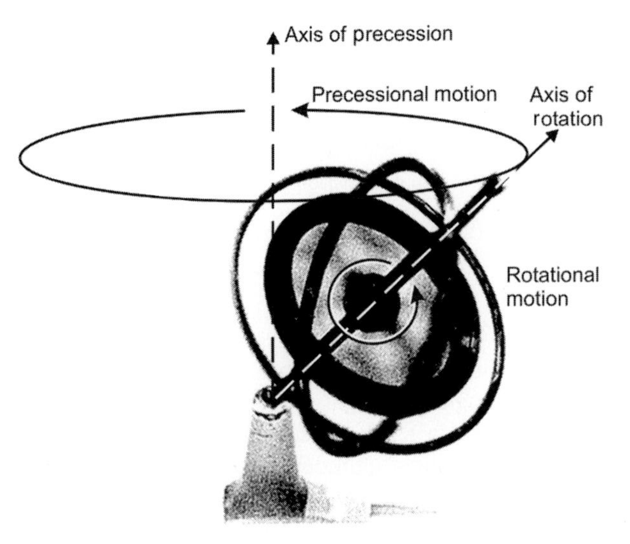

Figure 4.33 The motion of a toy gyroscope shows both rotation and precession. Both of these motions are circular and both have an axis of rotation.

The Earth's precessional motion

The Earth's gyroscopic motion is much more complex than the motion of this small gyroscope. The Earth's motion is also more difficult to analyze because its component motions are caused by its oblique shape (its equatorial bulge) and the gravitational force of many bodies (but mainly the Moon and Sun) acting on that oblique shape from various and changing directions.

The Earth is spinning like a gyroscope. However, the Earth is not solid and its rotational motion causes an equatorial bulge, or *polar flattening*. This is the reason we can state both a polar diameter and an equatorial diameter of the Earth. The Earth's polar radius is 6356.755 km and the equatorial radius is 6378.140 km (Lang, 1992). This difference causes the Earth to take the shape of an *oblate spheroid*. Jupiter and Saturn are very large planets with very high rotation rates. Their polar flattening can easily be seen in any small telescope. The Earth's polar flattening is small, but none the less present. The Sun's gravity and, more importantly, the Moon's gravity act on the Earth's equatorial bulge creating a torque (Appendix A, p. 245), which causes the Earth to precess.

The Earth is tilted by 23°.5. The meaning (measure) of this tilt is made clear in Section 5.4. This tilt angle remains essentially constant for the Earth. Some research has shown that the Moon is greatly responsible for this stability (Comins, 1993). Calculations show that the Earth's axial tilt would vary wildly without the stabilizing gravitational influence of the Moon. It has also been shown that the Earth's tilt does and has changed over long periods of time. Some interesting results of these calculations demonstrate how the Sahara region of Northern Africa was changed from lush tropical jungle to a desert by a one degree shift

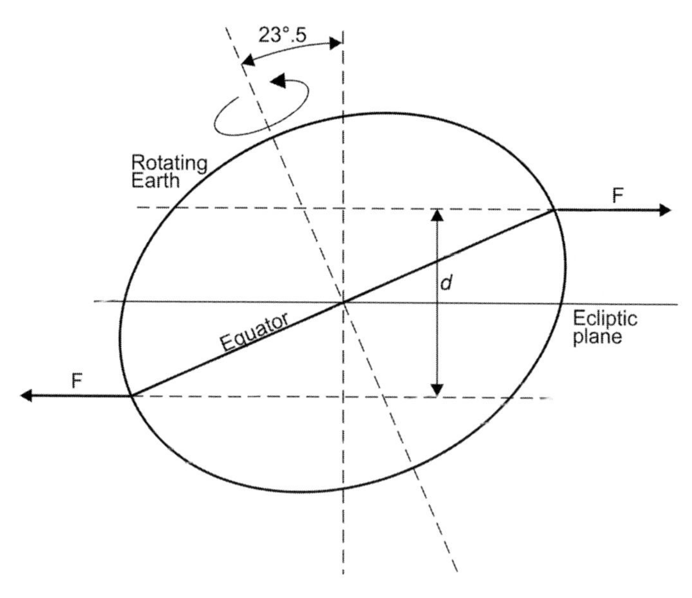

Figure 4.34 The tidal force pairs of the Moon or Sun act on the Earth's equatorial bulge producing torques that cause the Earth to precess. This figure shows only one tidal force pair (labeled F), perhaps the Sun's. Each object creating a tidal force on the Earth has a corresponding force pair. The oblateness of the Earth is greatly exaggerated here.

in the axial tilt. This might account for the presence of so much oil under the desert sands.

To simplify this discussion, we ignore other gyroscopic motions and complications for both the gyroscope and the Earth. The gyroscope precesses around its pedestal remaining tilted at the same angle with respect to the axis of precession. The Moon and the Sun produce tidal forces on the Earth. Figure 4.34 shows how the tidal forces (each force pair) acting on the Earth's equatorial bulge are not in a single line of action, due to the Earth's tilted rotational axis (Section 5.4). This is the torque that is causing the Earth's precessional motion. The toy gyroscope may take only a few seconds to precess once around its pedestal. The period of the Earth's precession is so long – 26 000 years – it is not noticeable within one lifetime. It is barely noticeable within many lifetimes. More precisely, it is 25 982 years (Lang, 1980, 1992).

Motion of the celestial poles One result of the Earth's precession is the change in position of both north and south celestial poles. As shown in Figures 4.35 and 4.36, the NCP and SCP move in a circle, or orbit, through the constellations. The center of the NCP orbit is the **north ecliptic pole** (NEP). The center of the SCP orbit is the **south ecliptic pole** (SEP). The north and south ecliptic poles are those points where the line perpendicular to the ecliptic plane (defined on p. 132) passes through the celestial sphere. These two points and the ecliptic plane form the basis for the ecliptic coordinate system (p. 35).

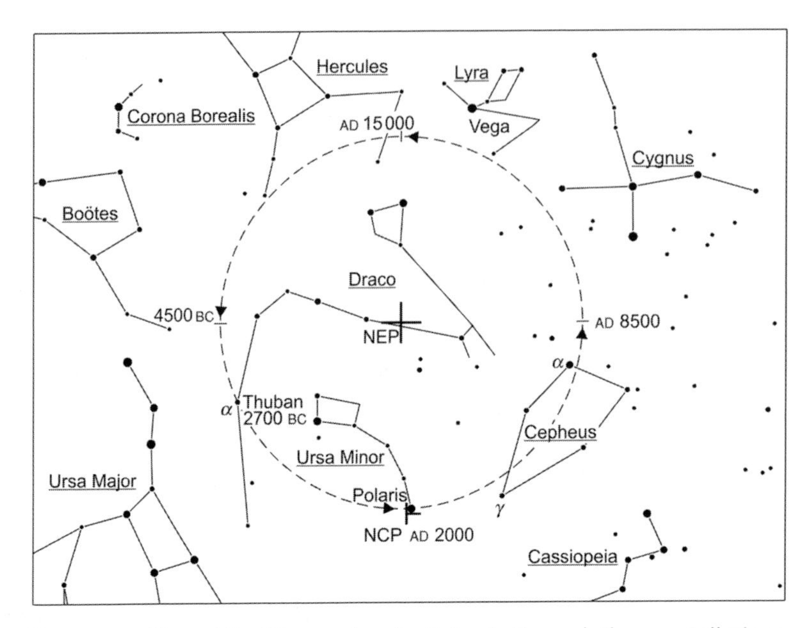

Figure 4.35 The orbit of the north celestial pole through the constellations is caused by the precessional motion of the Earth. The north ecliptic pole (NEP) is found at the center of this circle.

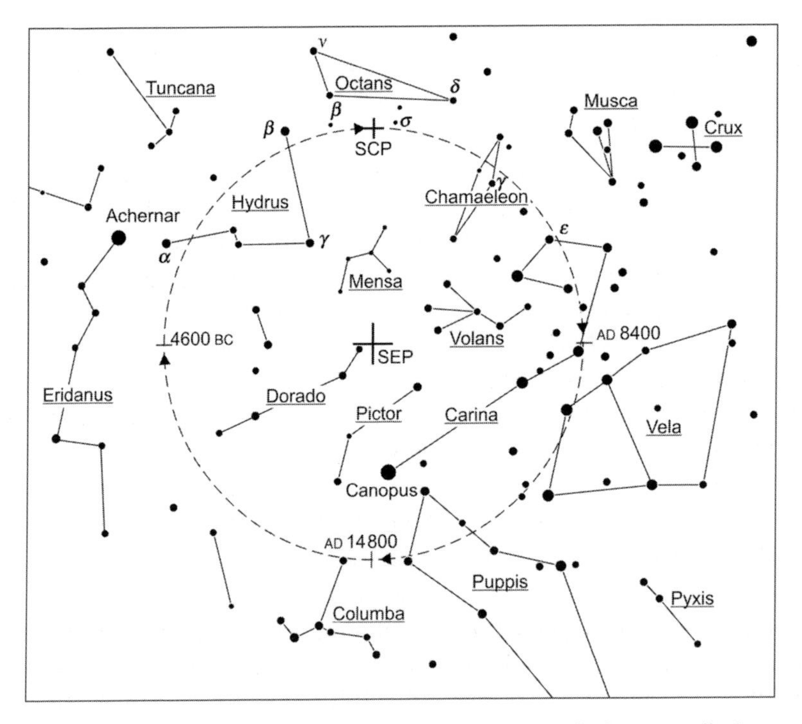

Figure 4.36 The orbit of the south celestial pole through the constellations is caused by the precessional motion of the Earth. The south ecliptic pole (SEP) is found at the center of this circle.

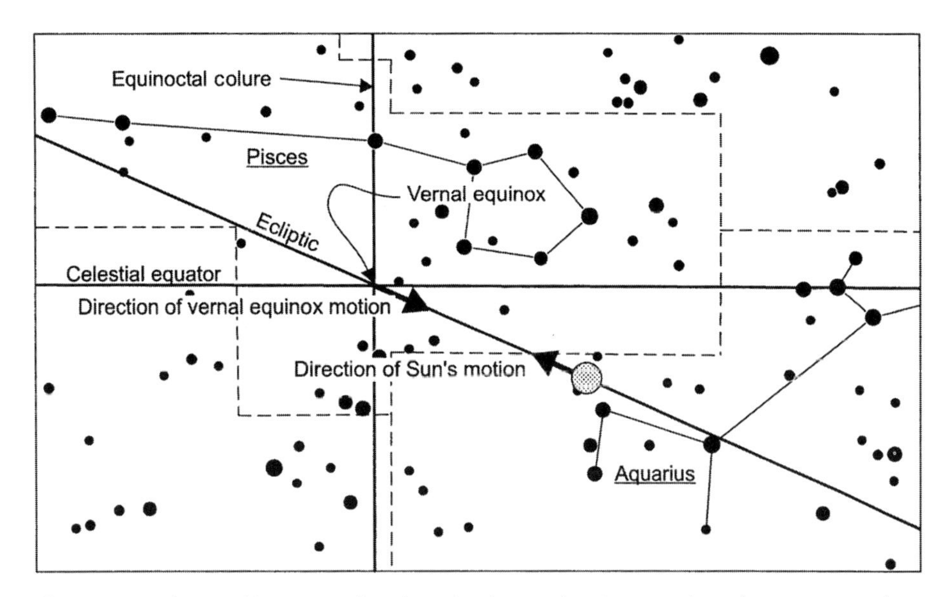

Figure 4.37 The Earth's precessional motion is causing the vernal equinox to creep along the ecliptic line in the opposite direction to the annual motion of the Sun. This has repercussions in terms of defining the year and in the measurement of the equatorial coordinates of celestial objects.

Polaris has not always been and will not always be the North Star. Thuban (α Draconis) was the North Star during the ancient Egyptian Empire (2700 BC). Vega will be the North Star in AD 12 000.

Currently, the declination of Polaris is +89° 15′ 51″. Its closest approach to the NCP occurs in AD 2102 at a distance from the NCP of 27′ 31″ (about the width of the full Moon). The next north stars will be γ Cephei and then α Cephei. Currently, the star closest to the south celestial pole is the very dim star σ Octantis. In the future, another dim star, β Chamaeleontis, will be closest to the SCP.

Because of the Earth's rotation, the NEP and the SEP make daily circular orbits around the NCP and SCP respectively. However, if you could watch the motion over a long period of time, the NCP would slowly move counterclockwise around the NEP. In the southern hemisphere, the SCP moves clockwise around the SEP. Compare the orientation of the orbits between the *North-polar Chart* and the *South-polar Chart*.

Precession of the equinoxes Another effect of precession is a shift in the position of the equinoxes. The equinoxes are moving westward along the ecliptic line at the rate of slightly more than 50″ of arc per year (Figure 4.37). The vernal equinox has not always been, and will not always be on 21 March, unless we make appropriate adjustments to our calendar. On the Roman (Julian) calendars, the beginning of spring was closer to the beginning of March. This changed with the Gregorian calendar. Because of the Earth's precession, the vernal equinox tries to move to the beginning of March but is unsuccessful because we adjust our calendars.

The vernal equinox is moving out of the constellation Pisces and into the constellation Aquarius. This is the celestial meaning of the popular 1960's phrase, "the dawn of the Age of Aquarius." With the constellation borders as defined now, this transition will occur around the year AD 2600. Written history spans the "Age of Aries" and, the "Age of Pisces" and thus, represents about 23% of one "wobble." Since the extinction of the dinosaurs about 65 million years ago, the Earth has "wobbled" on its axis about 2500 times. Dinosaurs ruled the Earth for 140 million years or 5425 "wobbles." In comparison, our written history is an insignificant length of time.

The motion of the vernal equinox also causes a problem for astronomers because it is the origin for the equatorial coordinate system (p. 31). Its motion requires the coordinates of every object to be periodically updated. Each star chart must show its epoch (p. 35).

Misalignment of zodiacal signs and constellations Precessional motion has caused the astrological signs of the zodiac to become misaligned from the zodiac constellations. Babylonian astrologers segmented the ecliptic into twelve equal lengths of 30°, starting at the "First Point of Aries," i.e. the vernal equinox. These ecliptic segments became the twelve signs and they shared their names with the nearest constellation.

When the zodiac signs were established nearly 3000 years ago, the constellation Leo was within the 30° ecliptic segment designated as the sign of Leo. The Earth's precessional motion has caused the constellation and sign to become misaligned. This is of no concern to most astrologers because it is the position of the planets within the signs (their position relative to the vernal equinox), not their position within the constellations, that formulates the astrological traits of an individual. There is a minority of astrologers who believe that precession must be included with the casting of charts.

Visible stars and constellations During the precessional cycle the declination of the stars and constellations varies by 47°. Some constellations now visible at your latitude will not be visible 12 000 years from now. On the other hand, some of those that are now not visible will become visible.

The zodiac constellations creep along the ecliptic to the east as the equinox and solstice points move to the west. This slow progression changes their declination, and the declination of all the other stars and constellations. Twelve thousand years from now Sagittarius will be where Taurus is now. Those constellations below Sagittarius and perhaps below your horizon, will be transiting the local meridian at about the altitude of the celestial equator. Other constellations, such as Orion, which are easily seen near the celestial equator will be 47° lower in declination and may not be visible at your latitude. Which constellations become visible or not visible is very latitude dependent.

Planetary precession Newton's law of universal gravity means that gravity acts between any two objects having mass. Gravity acts not only between the Earth and the Sun, but also between the Earth and all the other planets in the Solar System. The other planets have much less mass than the Sun and they are much farther away from the Earth, but their gravitational influences can still be measured. While their tidal effects are nowhere near as large as the Moon's and Sun's (because of their distance), they do cause changes to the Earth's orbital motion. These changes are not only changes in the shape of the orbit within the ecliptic plane, but also changes in the positioning of the ecliptic plane itself.

The precession caused by the gravitational influence of the Moon and Sun on the Earth's equatorial bulge is called *luni-solar precession*. The motion of the ecliptic plane caused by the gravitational influence of the other planets is called *planetary precession*. The observational effect of planetary precession is similar to that of the luni-solar precession. Planetary precession causes the equinoxes to move along the ecliptic in the opposite direction (eastward) from that of luni-solar precession (westward) and at a much slower rate: 0″.12 per year. The luni-solar precession is 50″.4 per year. We subtract the planetary precession from this to get the general precession rate of 50″.3 per year, which is the observed precessional rate of the equinoxes.

Planetary precession causes a slight change in the obliquity of the ecliptic (the tilt of the Earth, see p. 131). This periodic change takes about 41 000 years. The angle between the equator and the ecliptic varies between 24°.5 and 21°.5. The change occurs at the rate of 0″.47 per year. Its current value is, of course, 23°.5. Over the precessional period of 26 000 years these small gravitational influences cause the path of the celestial poles (Figures 4.35 and 4.36) to be an open path. That is, the path does not close on itself as it is drawn. The next time Polaris is the North Star it will be as far as 3° from the NCP rather than the 0′.5 distance encountered on this trip around.

The periodic changes in the Earth's orbit and the changes to the obliquity of the ecliptic may cause periodic changes in the Earth's climate. These cyclic changes are called the *Milankovitch cycles*. It is thought that they may be responsible for our ice ages, found throughout the geologic record. There is more about the Milankovitch cycles in Section 5.4.

4.7 Nutational motion

Another gyroscopic motion is called *nutation*. **Nutation** is an "up and down bobbing" motion of the gyroscope as it precesses. Nutation occurs with many forms or patterns to its motion. Generally, it is a very complex pattern of which we have little understanding. The Earth's rotational motion also has nutations – slightly over 100 of them are recognized and measurable! Nutational motion is another reason the celestial poles do not return to the same starting point after completing a "wobble."

While some of the small nutations are caused by the Earth's internal structure, the greatest nutational motions are caused by the Moon and the Sun. As they move along (or near, for the Moon) the ecliptic line on the celestial sphere, the Moon crosses the celestial equator twice a month and the Sun crosses it twice a year. As they move and cross the equator, they change the line of action of the forces creating the torques which create the precessional motion. Because the major torque is the one created by the Moon, the celestial poles do not actually orbit the ecliptic poles. Rather, they orbit the poles of the Moon's orbital plane, which are $5^\circ.09$ from the ecliptic poles. However, as discussed on p. 185, the Moon's orbital plane also precesses with an 18.6-year period, causing the Moon's orbital poles to revolve around the ecliptic poles.

The nutational motion causes the orbital speed of the celestial poles to increase and decrease, compared to their average speed around the ecliptic poles. The celestial poles also move slightly closer to and farther from (about $0''.9$) the ecliptic poles with a regular period. This causes another small periodic motion in the obliquity of the ecliptic. These nutations are caused by the variation of the precessional torques of the Moon and Sun. The precessional torques are strongest when the Moon or Sun is at the solstice points and zero when at the equinox points.

Questions for review and further thought

1. Why do the constellations not change their shape (stick figure) during a year's time?
2. List and briefly describe the four major motions of the Earth.
3. Describe the basic diurnal motion of the stars in the night sky, particularly for your latitude.
4. For regions north or south of the Earth's equator, there are some stars that never set (go below the horizon). What are these stars (as a class) called? Describe their motion.
5. Describe and compare the diurnal motion of the stars as observed when standing at latitudes 45° N and 45° S. Include a description of the appearance of the celestial equator.
6. What are we talking about when discussing an object's upper and lower culmination?
7. Describe the procedure for finding the upper and lower culminations of a circumpolar object for northern hemisphere observers.
8. Describe the procedure for finding the transit altitude of objects for an observer at the Earth's equator.
9. Describe the procedure for finding the upper and lower culminations of a circumpolar object for southern hemisphere observers.

10. Describe the annual motion of the stars. What direction do they move? How fast do they move?
11. Compare the annual motion of the Big Dipper with the annual motion of the Southern Cross.
12. Describe the diurnal motion of the Sun.
13. Define one *solar day*.
14. Why is there a 30 minute variation in the duration of the solar day during the year?
15. Describe the Sun's annual motion.
16. What is the apparent path of the Sun called?
17. At what rate does the Sun's position change as it moves along its apparent path?
18. Describe the location (on the celestial sphere) of the vernal equinox. On what date is the Sun found there?
19. What are those constellations through which the ecliptic passes called?
20. What is the difference between a zodiac sign and a zodiac constellation?
21. Why do the dates of the Sun's entry into the zodiac constellations not match with the dates of its entry into the zodiac signs?
22. How do we define our measure of the day or the year?
23. What is universal time? Describe.
24. When and why were time zones established?
25. Describe the analemma.
26. How is precession sometimes described?
27. What is *polar flattening*? How does it relate to equatorial bulge?
28. What is the period of the Earth's precessional motion?
29. Define the north and south ecliptic poles.
30. Why hasn't Polaris always been our North Star?
31. Along which line of the celestial sphere are the equinoxes moving?
32. List the effects of the motion of the vernal equinox.
33. Explain the term *epoch*.

Review problems

1. Determine if the following are circumpolar for latitudes higher than 40° N: Polaris, Thuban, Mizar, Capella, Abireo, Altair.
2. Determine if the following are circumpolar for latitudes lower than 40° S: β Octantis, Volans, Achernar, Canopus, α Centauri.
3. Calculate the upper (A_{UC}) and lower culmination (A_{LC}) of Altair for your latitude.
4. Calculate the upper (A_{UC}) and lower culmination (A_{LC}) of Sirius for your latitude.
5. Calculate the transit altitude (A_T) of the following objects for your latitude: Regulus, Arcturus, Great Square, Antares, Canopus, Achernar, Thuban,

Fomalhaut, Capricornus, Mizar. If the object is circumpolar, calculate its upper (A_{UC}) and lower culminations (A_{LC}).

6. Currently, it is 20 : 00 (solar) hours and the star Procyon is on the local meridian. What does the sidereal clock read? When (roughly) will Spica be on the local meridian?

There are more problems dealing with the motion of the Earth in Chapter 8, Observation Project 1.

5
The seasons

As we have seen in the previous chapter, the Sun has a daily motion across the sky and a yearly west–east motion among the zodiac constellations. In this chapter we discuss the effects of the second component of the Sun's annual motion – the north–south annual motion – which is due to the obliquity of the ecliptic. The most important result of this annual north–south motion is the seasons.

5.1 A description of the seasons

Many people have come to believe that we have seasons because of our changing distance from the Sun, caused by the Earth's slightly elliptical orbit. They think we have summer when we are closest to the Sun and winter when we are farthest away. We might call this the "solar distance hypothesis." There are two observations to consider when thinking of the seasons in this way. First, we are closest to the Sun on 4 January and farthest from the Sun on 4 July (within a day or two variation). Earth's perihelion (closest) distance is 1.4710×10^8 km (91 403 700 miles), and aphelion (farthest) distance is 1.5210×10^8 km (94 510 600 miles). See Table 5.1.

While this idea would still appear to work for the seasons in the southern hemisphere, the second observation about the seasons certainly eliminates the solar distance hypothesis. That is, the northern and southern hemispheres have opposite seasons. When it's summer in the northern hemisphere it's winter in the southern hemisphere. The solar distance hypothesis would require the same season over the entire Earth, with summer occurring in January and winter in July. Clearly, this hypothesis is wrong.

> *The seasons occur because of two effects – the efficiency with which the Sun heats the ground and the length of daylight at the observer's latitude. Both of these effects are ultimately caused by the tilt of the Earth's rotational axis and, so, the reason we have seasons is that the Earth's axis of rotation is tilted.*

Table 5.1 *Dates of the solstices and equinoxes*

Year	March equinox	June solstice	September equinox	December solstice	Earth–Sun distance	
					Min	Max
2004	20 Mar	21 Jun	22 Sep	21 Dec	4 Jan	5 Jul
2005	20 Mar	21 Jun	22 Sep	21 Dec	2 Jan	5 Jul
2006	20 Mar	21 Jun	23 Sep	22 Dec	4 Jan	3 Jul
2007	21 Mar	21 Jun	23 Sep	22 Dec	3 Jan	7 Jul
2008	20 Mar	20 Jun	22 Sep	21 Dec	3 Jan	4 Jul
2009	20 Mar	21 Jun	22 Sep	21 Dec	4 Jan	4 Jul
2010	20 Mar	21 Jun	23 Sep	21 Dec	3 Jan	6 Jul
2011	20 Mar	21 Jun	23 Sep	22 Dec	3 Jan	4 Jul
2012	20 Mar	20 Jun	22 Sep	21 Dec	5 Jan	5 Jul
2013	20 Mar	21 Jun	22 Sep	21 Dec	2 Jan	5 Jul

Table 5.2 *Names of the important solar or season dates*

Date	Northern hemisphere	Southern hemisphere
21 Mar	Spring equinox	Fall equinox
21 Jun	Summer solstice	Winter solstice
21 Sep	Fall equinox	Spring equinox
21 Dec	Winter solstice	Summer solstice

This tilt is also called the *obliquity of the ecliptic*. The real question becomes, what does "tilted" mean? Before we try to answer this question, let's first understand the seasons and the apparent motion of the Sun that creates them.

Important dates

There are four important dates for the seasons of the year. Any astronomer should certainly know each date and be able to find the position of the Sun on the celestial sphere (its right ascension and declination) and in the high noon sky (its local meridian transit altitude, or high noon altitude). The altitude of the Sun at noon depends on your latitude. We discuss the altitude of the Sun after the description of the general characteristics of each of these days. You can see from Table 5.1 that these days may occur anywhere between the 19th and 23rd of the month. This variation occurs because of things like leap years – artifacts of our calendars. For these descriptions I chose to simplify this variation by using the average date, the 21st of the month.

Each of the four dates has a name. However, the name depends on the observer's hemisphere and so they are shown in Table 5.2 for clarity. The Latin word **equinox** means that on these days the number of daytime hours and night-time hours is equal – 12 hours each. At least, this is what should happen.

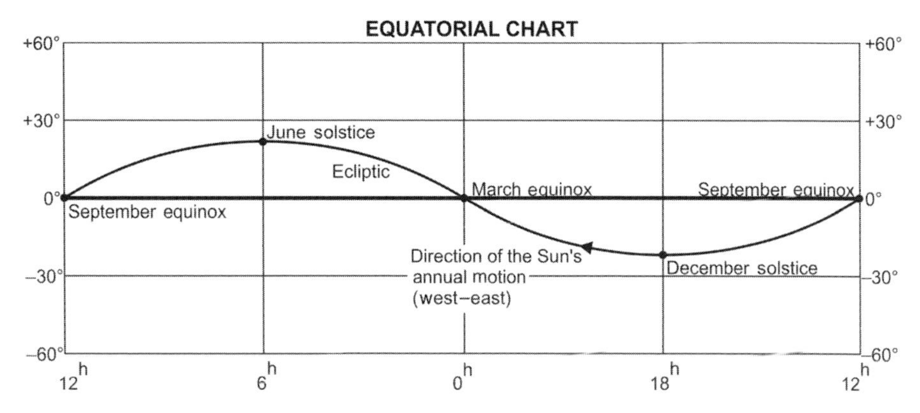

Figure 5.1 These are the locations of the four important solar positions along the ecliptic as found on the *Equatorial Chart*. Remember, the chart wraps around at the edges. The points are named with their months because the seasons are not the same for the Earth's northern and southern hemispheres.

The Earth's atmosphere causes a slight refraction of sunlight so an observer can see the Sun even though it is actually slightly below the horizon plane. Thus, the number of minutes of daylight on an equinox day is slightly greater than 720 (12 hours). The Latin word **solstice** comes from the words *Sol* and *sistere*, and means "sun stand still." On the solstice days, the Sun reverses its apparent annual north–south motion.

The ancient northern Europeans held a "Winter Festival" on 21 December, in an attempt to convince the sun god to return to the north and bring an end to winter. At the least, this tells us these people were well aware of the motion of the Sun and its connection to the seasons. Some historians claim that Christmas was set on 25 December both to compete with this winter festival and to give the pagan barbarians who converted to Christianity an equivalent festival so they wouldn't feel "left out."

In the northern hemisphere, winter is the shortest season of the year. Summer is 4.5 days longer, spring 3.8 days longer, and autumn is 0.9 days longer than winter. This is reversed for the southern hemisphere, where summer is the shortest season of the year.

Use Figure 5.1 in conjunction with your *Equatorial Chart* for the following sections. The dates are listed in their calendrical order.

21 March On 21 March, the Sun crosses the celestial equator and is thus directly above the Earth's equator. The Sun's declination is 0° and its right ascension is 0 hours. On this day the periods of daylight and darkness are equal everywhere on Earth. For all observers (except at the Earth's poles), the Sun rises due east and sets due west. For anyone standing on the Earth's equator, the Sun rises straight up in the east, passes through the zenith, and sets straight down in the west.

For anyone standing on the North Pole or the South Pole, the Sun circles the sky right along the horizon, never appearing to rise or set for the entire day. Actually, at the North Pole, the Sun is rising for a six months period of daylight and at the South Pole, the Sun is setting for a six months period of night (Figure 5.13).

21 June The declination of the Sun is $+23^\circ.5$ and its right ascension is 6 hours. The Sun is farthest north of the celestial equator and stays at nearly the same declination (and thus nearly the same transit altitude) for about two weeks. This places the Sun high in the southern sky in the northern hemisphere (summer) and low in the northern sky in the southern hemisphere (winter). During this day, the Sun is directly over the Tropic of Cancer, latitude $23^\circ.5$ N. In the northern hemisphere, this is the longest day of the year and in the southern hemisphere it is the shortest.

21 September The Sun is again crossing the celestial equator and is thus directly above the Earth's equator. Its diurnal path matches the celestial equator at any observation latitude. The declination of the Sun is 0° and its right ascension is 12 hours. The observed conditions on Earth are the same as described for 21 March except for a slight difference at the poles. At the North Pole, the Sun is setting for a six months period of night and at the South Pole, the Sun is rising for a six months period of daylight.

21 December The declination of the Sun is $-23^\circ.5$ and its right ascension is 18 hours. The Sun is farthest south of the celestial equator. It is low in the southern sky in the northern hemisphere (winter) and high in the northern sky in the southern hemisphere (summer). During this day, the Sun is directly over the Tropic of Capricorn, latitude $23^\circ.5$ S. In the northern hemisphere, this is the shortest day of the year and in the southern hemisphere it is the longest.

The cross quarter days While the dates listed above represent the beginning of the seasons in North America, not all countries and cultures recognized them in the same way. In some countries, these dates represent the midpoint of a season. For instance, Shakespeare's *A Midsummer's Night Dream* took place on the night of 21 June. In most cultures, the dates midway between the equinox and solstice days have some, if not the same, significance.

The second day of February is midwinter in the northern hemisphere and is also known as *Groundhog's Day* in the USA or as *Candlemas* in other countries. It is halfway between the beginning of winter and the beginning of spring. Groundhog's Day is a minor celebration to decide if winter will last another six weeks. Since all the seasons last about twelve weeks and Groundhog's Day is halfway through winter, is there any doubt winter will last six more weeks? (Yeah, okay, it's done just for the fun of it.) The other cross quarter days are

Hallowe'en (All Saints' Day Eve) on 31 October, *May Eve* on 30 April, and *Lammas Day* on 1 August.

Whether the dates listed above represent the beginning or the midpoint of a season is a local cultural preference, and therefore we refer to these important solar days simply by their dates. Table 5.3 describes and compares the characteristics of these season dates.

5.2 Determining the Sun's transit altitude

Like all other celestial objects, the altitude of the Sun when it transits your local meridian (high noon) depends on your latitude. The altitude can be found for any latitude on any day by following these three steps:

1. Given the latitude, find the altitude of the celestial equator. Following the results from Section 2.5 (pp. 28, 29), this is:

   $$A_{\mathrm{CE}} = 90^{\circ} - \lambda.$$

2. Given the date, find the declination coordinate of the Sun. This can be found from the *Equatorial Chart* or, if it is one of the four important dates listed above, you should know the Sun's declination. There are other sources for this information as well, such as computer planetarium programs.
3. In the northern hemisphere, add the Sun's declination to the altitude of the celestial equator, making sure to keep any minus signs. This is the Sun's transit altitude off the southern horizon. This is the procedure used on any celestial object (see p. 70 for review). If the result is greater than 90°, subtract it from 180° to obtain the Sun's transit altitude off the northern horizon.

   In the southern hemisphere, subtract the Sun's declination from the altitude of the celestial equator, making sure to handle all minus signs correctly. This is the Sun's transit altitude off the northern horizon. If the result is greater than 90°, subtract it from 180° to obtain the Sun's transit altitude off the southern horizon.

   At the Earth's equator, always subtract the Sun's declination from 90°, ignoring the declination sign (use the absolute value of the declination). When the declination is positive, the Sun transits off the northern horizon and when it is negative it transits off the southern horizon (Figure 5.8).

   At the poles, the celestial equator matches the horizon. The altitude of the Sun is the same all day. That is, it travels on an almucantar. At the North Pole, the altitude of the Sun's almucantar is equal to its declination. At the South Pole, the altitude of the Sun's almucantar is equal to the negative of its declination (reverse the declination sign).

In any case, the Sun's altitude can never be greater than 90°. Table 5.4 is a list of example results when these steps are applied to various latitudes.

Table 5.3 *A comparison of the equinox and solstice days*

Observation	March equinox (March 21)	June solstice (June 21)	September equinox (September 21)	December solstice (December 21)
Dec. of the Sun	0°	$+23^\circ.5$	0°	$-23^\circ.5$
RA of the Sun	0^h	6^h	12^h	18^h
Sun passes through zenith:	At the equator	At the Tropic of Cancer	At the equator	At the Tropic of Capricorn
Sunrise:	Due east	Northeast	Due east	Southeast
Sunset:	Due west	Northwest	Due west	Southwest
Northern hemisphere	Spring equinox: equal length day and night	Summer solstice: long days, short nights	Fall equinox: equal length day and night	Winter solstice: short days, long nights
Southern hemisphere	Fall equinox: equal length day and night	Winter solstice: short days, long nights	Spring equinox: equal length day and night	Summer solstice: long days, short nights
North Pole (and Arctic):	Sunrise all day	24 hours of day above Arctic Circle	Sunset all day	24 hours of night above Arctic Circle
South Pole (and Antarctic):	Sunset all day	24 hours of night below Antarctic Circle	Sunrise all day	24 hours of day below Antarctic Circle

Table 5.4 *The Sun's transit (high noon) altitude for various locations*[d] *on the Earth*

Location	Latitude of location	Altitude of NCP	Altitude of SCP	Altitude of celestial equator	Position of the Sun (degrees)							
					21 Mar		21 Jun		21 Sep		21 Dec	
					δ	Alt	δ	Alt	δ	Alt	δ	Alt
North Pole	90.0 N	90.0[a]	−90.0[a]	0.0[a]	0.0	0.0[a]	+23.5	23.5[a]	0.0	0.0[a]	−23.5	−23.5[a]
Arctic Circle	66.5 N	66.5[b]	−23.5[c]	23.5[c]	0.0	23.5[c]	+23.5	47.0[c]	0.0	23.5[c]	−23.5	0.0[c]
Chicago, IL	42.0 N	42.0[b]	−42.0[c]	48.0[c]	0.0	48.0[c]	+23.5	71.5[c]	0.0	48.0[c]	−23.5	24.5c
Miami, FL	26.0 N	26.0[b]	−26.0[c]	64.0[c]	0.0	64.0[c]	+23.5	87.5[c]	0.0	64.0[c]	−23.5	40.5[c]
Tropic of Cancer	23.5 N	23.5[b]	−23.5[c]	66.5[c]	0.0	66.5[c]	+23.5	90.0[a]	0.0	66.5[c]	−23.5	43.0[c]
Equator	0.0	0.0[b]	0.0[c]	90.0[b]	0.0	90.0[a]	+23.5	66.5[b]	0.0	90.0[a]	−23.5	66.5[c]
Tropic of Capricorn	23.5 S	−23.5[b]	23.5[c]	66.5[b]	0.0	66.5[b]	+23.5	43.0[b]	0.0	66.5[b]	−23.5	90.0[b]
Sydney, Australia	35.0 S	−35.0[b]	35.0[c]	55.0[b]	0.0	55.0[b]	+23.5	31.5[b]	0.0	55.0[b]	−23.5	78.5[b]
Antarctic Circle	66.5 S	−66.5[b]	66.5[c]	23.5[b]	0.0	23.5[b]	+23.5	0.0[b]	0.0	23.5[b]	−23.5	47.0[b]
South Pole	90.0 S	−90.0[a]	90.0[a]	0.0[a]	0.0	0.0[a]	+23.5	−23.5[a]	0.0	0.0[a]	−23.5	23.5[a]

[a] Measurement made from any horizon.
[b] Measurement made from the northern horizon.
[c] Measurement made from the southern horizon.
[d] The first two columns select a location on Earth with its latitude coordinate. The altitude of the CP at any location is equal to the location's latitude (see p. 20). Notice the altitude of the NCP is negative at locations south of the Earth's equator. The altitude of the celestial equator is equal to 90° minus the location's latitude (see p. 28). For each date, the δ column is the declination coordinate of the Sun on that date. Notice this declination is independent of the Earthly locations. The "Alt" column is the high noon (transit) altitude of the Sun on that date, at that location. Make a careful comparison of the arithmetic for the northern and southern hemispheres.

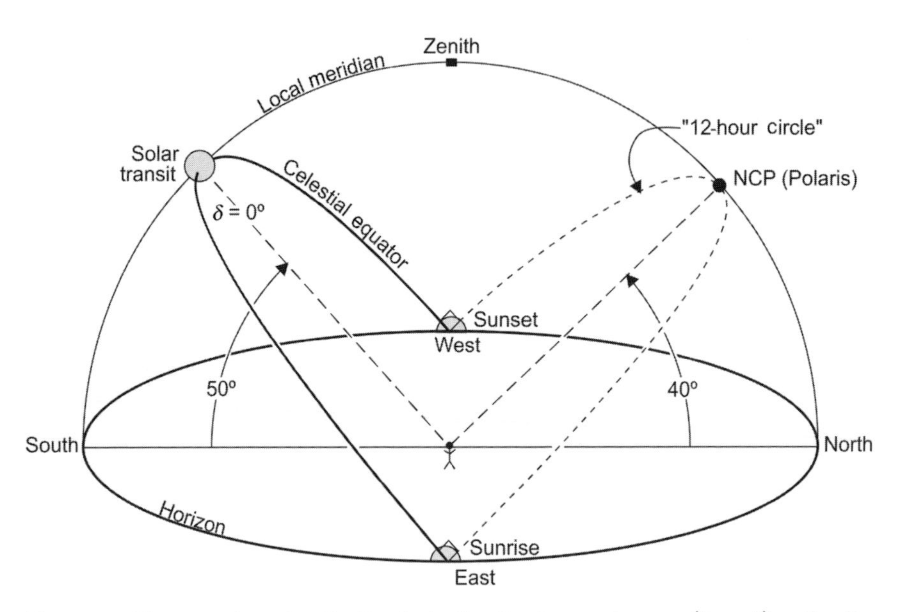

Figure 5.2 On an equinox day, the Sun's declination is zero degrees ($\delta = 0^\circ$) so the diurnal path of the Sun for these days matches the celestial equator. It rises due east and sets due west. At 40° N, the Sun transits the local meridian at $90^\circ - 40^\circ + 0^\circ = 50^\circ$ off the southern horizon. Daylight lasts 12 hours. Compare this figure to Figures 5.3, 5.4, and 5.5.

At the equinoxes 21 March and 21 September

Because the Sun is on the celestial equator, its altitude is always the same as the celestial equator's altitude at your latitude.

> *Because the Sun is on the celestial equator and the celestial equator's transit altitude is equal to 90° minus the observer's latitude, the Sun's transit (high noon) altitude equals 90° minus the observer's latitude.*

Example: *Northern hemisphere*: at 40° N, we get $90^\circ - 40^\circ = 50^\circ$ for the celestial equator. Thus, on the equinox days (Figures 5.2 and 5.3), the Sun's high noon altitude is 50° off the southern horizon. Make a comparison between the two northern hemisphere figures.

Example: *Earth's equator*: the latitude is 0° and we get $90^\circ - 0^\circ = 90^\circ$ for the celestial equator (Figure 5.4). For observers at the Earth's equator on an equinox day, the Sun passes through the zenith.

Example: *Southern hemisphere*: at 40° S, we get $90^\circ - 40^\circ = 50^\circ$ for the celestial equator (Figure 5.5). The Sun's high noon altitude is 50° off the northern horizon.

Example: At the poles, the altitude of the celestial equator is zero degrees and thus the Sun's altitude is also zero degrees (Figure 5.13).

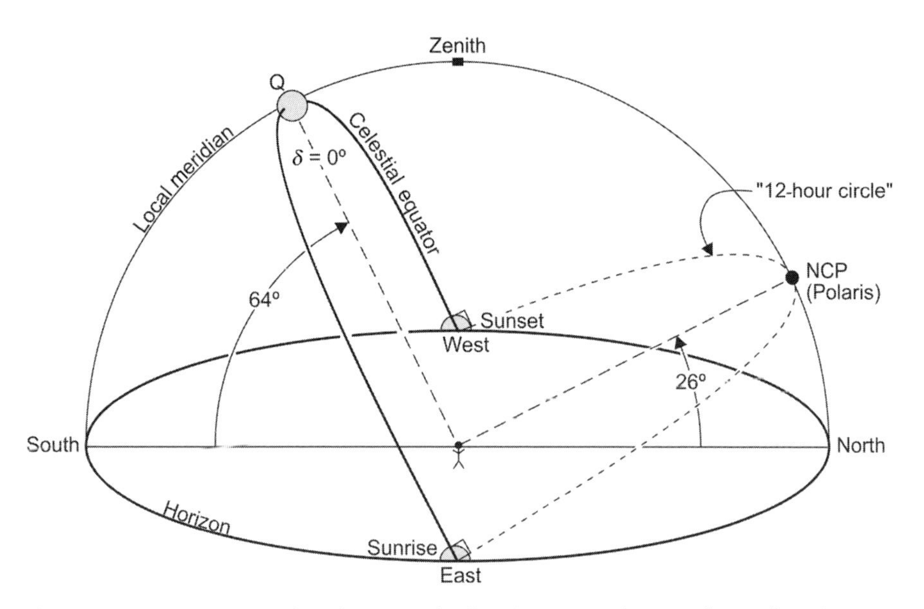

Figure 5.3 On an equinox day, the Sun's declination is zero degrees ($\delta = 0°$) so the diurnal path of the Sun for these days matches the celestial equator. It rises due east and sets due west. At 26° N, the Sun transits the local meridian at $90° - 26° + 0° = 64°$ off the southern horizon. Daylight lasts 12 hours. Compare this figure to Figures 5.2, 5.4, and 5.5.

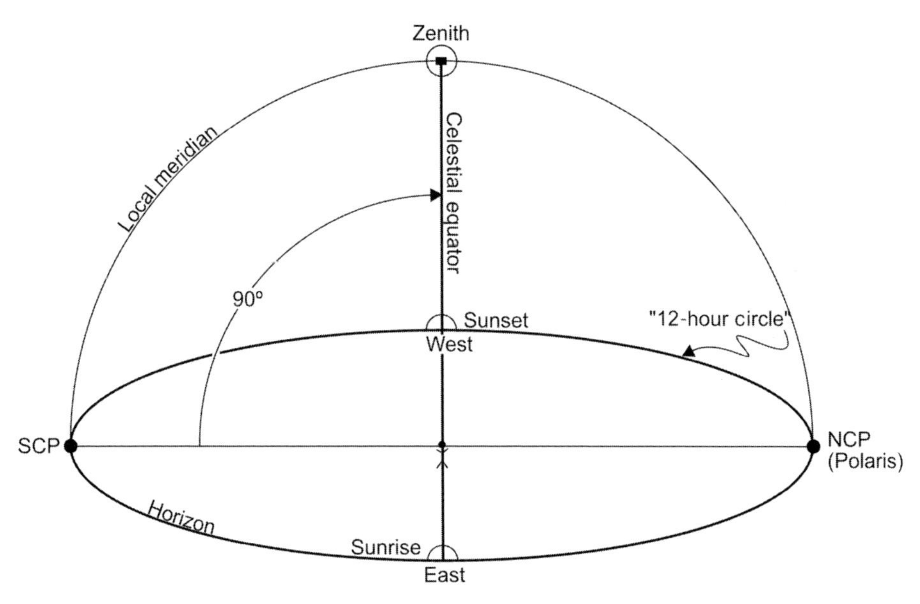

Figure 5.4 On an equinox day at the Earth's equator, the Sun transits the zenith. It rises straight up, due east, and sets straight down, due west. Because the 12-hour circle matches the horizon at this latitude (0°), the Sun is always above the horizon for 12 hours. Compare this figure to Figures 5.2, 5.3, and 5.5.

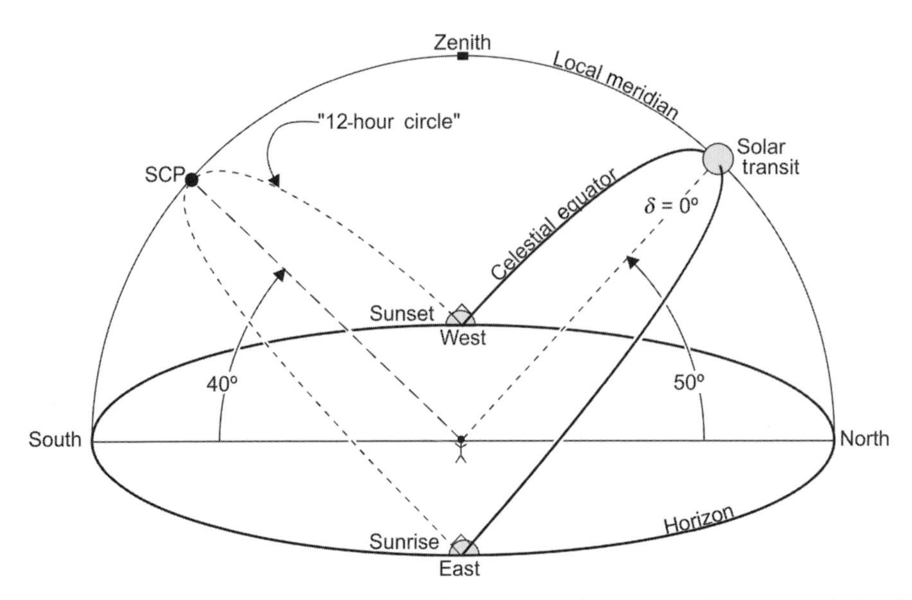

Figure 5.5 This is the apparent motion of the Sun on the equinox days at 40° S. On the equinox days the Sun transits at the same altitude as the celestial equator. The celestial equator represents the diurnal path of the Sun on this day. The Sun rises due east and sets due west with daylight lasting for 12 hours. Compare this figure to Figures 5.2, 5.3, and 5.4.

At the June solstice

The Sun is at $+23^\circ.5$ declination. The declination of Sol stops increasing (solstice) and begins to decrease.

Example: In the *northern hemisphere* (Figures 5.6 and 5.7), add $23^\circ.5$ to the celestial equator's transit altitude to get the Sun's transit altitude. For 40° N, we get $50^\circ + 23^\circ.5 = 73^\circ.5$. So, on the June (summer) solstice, the maximum (high noon) altitude of the Sun is $73^\circ.5$ above the southern horizon. Notice the Sun can *never* be directly overhead at this latitude.

Be careful to note that to an observer in the northern hemisphere, but south of the Tropic of Cancer, the Sun actually transits the local meridian off the northern horizon for this time of year, while it is in the southern sky for most of the rest of the year (see Figure 5.8).

Example: For an observer at 15° N, the altitude of the celestial equator is $90^\circ - 15^\circ = 75^\circ$. Adding the declination of the Sun to this value gives the transit altitude of the Sun as $75^\circ + 23^\circ.5 = 98^\circ.5$. Following our rules for such cases (in step three of the above procedure), the Sun transits the local meridian at $81^\circ.5$ off the northern horizon.

Example: At the *Earth's equator*, the Sun is $90^\circ - |+23^\circ.5| = 66^\circ.5$ off the northern horizon (Figure 5.8).

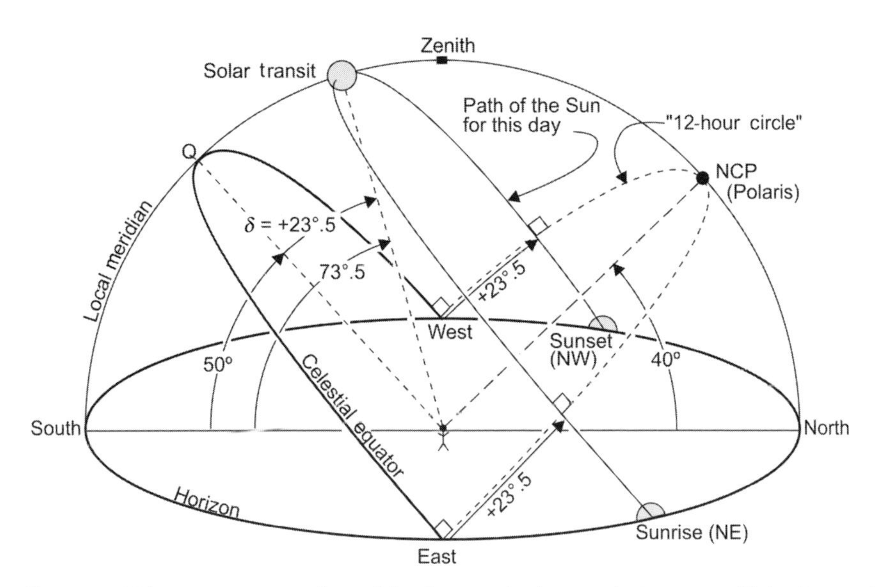

Figure 5.6 The apparent motion of the Sun on the June solstice at 40° N. In the northern hemisphere, the June solstice is the summer solstice. The Sun rises in the northeast, transits high in the southern sky at $90^\circ - 40^\circ + 23^\circ.5 = 73^\circ.5$, and sets in the northwest. Its path for the day is parallel to the celestial equator, with the Sun at a constant declination of $+23^\circ.5$. Daylight lasts for more than 12 hours. Compare this figure to Figures 5.7, 5.8, and 5.9.

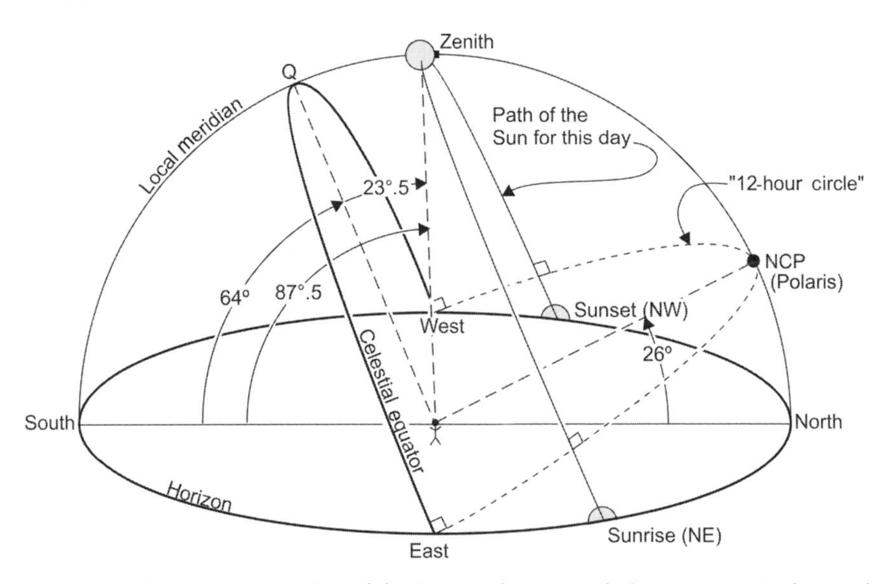

Figure 5.7 The apparent motion of the Sun on the June solstice at 26° N. In the northern hemisphere, the June solstice is the summer solstice. The Sun rises in the northeast, transits high in the southern sky at $90^\circ - 26^\circ + 23^\circ.5 = 87^\circ.5$, and sets in the northwest. Daylight lasts for more than 12 hours. Compare this figure to Figures 5.6, 5.8, and 5.9. Note especially, as one gets closer to the equator, the variation in the length of daylight during the year decreases. Sunrise and sunset occur closer to the 12-hour circle.

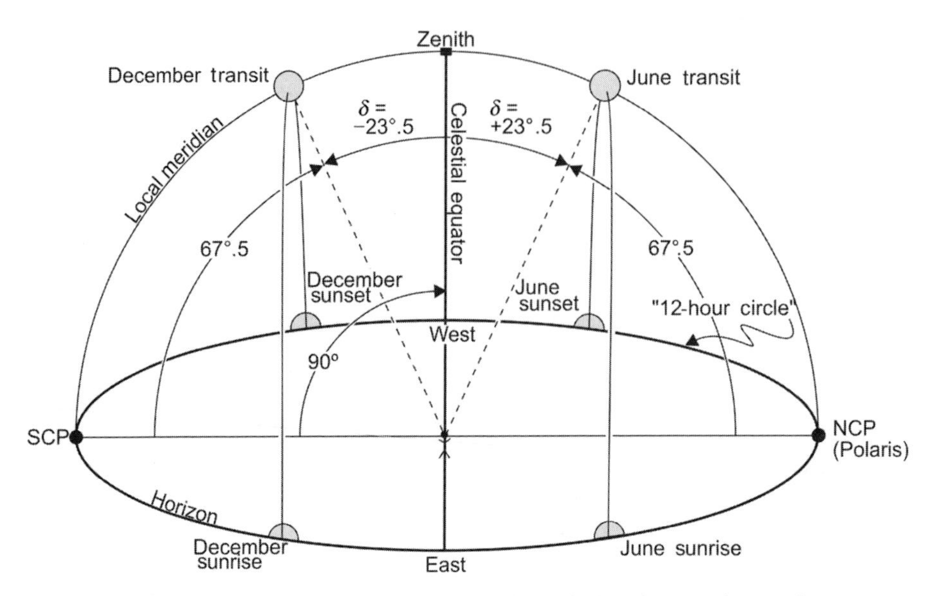

Figure 5.8 The apparent motion of the Sun on the solstice days at the Earth's equator. On the solstice days at the equator, the Sun transits at 67°.5. It rises straight up and sets straight down. Because the 12-hour circle matches the horizon, the Sun is above the horizon for 12 hours – every day – for observers at the equator. Compare this figure with the other solstice figures, Figures 5.6, 5.7, 5.9, 5.10, 5.11, and 5.12.

Example: In the *southern latitudes* (Figure 5.9), subtract 23°.5 from the celestial equator's altitude. For 40° S, we get 50° − (+23°.5) = 26°.5. So, on the June (winter) solstice, the maximum (high noon) altitude of the Sun is 26°.5 above the northern horizon. With the Sun so low in the sky, winter weather dominates.

Example: At the North Pole (Figure 5.13), the Sun circles the sky 23°.5 above the horizon, so it is daytime for 24 hours! The North Pole sees daylight 24 hours a day, for the days between 21 March and 21 September. During this six month period, the South Pole is in 24-hour darkness.

At the December solstice

The Sun is at −23°.5 declination. The declination of Sol stops decreasing (solstice) and begins to increase.

Example: At *northern latitudes* (Figures 5.10 and 5.11), add the negative declination to the celestial equator's altitude. For 40° N, we get 50° + (−23°.5) = 26°.5. So, on the December (winter) solstice, the maximum (high noon) altitude of the Sun is 26°.5 above the southern horizon. Compare this with the altitude of the Sun on 21 June in the southern hemisphere as calculated above.

Example: At *southern latitudes* (Figure 5.12), subtract the negative declination from the celestial equator's altitude. For 40° S, the celestial equator's altitude is

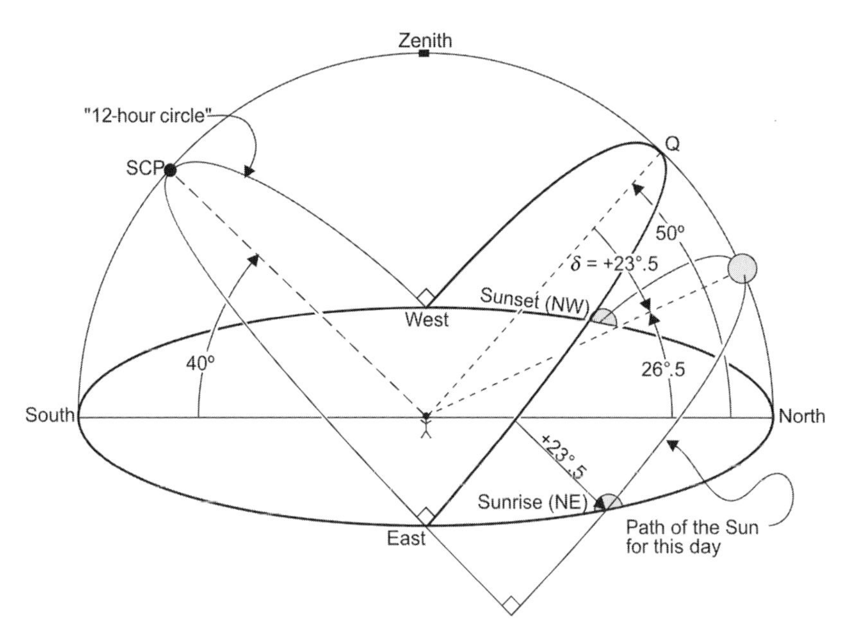

Figure 5.9 The apparent motion of the Sun on the June solstice at 40° S. For the southern hemisphere, the June solstice is the winter solstice. The Sun rises in the northeast, transits the local meridian low in the northern sky and sets in the northwest. The Sun crosses the 12-hour circle before it rises in the northeast and after it sets in the northwest. Daylight lasts less than 12 hours. Compare this figure with Figures 5.6, 5.7, and 5.8.

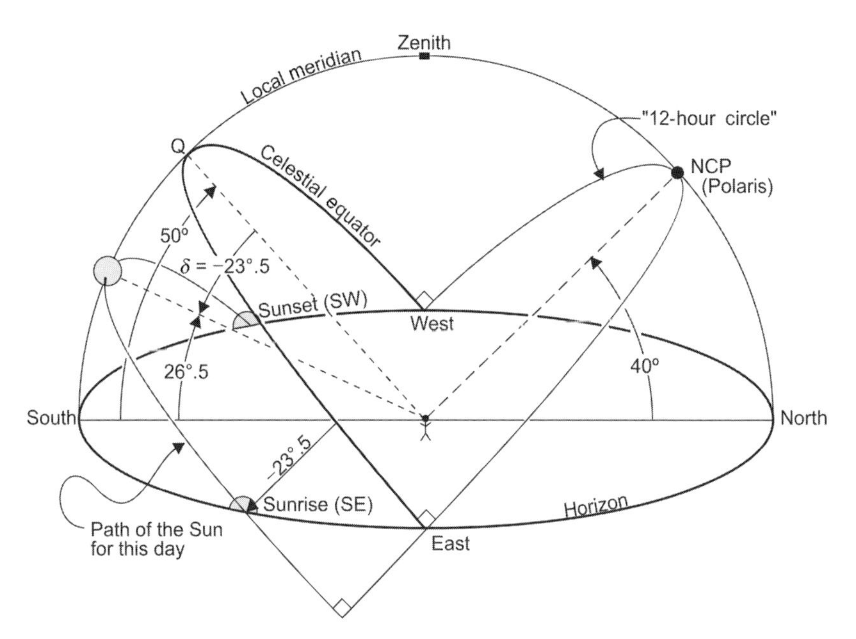

Figure 5.10 The apparent motion of the Sun on the December solstice at 40° N. In the northern hemisphere, the December solstice is the winter solstice. The Sun rises in the southeast, transits low in the southern sky at $90^\circ - 40^\circ - 23^\circ.5 = 26^\circ.5$, and sets in the southwest. Its path for the day is parallel to the celestial equator, with the Sun at a constant declination of $-23^\circ.5$. Daylight lasts for less than 12 hours. Compare this figure to Figures 5.8, 5.11, and 5.12.

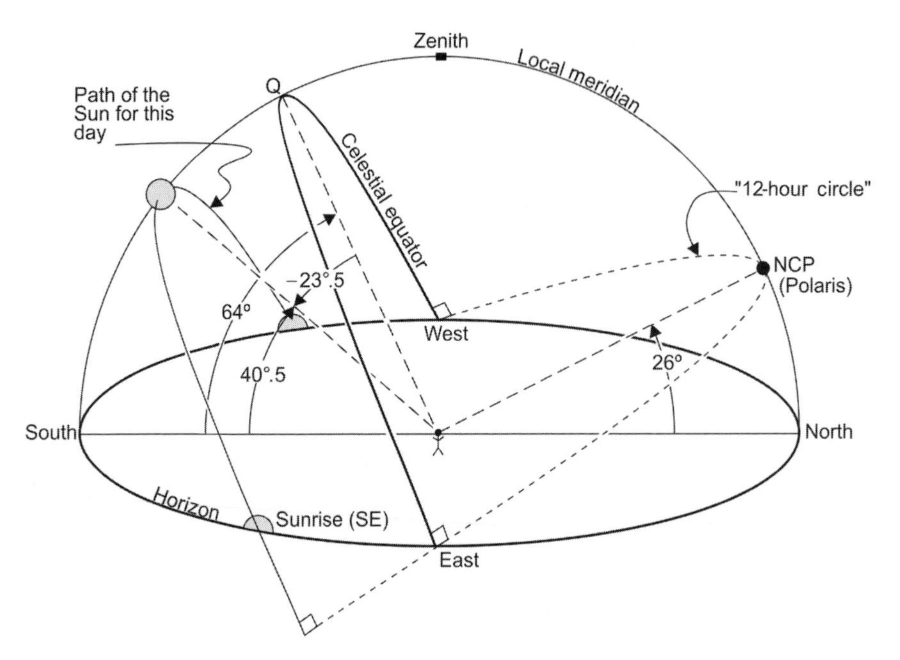

Figure 5.11 The apparent motion of the Sun on the December solstice at 26° N. Here again the December solstice is the winter solstice. The Sun transits low in the southern sky. It rises in the southeast, sets in the southwest, and it is above the horizon for less than 12 hours. Compare this figure with Figures 5.8, 5.10, and 5.12. Note especially in comparison to Figure 5.10 that at this latitude (Miami, Florida), the Sun remains high enough during winter that snow is not a concern.

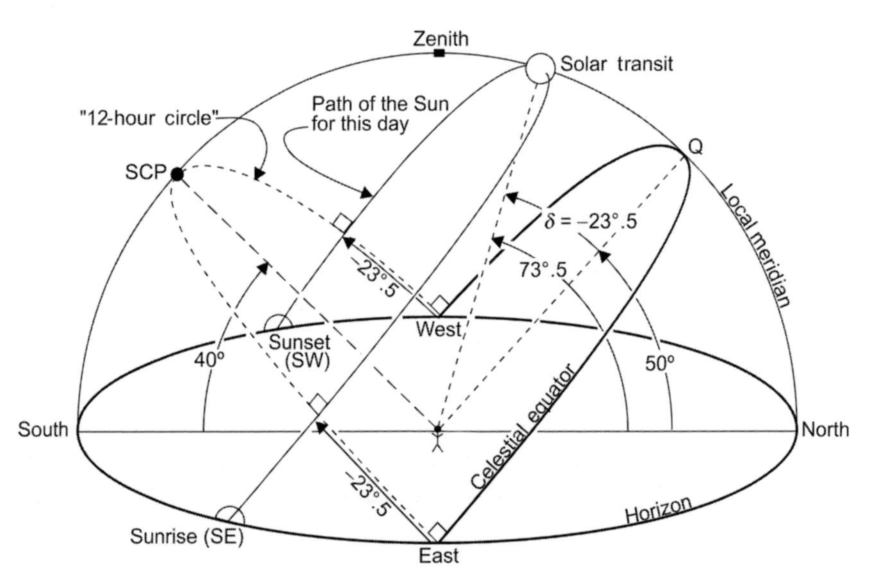

Figure 5.12 The apparent motion of the Sun on the December solstice at 40° S. For the southern hemisphere, the December solstice is the summer solstice. The Sun rises in the southeast, transits the local meridian high in the northern sky, and sets in the southwest. The Sun follows a path parallel to the celestial equator, bringing it above the horizon before it gets to the 12-hour circle. Daylight lasts longer than 12 hours. Compare this figure with Figures 5.8, 5.10, and 5.11.

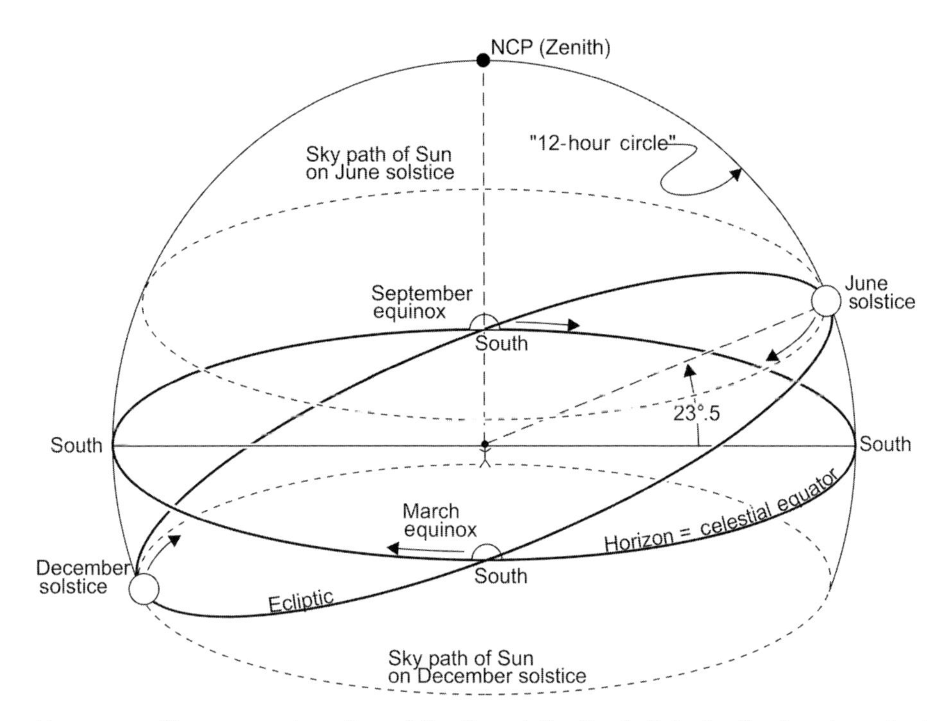

Figure 5.13 The apparent motion of the Sun at the North Pole for the four important seasonal dates. Like everything else, the Sun's daily path must be parallel to the celestial equator. For the June solstice the Sun is 23°.5 above the horizon all day. On the equinox days there is 24 hours of sunrise or sunset. On the December solstice the Sun never rises – 24 hours of night. For the South Pole, reverse the results for June and December and reverse the direction of the motion arrows along their respective paths.

50° and we get $50^\circ - (-23^\circ.5) = 73^\circ.5$. So, on the December (summer) solstice, the maximum (high noon) altitude of the Sun is 73°.5 above the northern horizon.

Note that the situation for observers near the Earth's equator described in the previous paragraph and example occurs for observers near the equator in both hemispheres. For observers in the southern hemisphere, but north of the Tropic of Capricorn, the Sun's altitude may need to be adjusted for azimuth.

Example: The *North Pole* is in darkness for 24 hours a day with the Sun circling at $-(+23^\circ.5)$ below the horizon (negative altitude). The *South Pole* is now enjoying 24 hours of daylight with the Sun circling $-(-23^\circ.5) = +23^\circ.5$ above the horizon (see Figure 5.13).

5.3 The changing solar declination

Because of the Sun's changing declination (+23°.5 to −23°.5), the Sun's transit altitude varies by 47° during the year. This variation of the Sun's altitude creates heating and cooling effects which in turn create the seasons. We now look at these effects and how they create the seasons.

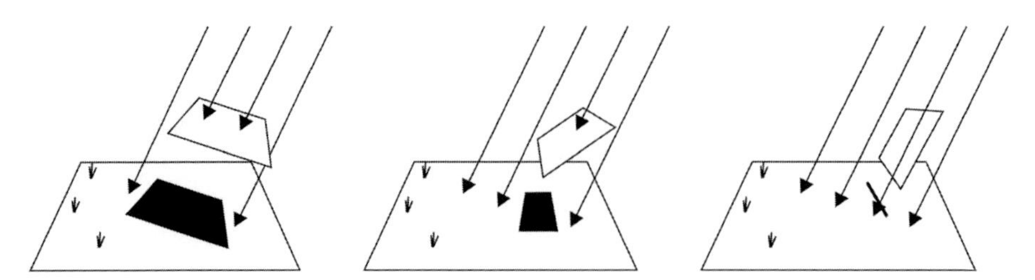

Figure 5.14 The shadow cast by a metal plate gives an indication of how well the sunlight is heating the plate. In the left-hand drawing the plate is casting a maximum shadow and the plate heating rate is maximum. In the center drawing the plate is angled, casting a smaller shadow, blocking a smaller amount of sunlight and thus the plate is not heated as efficiently. In the right-hand drawing the plate casts a minimum shadow. The sunlight passes over the plate, not heating it at all.

The Sun's altitude and ground heating

The gases of the atmosphere absorb very little sunlight directly. They are mainly heated indirectly by the ground. Sunlight passes through the air to heat the ground, which in turn, heats the air. To understand seasons, we must understand ground (earth) heating. Here, I want to point out the use of "Earth" to represent the planet as a whole, in the astronomical sense, and "earth" to represent the ground we walk on.

Light is electromagnetic energy – one of the many forms of energy. Classically, it is energy transported by waves passing through a combination of linked electric and magnetic fields. The intensity of the light and the energy transported is proportional to the amplitude of the wave. In modern physics, light energy is carried by particles called *photons*. The energy carried by a photon is proportional to its frequency.

Because sunlight is a form of energy, it warms any surface it strikes. The rate (or efficiency) of warming depends on the angle at which the sunlight hits the surface of the object. The warming rate is highest when light hits the surface straight on, at what most people would call a right angle. However, in optics, angles are measured from the normal line, which is a line drawn perpendicular to the surface (interface) between two materials, or a material and vacuum. Thus, maximum heating occurs when the sunlight strikes the surface at zero degrees to the normal.

As the surface is tilted away from the sunlight, the rate of warming decreases. Look at Figure 5.14. A thin metal plate, held so it casts a thin line shadow on the ground, is not warmed by the Sun. If it is held so it casts a large shadow, it is easily warmed by the sunlight. Now look at Figure 5.15. If we leave the metal plate on the ground for a year, the changing altitude of the Sun has the same effect as changing the tilt of the metal plate to the Sun's rays. During the summer, sunlight hits the metal plate (and the ground) more directly than during the winter. Like the plate, the earth is warmed in the summer, and allowed to cool in the winter.

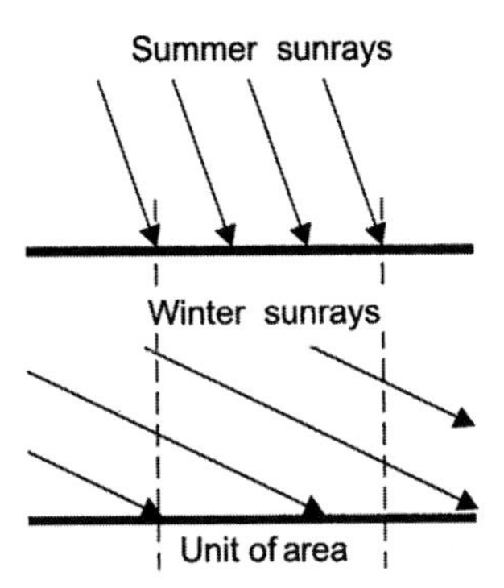

Figure 5.15 When the Sun is high in the sky during the summer, the amount of sunlight landing on a unit of area is larger than the amount landing on the same unit of area during the winter. The sunrays drawn here are all the same distance apart, but because they hit the earth at a lower angle during the winter, they land much farther apart than during the summer. Thus, the unit area of ground does not receive as much heat energy from the Sun during the winter as during the summer. This is exactly the same effect as tilting the metal plate, shown in Figure 5.14. The dashed lines mark an equal amount of area on each surface and also show the normal to each surface.

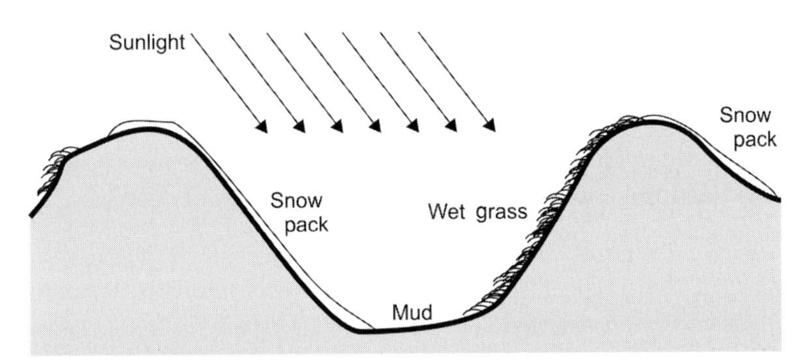

Figure 5.16 The efficiency of solar heating is determined by the angle at which the sunlight strikes the ground. Maximum heating occurs when the sunlight strikes the ground at high angles (or angles near the surface normal), such as on the Sun-facing slope. Minimum heating occurs when the sunlight passes over a surface rather than landing on it, such as the opposite-facing slope. The snow remains on this slope because the sunlight "skims over" the surface of the snow pack. Compare what's happening here with the views presented in Figures 5.14 and 5.15.

Figure 5.16 shows another view of this concept. When the snow melts in the spring, those (sloped) parts of the ground facing the Sun lose the snow first, because the sunlight strikes this ground straight on. The sloped parts that are not facing the Sun, must wait for the warmed air to melt the snow.

The Sun's rays always strike the equatorial regions of the Earth almost straight on. However, at the poles of the Earth, the Sun's rays always strike the ground at a very shallow angle. Look at Figure 5.17. At the Earth's equator, the metal

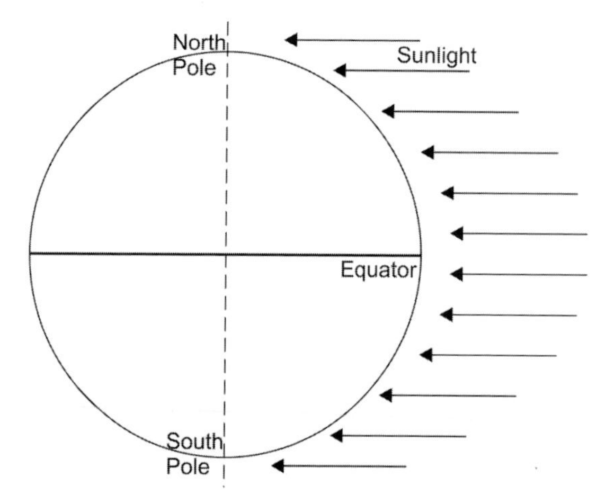

Figure 5.17 Solar heating of terrestrial regions. The Sun heats the equatorial regions with greater efficiency than the polar regions because of the angle at which the sunlight strikes the surface of the Earth in these regions.

plate (the earth) is face on to the Sun's rays, and at the poles the metal plate (the earth) is edge on. The equatorial region is warm and the polar regions are cold because of the angle of the Sun's rays as they strike the surface of the Earth in those regions. While the temperature difference between these regions may seem extreme to us humans, it is moderated by the Earth's atmosphere. On planets without an atmosphere (Mercury), the ground temperature difference between equatorial and polar regions can be hundreds of degrees.

The seasons are reversed north and south of the Earth's equator. When it is winter in the northern hemisphere, it is summer in the southern hemisphere. When the Sun is low in the northern hemisphere's southern sky, it is high in the southern hemisphere's northern sky.

To see this effect clearly, compare Figures 5.10 and 5.12. Ground heating is the effect that causes the northern and southern hemispheres to have opposite seasons. There is an unequal effectiveness of ground heating between the two hemispheres.

The Sun's declination and the length of daylight

The change in the Sun's declination also changes the length of daylight. Looking at the drawings for the northern hemisphere, you should see how the sunrise point along the eastern horizon (its azimuth) changes during the year. On an equinox day (Figures 5.2 and 5.3), the Sun rises due east and sets due west. During the summer (Figures 5.6 and 5.7), the Sun rises in the northeast and sets in the northwest. During the winter months (Figures 5.10 and 5.11), the Sun rises in the southeast and sets in the southwest. In the southern hemisphere, the Sun rises in the northeast and sets in the northwest during the winter months

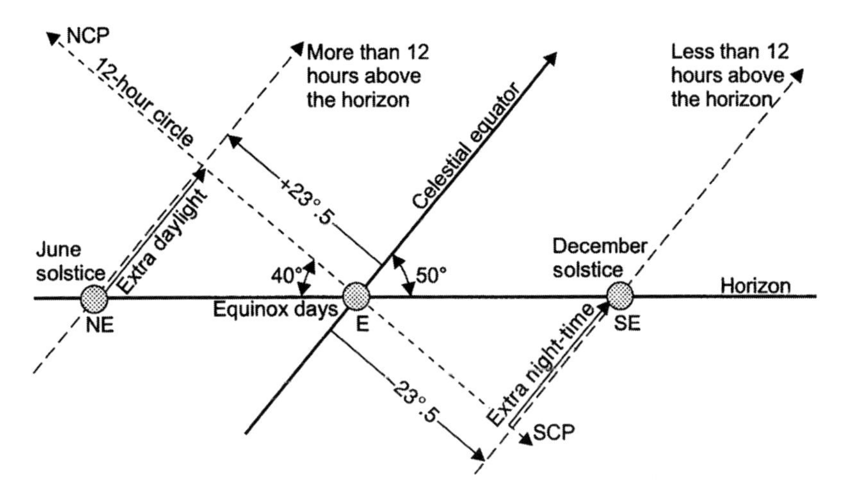

Figure 5.18 Sunrise as seen by an observer at 40° N from "inside the bowl" of Figures 5.2, 5.6, and 5.10. Curvature in the lines that would occur from projecting a sphere onto the page is not taken into account in this diagram. However, you should see a better understanding of why the Sun's rising point moves along the horizon and why the amount of time the Sun spends above the horizon changes with the seasons. Notice the relationship between the angle of the 12-hour circle and the horizon, the celestial equator and the horizon, and the observation latitude. Compare this figure to Figure 5.19.

(Figure 5.9). During the summer, the Sun rises in the southeast and sets in the southwest (Figure 5.12).

In Figures 5.2 through 5.12, the *12-hour circle* is drawn from the NCP to the SCP such that it crosses the celestial equator on the observer's due-east and due-west horizon points. It always takes the Sun 12 hours to travel from the east side of this circle to the west side, independent of the season or observer latitude. Figures 5.18 and 5.19 show the sunrise from "inside the bowl." Using the equations for converting equatorial to horizon coordinates (Kaler, 1996) we can find the azimuth of sunrise as a function of the observer's latitude, λ. For sunrise on the June solstice we have

$$A = \arccos\left[\frac{\sin 23^\circ.5}{\cos\lambda}\right].$$

For sunrise on the December solstice we have

$$A = \arccos\left[\frac{\sin(-23^\circ.5)}{\cos\lambda}\right].$$

In both of these formulas, A is the azimuth of sunrise measured from due north. Because the Sun does not rise or set at latitudes above 66°.5 N or below 66°.5 S, the latitude λ is restricted between 66°.5 (maximum north latitude) and −66°.5 (maximum south latitude). They work for both the northern and southern hemisphere.

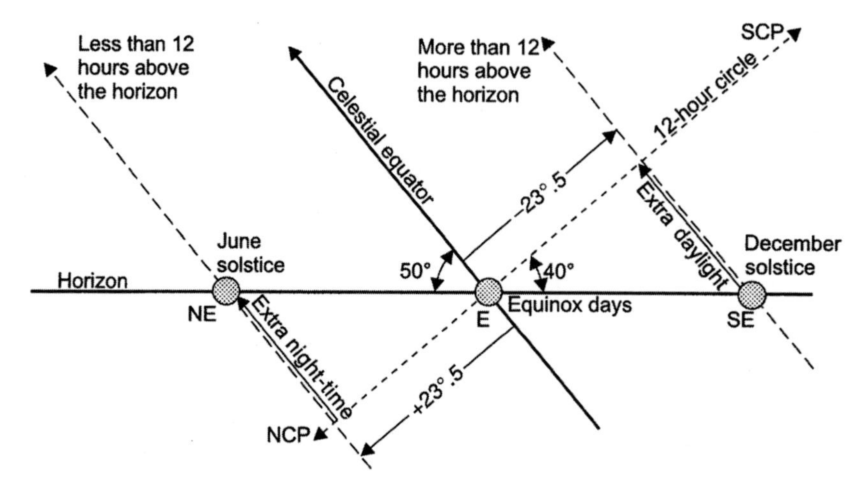

Figure 5.19 Sunrise as seen by an observer at 40° S from "inside the bowl" of Figures 5.5, 5.9, and 5.12. Curvature in the lines that would occur from projecting a sphere onto the page is not taken into account in this diagram. However, you should see a better understanding of why the Sun's rising point moves along the horizon and why the amount of time the Sun spends above the horizon changes with the seasons. Notice the relationship between the angle of the 12-hour circle and the horizon, the celestial equator and the horizon, and the observation latitude. Compare this figure to Figure 5.18.

In either hemisphere, in the summer, the Sun rises before it gets to the 12-hour circle in the east and sets after it crosses the 12-hour circle in the west. Thus, summer daylight is longer than 12 hours. In winter, the Sun crosses the 12-hour circle before it rises and after it sets. Thus, winter daylight is shorter than 12 hours.

Note that as an observer gets closer to the Earth's equator, the variation in the length of the day decreases. At 40° N we see about a six hour change in the length of the day during the year. Miami, Florida, at 26° N sees only about a three hour change in the length of the day. For people living on the Earth's equator, the 12-hour circle is on the horizon. Thus, at the equator, the days and nights are always 12 hours, all year long.

Creating the seasons

The seasonal weather is created by the combination of the efficiency with which the earth (and thus the atmosphere) is heated and the length of time (daylight) given to heat it.

During the summer months of either hemisphere, the duration of daylight is long and the Sun is high in the sky. This combination warms the earth efficiently and causes the long, warm summer days. During the winter months, the Sun is low in the sky and the duration of daylight is shorter. This combination does not heat the ground very well, creating the colder winter weather.

You may have noticed the coldest winter days come about a month after the winter solstice and the warmest days of summer come about a month after the summer solstice. It takes about one month for the cooling or warming process to affect our weather. This effect is called, *seasonal lag*. After all, it takes time for a pan of water to reach boiling after placing it on the burner. Also, hopefully, you would not place your hand on the burner only one minute after turning it off. In any physical system, it takes time for heat energy to build up, or to dissipate.

Now we have seen how the change in the Sun's declination – and thus its altitude – changes the length of daylight and the sunlight's efficiency in ground heating. Because the changing altitude is caused by the Sun's changing declination, the next step to understanding seasons is to understand why the Sun's declination changes during the year. So we now look at why the Sun has an annual north–south motion.

5.4 The Earth's tilted axis of rotation

Look at the *Equatorial Chart*. The celestial equator is the straight line across the middle of the map. The 23°.5 tilt of the Earth's axis of rotation (Figure 5.21) causes the ecliptic to snake 23°.5 N and S of the celestial equator, thus changing the Sun's declination, which in turn causes the seasonal variation in the Sun's local meridian (high noon) altitude and the change in the length of daylight. What does a "23°.5 tilt" really mean?

The Earth's axial tilt, also known as the **obliquity of the ecliptic**, is currently 23° 16′ 21″.448 or 23°.273 (Epoch 2000.0; Lang, 1992). Recent research suggests that the stability of the Earth's rotational axis is due to the presence of the Moon. In other words, without the Moon, the Earth's axial tilt would change drastically and somewhat randomly (Comins, 1999).

Measuring the Earth's tilt

Humans determine "up" and "down" from their sensation of the Earth's gravity. However, in interplanetary space there is no sensation of gravity to indicate these directions – there is no sense of "up" and "down." (In interplanetary space we are actually in the gravitational field of the Sun. It is possible to define an up and down relative to the Sun's field. However, because the field is spherical and each planet has its own orbital path around the Sun, this definition is not convenient for measuring the rotational axis tilt of the planets.) We must be careful to specify a reference for measuring the Earth's tilt. We need a reference which is independent of the position of the observer and the position of other bodies within the solar system.

As the Earth travels around the Sun, it follows an orbital path. We now make the following general statement:

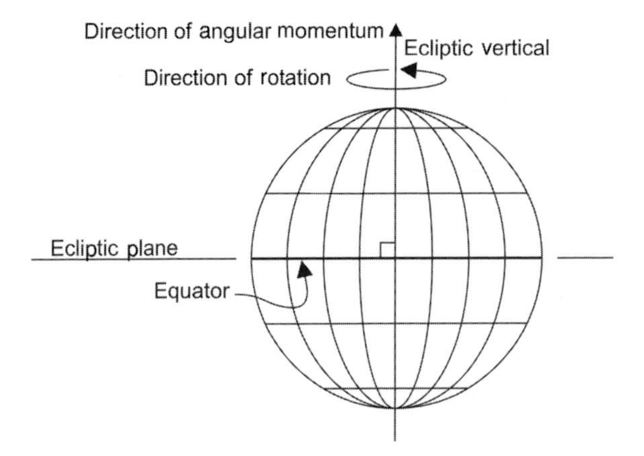

Figure 5.20 If the Earth's axis were not tilted, the axis would be perpendicular to the ecliptic plane. The ecliptic plane would pass through the Earth at the Earth's equator, causing the celestial equator and the ecliptic line to be the same. In this case, there would be no variation in the Sun's declination and thus, no seasons. There would perhaps be climate zones with frozen polar regions, temperate latitudes, and tropical equatorial zones.

Any two objects in orbit around each other must orbit each other in a mathematical plane. The plane passes through the center of both objects. In general, this plane is called an **orbital plane** *and it contains the orbital paths of both bodies.*

When dealing with the planetary orbits we sometimes forget that two objects orbit each other. That is, the Earth orbits the Sun *and* the Sun orbits the Earth. Strictly speaking, two objects orbiting each other actually orbit their *center of mass*, which is also called their *barycenter*. In the case of the planets, the Sun is so overwhelmingly massive, its motion about the barycenter is not easily perceived. (Measuring this motion of a star is how we've found planets outside our own Solar System.)

The Earth's orbital plane has a special name. It is called the **ecliptic plane**. We use the ecliptic plane to define "up" and "down" in the interplanetary space of the Solar System. If we are located above the ecliptic plane then we can look "down" to see the Earth's north pole. If we are below the ecliptic plane, we can look "up" to see the Earth's south pole. The line *perpendicular* to the ecliptic plane defines the vertical (up and down) of the Solar System. This perpendicular line passes through (and defines) the north ecliptic pole and south ecliptic pole (the NEP and SEP; see Figures 4.35, 4.36, the *North-polar Chart* or the *South-polar Chart*).

The 23°.5 tilt of the Earth is measured with respect to the ecliptic plane's definition of vertical. If the Earth were not tilted on its axis (Figure 5.20), its axis would be vertical, or perpendicular to the ecliptic plane. The celestial poles and the ecliptic poles would be the same point on the celestial sphere. The ecliptic line (the apparent path of the Sun through the constellations) represents the location where the ecliptic plane passes through the celestial sphere. If the Earth were

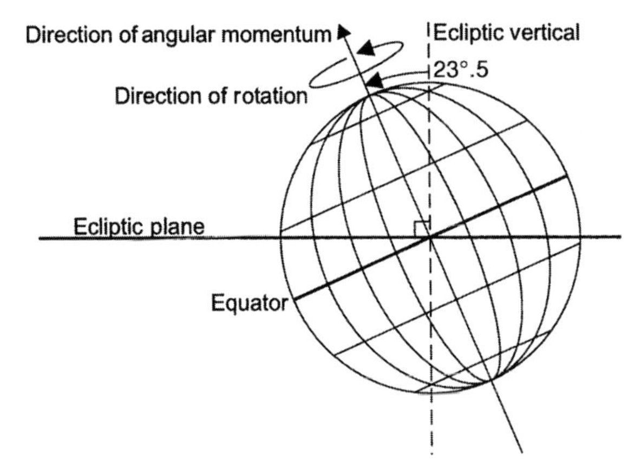

Figure 5.21 The Earth's rotational axis is tilted by 23°.5 to the perpendicular to the ecliptic plane. The ecliptic line (as seen on the *Equatorial Chart*) is created by the intersection of the ecliptic plane with the celestial sphere. The ecliptic line (and thus the Sun's) declination varies from the celestial equator by the amount of the Earth's axial tilt. The variation in the Sun's declination causes the effects which bring about the seasons on Earth.

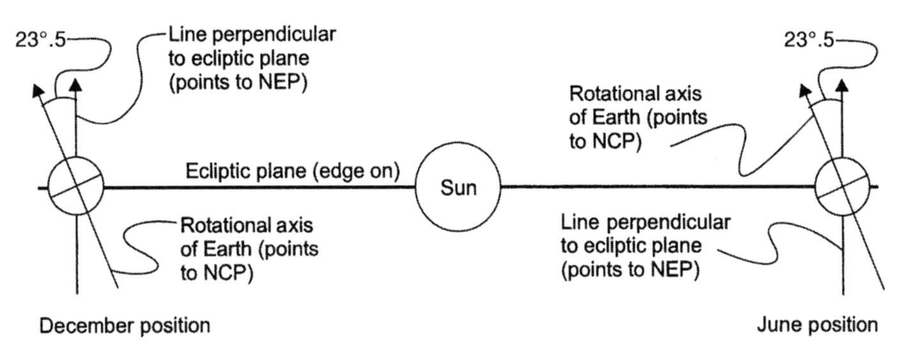

Figure 5.22 No matter where the Earth is located in its orbit about the Sun, the axis of rotation always points to the north celestial pole (Polaris, ignoring the precessional motion covered in Section 4.6). The short line shown across the diameter of the Earth is the equator. Notice the equator is also tilted by 23°.5. The two positions shown here are six months apart. The hemisphere having winter weather is "tilted away from the Sun," causing the sunrays to strike the ground at a lower angle. Ultimately, the seasons are caused by the tilt of the Earth's rotational axis.

not tilted, the ecliptic line would match the celestial equator because the ecliptic plane would pass through the Earth at its equator. The Sun would remain on the celestial equator (directly above the Earth's equator) all year long. The Earth would not have any seasons. Instead, there might be "climate zones."

However, the Earth's axis *is* tilted. The Earth's axis makes a 23°.5 angle with the vertical line of the ecliptic plane (Figures 5.21, 5.22, and 5.23). Notice (by looking at the *Equatorial Chart*) that the minimum and maximum solar declination is equal to the Earth's axial tilt.

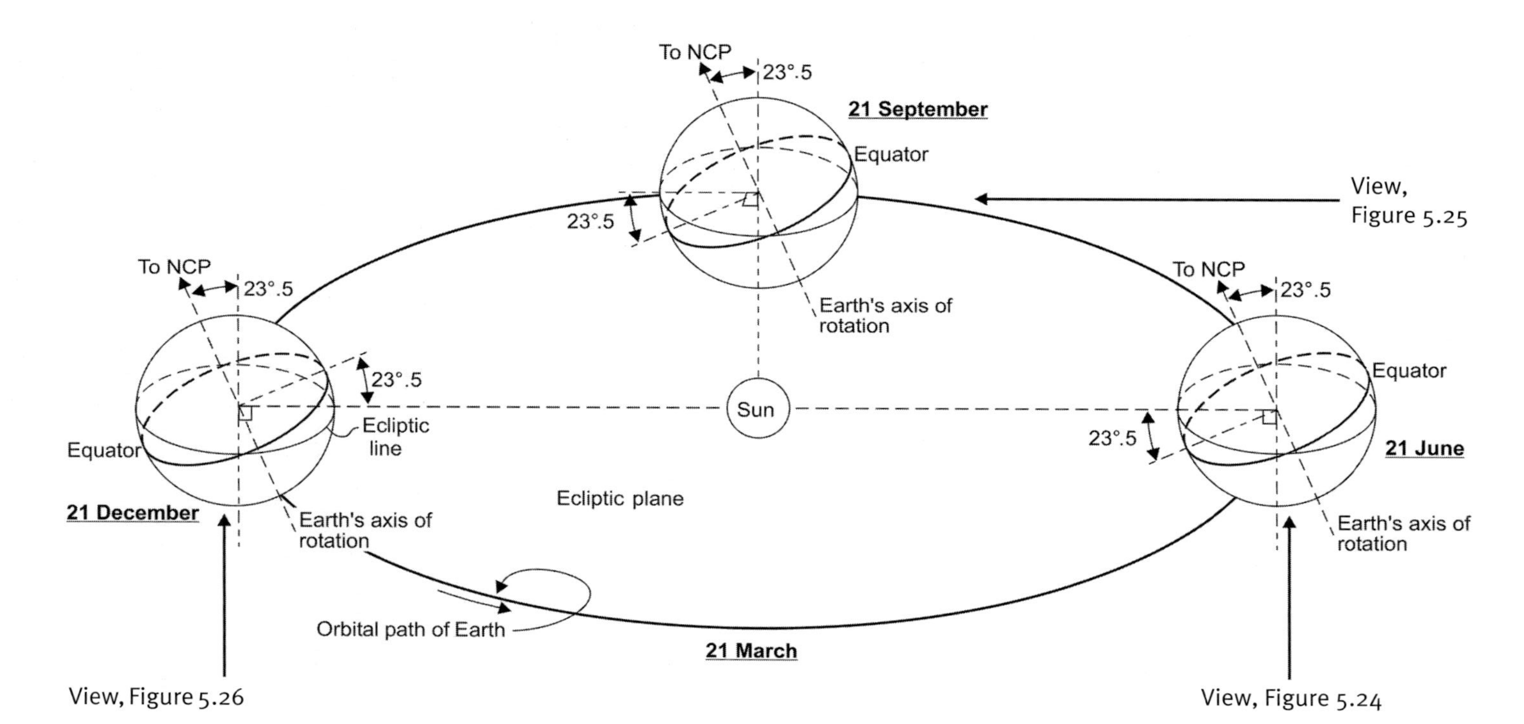

Figure 5.23 The seasons are caused by the tilt of the Earth's rotational axis with respect to the ecliptic plane. For each position shown, the darker line is the celestial equator and the lighter line is the ecliptic, as they would be seen on the celestial sphere. In June, the northern hemisphere is "tilted toward the Sun" and experiences summer, while the southern hemisphere is having winter weather. In December, the southern hemisphere is "tilted toward the Sun" and is in its summer season, while the northern hemisphere is in winter. Use this figure in combination with Figures 5.24 through 5.26 to understand how the sunlight hits the earth during each of the seasons. Imagine looking at the ecliptic plane edgewise in the respective direction for each point of view shown in these figures. The Earth must keep its axis of rotation pointed near Polaris all year because of *conservation of angular momentum* (see Appendix A, p. 244).

Because of the tilt of the Earth's rotational axis, the Sun's transit altitude varies during the year. This variation causes increasing or decreasing efficiency in the Sun's ability to warm the earth. This, along with the variation in the length of daylight also caused by the Earth's tilt, causes the seasons of the year. The obliquity of the ecliptic is the sole reason for the seasons.

Milankovitch cycles

Milutin Milankovitch (1879–1958) was a Serbian engineer and Professor of Applied Mathematics at the University of Belgrade. He worked out a mathematical model for the creation of the ice ages. This model was built on the small perturbations of the other planets in the Solar System and the small gravitational effects of neighboring stars. The result is a set of cyclic variations in the Earth's orbit around the Sun. These variations are:

1. Variations in the Earth's orbital eccentricity. The shape of the Earth's orbit around the Sun changes with a 21 000-year period.
2. Variation in the obliquity of the ecliptic. The angle between the Earth's rotational axis and the ecliptic varies between 24°.5 and 21°.5 with a period of about 41 000 years.
3. Precession: is the 26 000-year orbit of the celestial poles discussed in Section 4.6.

Collectively these cycles are known as the Milankovitch cycles.

Milankovitch proposed that changes in the amount of solar radiation received at some latitudes plays a greater role in the creation and decline of ice sheets than at other latitudes. Based on the results and suggestions of others working on the problem of the ice ages, he chose 65° N as the most important region to model and suggested these astronomical cycles as a major cause of the ice ages. Other factors play a role too. If the snow does not melt, it raises the reflectiveness (the *albedo*) of the Earth, making a large contribution to cooling the Earth. Also, with increased ice levels the atmosphere's greenhouse gases are reduced, which contributes to cooling the Earth.

The theory was ignored for 50 years or so, until new data about the ice ages pointed to a stronger connection between the astronomical cycles and the cycle of ice ages. These data show that there have been dozens of ice ages over the past few millions of years and that we are currently headed for another one. However, humankind's industrial revolution may cause a delay in the next ice age.

5.5 Tropics and circles

The tilt of the Earth's rotational axis allows the Sun to warm certain regions more efficiently than others. The heavier sunlight line on each of Figures 5.24 through 5.26 denotes the ray striking the earth at a right angle (parallel to the surface normal). That ray warms the earth most efficiently. During the June solstice, the Sun warms the northern hemisphere because it is "tilted toward"

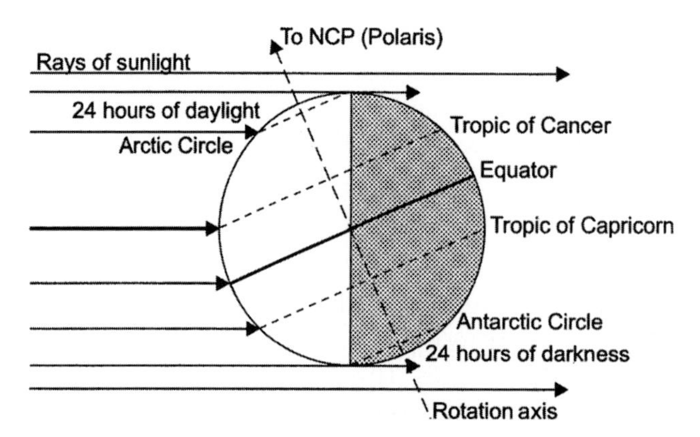

Figure 5.24 At the June solstice sunlight hits the Earth's surface directly on the Tropic of Cancer (23°.5 N). Anyone standing at this latitude sees the Sun directly overhead at high noon. Anyone below the Antarctic Circle has 24 hours of darkness. Anyone above the Arctic Circle has 24 hours of daylight. This is summertime for the northern hemisphere.

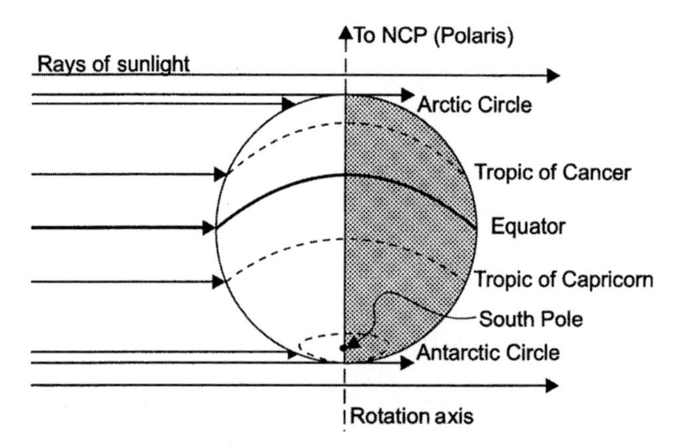

Figure 5.25 At the September equinox sunlight hits the Earth's surface directly on the equator. Anyone standing on the Earth's equator sees the Sun directly overhead at high noon. Everyone on Earth sees 12 hours of daylight and 12 hours of night, except at the poles, where they sees 24 hours of sunrise/sunset.

the Sun. During the December solstice, the northern hemisphere receives less heat because it is "tilted away" from the Sun. Of course, the seasons are reversed for the southern hemisphere. Use these three figures in combination with Figure 5.23 to understand how the northern hemisphere first tilts toward the Sun, then away.

On the June solstice, the declination of the Sun is +23°.5 and the Sun is directly over the *Tropic of Cancer*. The **Tropic of Cancer** is a line drawn parallel to the Earth's equator at latitude 23°.5 N. The **Tropic of Capricorn** is a line drawn parallel to the Earth's equator at latitude 23°.5 S. *Tropic* is a Latin word meaning "turning point." These lines represent the farthest latitude, north or south, at

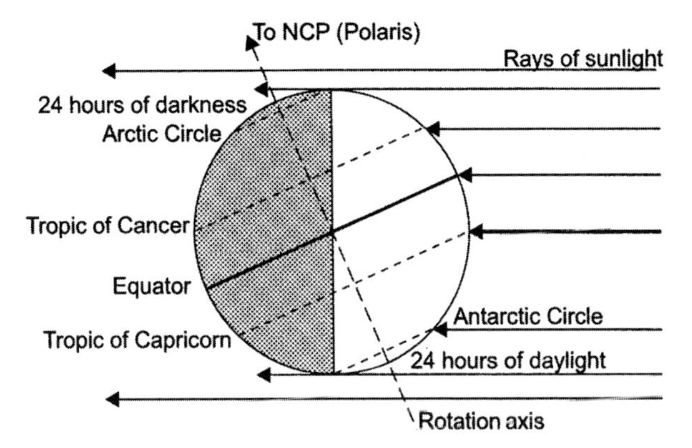

Figure 5.26 At the December solstice sunlight hits the Earth's surface directly on the Tropic of Capricorn (23°.5 S). Anyone standing at this latitude sees the Sun directly overhead at high noon. Anyone below the Antarctic Circle has 24 hours of daylight. Anyone above the Arctic Circle has 24 hours of darkness. This is summertime for the southern hemisphere.

which the Sun will pass through the zenith at least once (twice for latitudes between the tropic lines) during the year.

A few thousand years ago when these lines were named, the June solstice occurred when the Sun was in the constellation Cancer and the December solstice occurred when the Sun was in Capricornus. Because the Earth's rotational axis precesses (see Section 4.6) the June solstice is now in Gemini and the December solstice is located in Sagittarius. The latitudes between the tropic circles are called the *tropics*. For any latitude outside of the tropics, the Sun never passes through the zenith.

The **Arctic Circle** and the **Antarctic Circle** are located at 66°.5 N and 66°.5 S, respectively. The Arctic Circle designates the latitude above which the Sun never sets on the June solstice – it is above (or on) the horizon at midnight. This region has also been called "the land of the midnight Sun." During the December solstice this same region is in 24 hours of darkness – the Sun never rises. The Antarctic Circle represents the same events for the south-polar regions, but for the opposite solstice dates.

As one moves from the Arctic (or Antarctic) circle to the pole, the number of days of 24-hour daylight during the summer months changes. If standing on the circle, only the day of the summer solstice (for that particular hemisphere) has 24-hours of daylight. When standing on the Arctic Circle on 21 June, the Sun brushes the northern horizon at "midnight." When standing on the Antarctic Circle, the Sun brushes the southern horizon at "midnight." After the solstice, the Sun's declination is decreasing and so it goes below the horizon again. Moving a few degrees closer to the pole, increases the day count (of 24-hour daylight) by a few days. Only when standing directly on the Earth's pole, will the observer have six months (equinox to equinox) of 24-hour daylight.

The latitudes above the Arctic Circle are called the Arctic. The latitudes below the Antarctic Circle are called the Antarctic. The latitudes (both northern and southern hemispheres) between the tropics and polar circles are called the temperate zones. The temperate zones have seasonal weather patterns.

5.6 Seasonal last thoughts

There is another effect caused by the Earth's precessional motion (see Section 4.6). In 13 000 years, when Vega is the North Star, the seasons of the year will be slightly different. Now, the Earth is closest to the Sun on 4 January. Given that we adjust our calendar to keep the March equinox at 21 March, when Vega becomes the North Star, the Earth will be at its farthest distance from the Sun on 4 January. This will make the northern hemisphere winters a little cooler and the northern hemisphere summers a little warmer, with the lengths of the seasons swapping as well. The southern hemisphere's seasons will be affected in the opposite sense, with the winters getting slightly warmer and the summers getting slightly cooler. The overall character of our winters and summers is by far more affected by other climactic changes and conditions, such as the El Niño and La Niña effects of the interaction of the Pacific Ocean with the atmosphere.

If we choose *not* to update our calendars, then in 13 000 years the northern hemisphere's winter will be in July and summer will be in January. This is, of course, currently the case for the southern hemisphere. The hemispheres will just swap seasons. In 26 000 years they'd be back to "normal."

Questions for review and further thought

1. On what date (approximately) is the Earth farthest from the Sun?
2. In one sentence, why do we have seasons?
3. List the four important dates for the seasons.
4. What is the meaning of the term *equinox*?
5. Why does it happen that the date on which there is 12 hours of daylight and 12 hours of night (ignoring refraction) occur on the same date that the Sun is on the celestial equator?
6. What is the origin and meaning of the word *solstice*?
7. Describe the main process by which the atmosphere is heated.
8. At what (optical) angle does sunlight heat the ground most efficiently?
9. Explain why the Earth's equatorial regions are always warm and its polar regions are always cold.
10. Explain how ground heating causes the northern and southern hemispheres to have opposite seasons.
11. What role does the duration of daylight play in the seasons?
12. What other factors, besides the angle of the sunlight, affect the rate at which the Sun may heat the ground?
13. Explain why people living at the Earth's equator have 12-hour days throughout the year.

14. Explain how solar altitude and daylight duration are related to the seasons.
15. Explain the effect called *seasonal lag*.
16. What is an *orbital plane*?
17. What is the *ecliptic plane*?
18. Explain how the ecliptic plane is used to define "up" and "down" in the Solar System.
19. Explain how the Earth's axial tilt is related to (measured with reference to) the ecliptic plane.
20. Describe and explain the Tropic of Cancer and the Tropic of Capricorn.
21. Describe and explain the Arctic Circle and the Antarctic Circle.
22. Explain the phrase "land of the midnight Sun."
23. Describe possible climate zones of the Earth, if the Earth's axis were not tilted.

Review problems

1. Calculate the transit altitude of the Sun on the four seasonal dates for an observer at 75° N latitude.
2. Calculate the transit altitude of the Sun on the four seasonal dates for an observer at 35° N latitude.
3. Calculate the transit altitude of the Sun on the four seasonal dates for an observer at 15° N latitude.
4. Calculate the transit altitude of the Sun on the four seasonal dates for an observer at 15° S latitude.
5. Calculate the transit altitude of the Sun on the four seasonal dates for an observer at 35° S latitude.
6. Calculate the transit altitude of the Sun on the four seasonal dates for an observer at 75° S latitude.
7. Calculate the transit altitude of the Sun on the four seasonal dates for your latitude.
8. Make a set of diagrams similar to Figures 5.2 through 5.12 for your city's latitude.
9. Make a drawing similar to Figures 5.18 and 5.19 for the western horizon.

There are more problems dealing with the motion of the Sun in Chapter 8, Observation Project 1.

6
The phases of the Moon

When we look at the Moon, we might first notice its size, brightness, and surface features. Many people have heard the phrase "the man in the Moon." This is a folklore image formed by the pattern of light and dark areas on the surface. The lighter colored areas are rocky, mountainous regions called **terra** – the Latin word for earth. The dark areas are flat-plains regions called **maria**. *Mare* is a Latin word for "sea" (plural *maria*; see Figure 6.1). These misnomers are carried over from Galileo's lunar observations. We can also see many craters, caused mostly by the impact of Solar System debris. We find craters in both the mountainous and maria regions. Craters are named for prominent philosophers and scientists of the past.

Because the Moon is in a slightly elliptical orbit around the Earth, its apparent size varies from 29″.4 to 33″.5. It is about the same apparent size as the Sun. Sometimes we see a huge, orange, full Moon, rising over the eastern horizon. The size is an illusion – a trick of the mind. When the Moon is near the horizon we can compare its size to familiar objects, such as trees and buildings. Because we think of these objects as large, we also think of the Moon as being large. As it climbs into the sky, however, it moves away from these familiar objects and we can no longer compare it to them. When we view the high, isolated Moon, it appears small, simply because there is nothing next to it to which to compare it. A multiple exposure photograph of the rising Moon demonstrates its apparent size does not change as it moves away from the horizon.

The Moon's orange color when it is near the horizon is due to the same atmospheric effect that causes the red sunrise and sunset – Rayleigh scattering. This effect removes most of the blue light coming from the Moon, leaving the Moon with a reddish tint. The effect decreases as the Moon gets higher in the sky and the moonlight passes through less atmosphere. The effectiveness of Rayleigh scattering is influenced by the amount of particulate material in the atmosphere during the period of the observation.

Figure 6.1 The full Moon rises in the east at sunset and sets in the west at sunrise. The dark regions are the maria and the light colored regions are the terra. (Photo courtesy of Kevin S. Jung.)

6.1 The motion of the Moon

The Moon has two motions that we consider here – diurnal motion and synodic (monthly) motion. Of course its motion is far more complicated than the combination of just these two, but these are the only ones needed to understand its overall celestial sphere motion and lunar phases.

The Moon's diurnal motion

Because it is due to the Earth's rotation, the Moon has the same diurnal motion as all the other objects in the sky. It rises in the east and sets in the west.

The amount of time we can see the Moon in the daytime sky is a little less than the amount of time we can see it in the night-time sky, because as the Moon comes closer to the Sun it becomes a narrow crescent that is easily out shined by the glaring Sun. Also, the new Moon phase – when the Moon is in the same position as the Sun – cannot be seen at all. Therefore, although the Moon is in the daytime sky as well as in the night-time sky, its proximity to the Sun will not allow us to view the Moon during this entire time.

The Moon's synodic (orbital) motion

We always see the same face of the Moon. The Moon has a "near side" and a "far side." The far side can never be seen from Earth. This behavior does not mean the Moon does not rotate on its axis. To keep the same face always toward

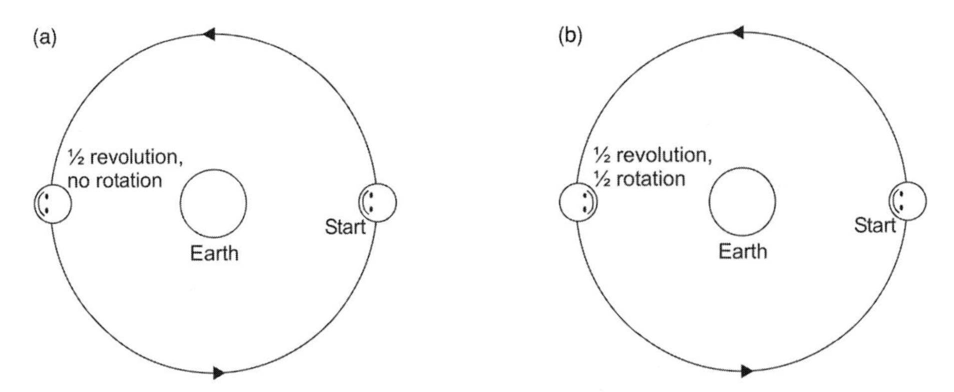

Figure 6.2 The rotation of the Moon on its axis is synchronized to its revolution around the Earth. This is called a synchronous orbit and it ensures the same lunar face is always toward the Earth. (a) If the Moon did not rotate on its axis we would not always see "the face." (b) The Moon rotates as it revolves around the Earth such that we always see the same face.

Earth, the Moon's rotation is synchronized to its orbit around the Earth. There are some places where reference is made to the "dark side" of the Moon. This usually refers to the far side. Because the Moon rotates on its axis, there is no permanent dark side of the Moon. Any given location on the Moon has day and night, just like the Earth. The Moon's rotation allows every part of its surface to be lit for half of the lunar day. Anyone living on the Moon experiences day and night, just like on Earth. However, lunar days and nights are both (roughly) 14 Earth days long!

Any object (usually a moon) whose rotation is synchronized to its orbit is in a **synchronous orbit**. The Moon is in a 1 : 1 (one-to-one) synchronous orbit. This means it rotates on its axis once for each orbit around the Earth. Mercury happens to be in a 3 : 2 synchronous orbit around the Sun – it rotates on its axis three times for every two orbits. The Moon's synchronous orbit means "the man in the Moon" always faces the Earth (Figure 6.2). Synchronous orbits are caused by tidal forces. In the distant past, the Moon rotated faster and showed all sides to the Earth. But the Earth's tidal forces acting on the Moon caused its rotation to slow to the point where its rotation is now synchronized to its revolution around the Earth.

The Moon also has *libration* motions which allow us to see more than half of its surface. In other words, the man in the Moon has a slight "head turning" motion that allows us to see around the "edge." We can actually see 59% of the Moon's surface from the Earth.

To keep the same face always toward Earth, the Moon's rotation is synchronized to its orbit around the Earth. The Moon is in a 1 : 1 *(one-to-one) synchronous orbit. This means it rotates on its axis once for each orbit around the Earth. The Moon's rotational motion allows anyone living on the Moon to experience day and night, just like on Earth.*

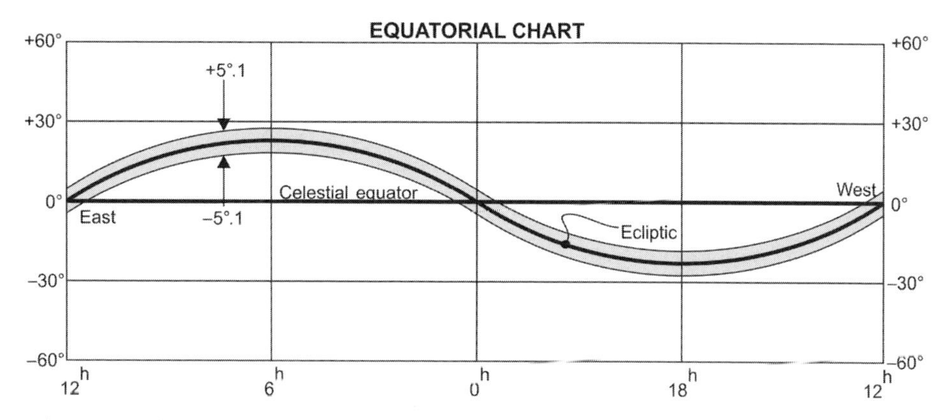

Figure 6.3 The Moon moves along a path near the ecliptic from west to east in 27.32 days (with reference to the stars), while the Sun moves along the ecliptic from west to east in 365.25 days. The Moon's orbital path lies within the shadowed region shown here. The Moon is never farther than 5°.1 (5° 8′ 42″) away from the ecliptic.

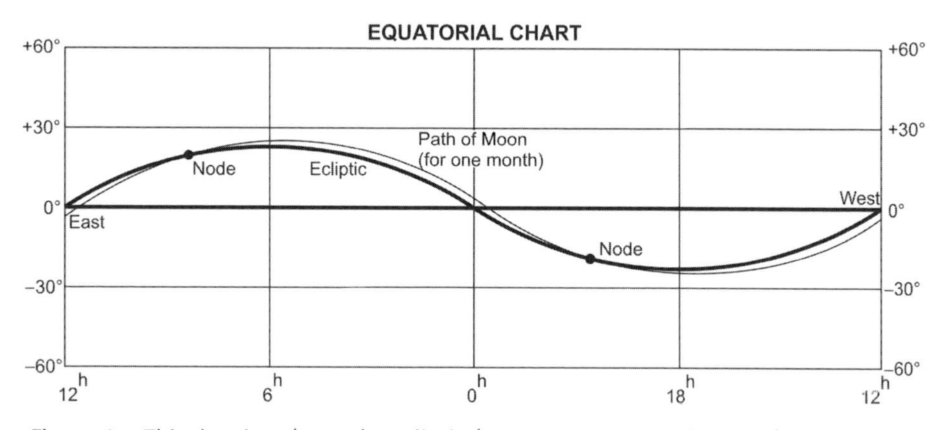

Figure 6.4 This drawing shows the ecliptic (one year's motion of the Sun) as the darker line and one month's motion of the Moon as the lighter line. The two small dots represent the nodes in the Moon's orbit for this month. The nodes are 180° (or 12 hours, right ascension) apart.

The Moon's orbital motion makes it appear to move in our sky from west to east, close to the ecliptic line. Its orbital path carries it off the ecliptic by at most 5°.1 (Figure 6.3). Because the Moon is always so close to the ecliptic, we see it only within the zodiac constellations, taking (on average) slightly more than two days to pass through each constellation. The Moon crosses the ecliptic twice during each orbit around the Earth. The points where the Moon's path crosses the ecliptic are called **nodes** (from the Latin word for "knot"). They are 180° apart, on opposite sides of the sky. The nodes move slowly along the ecliptic with a period of 18.61 years (Figures 6.4 and 7.21).

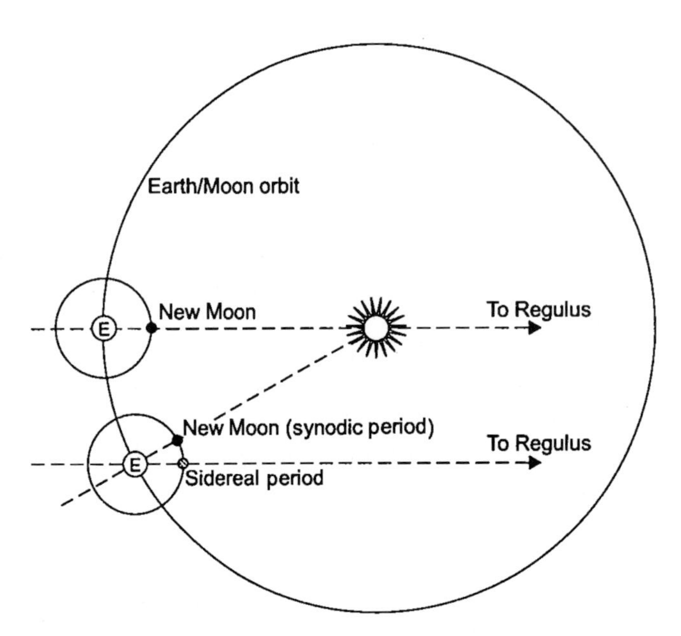

Figure 6.5 The sidereal period of the Moon is the time it requires for the Moon to orbit the Earth with reference to a star. The synodic period is the time between two successive new Moon phases. The difference is caused by the orbital motion of the Earth–Moon system around the Sun.

With reference to the stars, it takes the Moon 27.32 days to orbit the Earth. This is the Moon's *sidereal period*. The **sidereal period** of any object in orbit is the length of time it takes to orbit the parent body with reference to the stars. Because of the motion of the Earth–Moon system around the Sun, it takes 29.53 days for the Moon to complete an orbit with reference to the Sun. This is the Moon's *synodic period* (from the Greek word *synodos* meaning "meeting"). The Moon's **synodic period** is the time between two successive new Moon phases (Figure 6.5). Notice the synodic period is about one calendar month.

The Moon's orbital motion makes it appear to move in our sky from west to east, close to the ecliptic line. Its orbital path carries it off the ecliptic by $5^\circ.1$ at most. The Moon crosses the ecliptic twice during each orbit around the Earth. The points where the Moon's path crosses the ecliptic are called nodes.

6.2 The phases of the Moon

There are a few cases where I have run into students who were taught that the Moon has phases because it moves through the shadow of the Earth. While it is true that the Moon does moves through the Earth's shadow from time to time, this is not the reason for the phases of the Moon. The **lunar phases** are caused by the position of the Moon relative to the Earth and Sun, and our view of the Moon from here on Earth. The Moon revolves around (orbits) the Earth, and the Earth–Moon system revolves around the Sun. Because of the motion of the Earth–Moon system around the Sun, it takes 29.53 days for a complete cycle of lunar phases. This is again, the Moon's synodic period. A *lunar phase* refers to the apparent shape of the Moon as seen from Earth. Everyone on Earth sees the same lunar phase on any given night.

Given the current lunar phase and date, the Moon can be plotted on the *Equatorial Chart* along with the Sun. The dates shown along the bottom of the ecliptic line on this chart indicate the position of the Sun on any particular date. The degree markings along the top show the Sun's *celestial longitude* (λ) in degrees (p. 35). The degree markings along the ecliptic also give the celestial longitude of the Moon, even though its path actually carries it off the ecliptic line. A description of plotting the Moon's position (for use in Observation Project 1, Part V) is given in each of the following sections describing the phases.

In Figures 6.9, 6.10, and 6.12 through 6.17, the number of days given for each phase is the number of days past the new Moon phase. The dashed line labeled "L.M." represents the local meridian. As can be seen from these figures, the appearance of the Moon during each of its phases depends on the location (latitude) of the observer. The description of which portion of the Moon is lit is best made with compass directions, as opposed to "left" or "right" side. For instance, at the equator the Moon passes through or nearly through the zenith. Thus "left" and "right" depend only on whether the observer is facing north or south.

> *The lunar phases are caused by the position of the Moon relative to the Earth and Sun, and our view of the Moon from here on Earth. A lunar phase refers to the apparent shape of the Moon as seen from Earth. Because of the motion of the Earth–Moon system around the Sun, it takes* 29.53 *days for a complete cycle of lunar phases. This is also the Moon's synodic period.*

New Moon

The **new Moon** occurs when the Moon is between the Earth and the Sun. As seen from Earth, the Sun and the Moon are in the same direction in the sky. This phase cannot normally be seen for two reasons. First, because the lunar "face" is unlit and, second, the glare of the Sun prevents us from seeing much of any celestial object, let alone one that appears unlit. The far side of the Moon is lit by the Sun. The "man in the Moon" has the backside of his head lit by the Sun, not his face.

However, if the Moon happens to pass directly in front of the Sun, which happens when both bodies are at the same node simultaneously (see Section 7.4 and Figures 7.9 and 7.24) we see a *solar eclipse*. At this time, we can see a "true" new Moon. The Moon appears as a dark disk in front of the Sun. Technically, the new Moon phase lasts only a few hours – while it is positioned within a degree of the Sun. In practical (observational) terms, it may last a couple of days because it is so difficult to see the narrow crescent phases. We can see from Figures 6.6

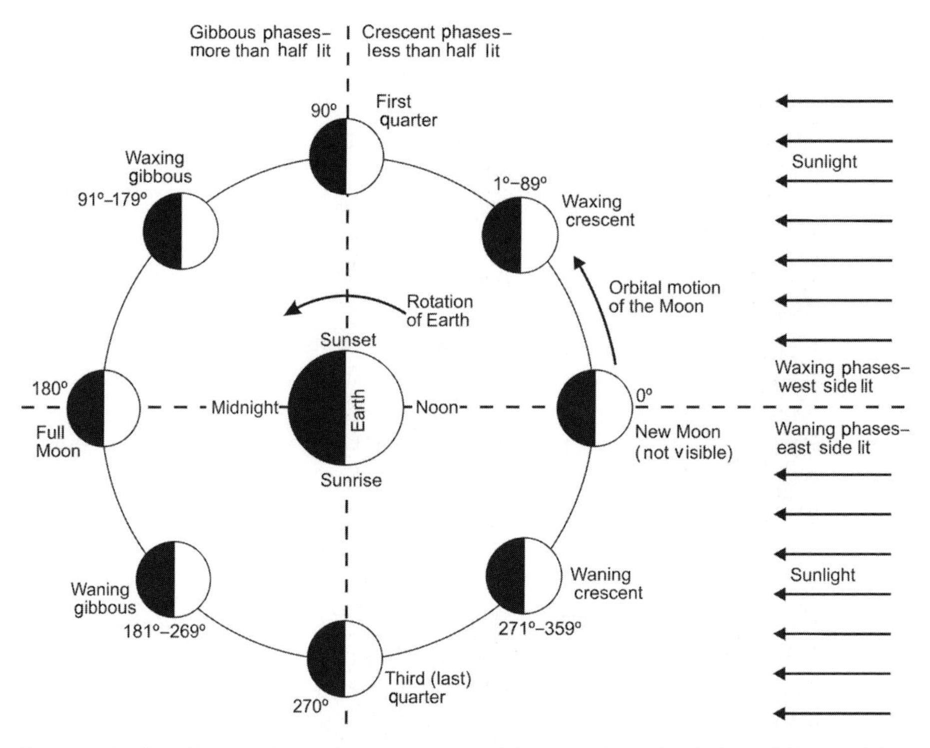

Figure 6.6 The phases of the Moon are caused by our view of it in its orbital position relative to the Earth and Sun. The new, first quarter, full, and last quarter phases last only a few hours. Therefore the date on which they occur can be printed on calendars. The crescent and gibbous phases last about one week each. It takes the Moon 29.5 days to complete one cycle of phases (its synodic period). The appearance of the Moon for each phase, as seen from Earth is shown in Figures 6.7 and 6.8. The lit halves of both the Earth and the Moon always point toward the Sun. The angles given for each phase are the difference in celestial longitude in the positions of the Moon and Sun. The angles are measured from the Sun's position eastward to the Moon's position.

to 6.10 and Table 6.1 (and Figures 6.14 and 6.15) the new Moon rises in the east at sunrise and sets in the west at sunset – as it must, because it is at the same position as the Sun.

Example: When plotting the new Moon on the *Equatorial Chart*, it is positioned in the same place as the Sun. As seen in Figure 6.6, the angle between the Sun and the Moon is zero degrees. On 1 July, the Sun's celestial longitude is 99°. If a new Moon occurs on 1 July, it is also positioned at longitude 99°, but its declination is probably not the same as the Sun's. If it is the same (or very nearly the same), then the Sun and the Moon are near a lunar node and a solar eclipse occurs.

Waxing phases

The **waxing phases** are all the lunar phases where the western side of the Moon is sunlit. "We are 'waxing up' the Moon to make it look good when it is full!"

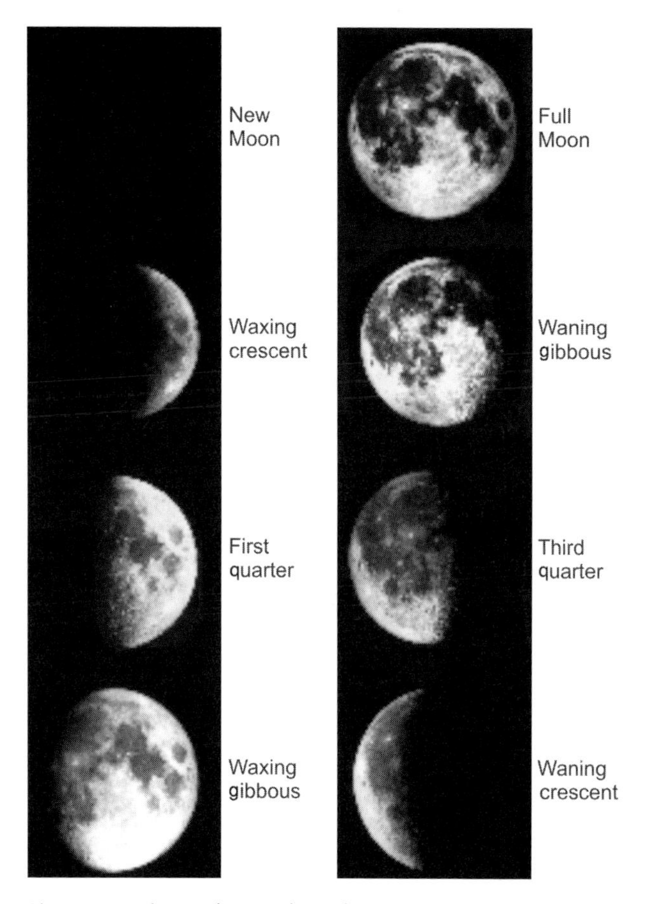

Figure 6.7 These photos show the appearance of the Moon during its phases for observers in the northern latitudes.

"Waxing" comes from the German word, *wachsen* – "to grow." The waxing phases are best seen in the late afternoon and evening skies.

Waxing crescent Moon The **waxing crescent Moon** (Figure 6.11) is crescent shaped and lit on the west side. The "horns" (also called the "cusp") of the crescent always point away from the Sun. Although this phase of the Moon can be seen during the daytime (the Moon is just to the east of the Sun), a young crescent moon is usually washed out by the glare of the Sun. The waxing crescent rises just after sunrise and it sets just after sunset, but it is most easily seen just after sunset in the west. See Figures 6.6, 6.9, and 6.10, and Figures 6.16 and 6.17. This is the lunar phase between the new and first quarter phases, lasting about seven days. "Crescent" comes from the Latin word, *cresco* – "to grow." On the *Equatorial Chart*, this lunar phase is plotted anywhere from 1° to 89° to the left of the Sun's position for the given date.

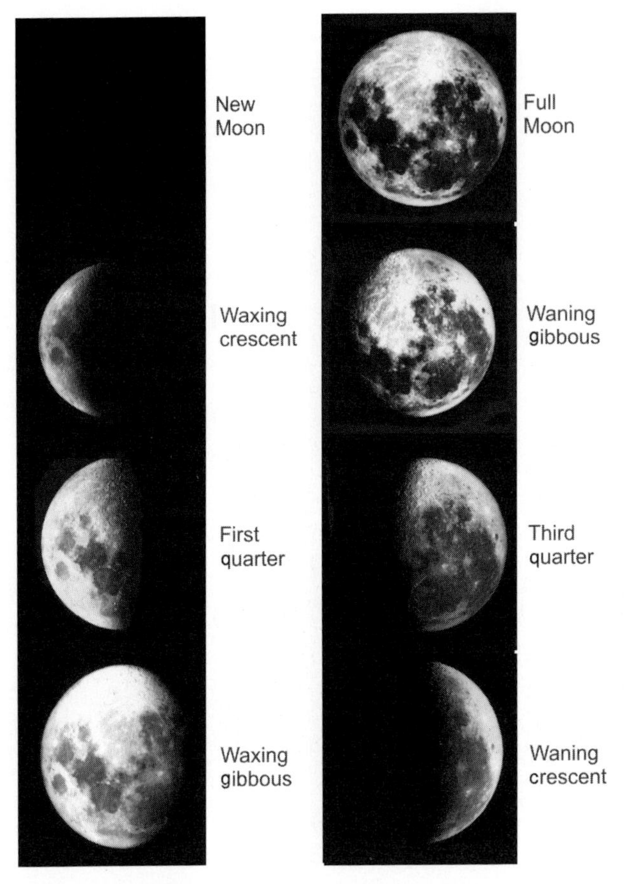

Figure 6.8 These photos show the appearance of the Moon during its phases for observers in the southern latitudes.

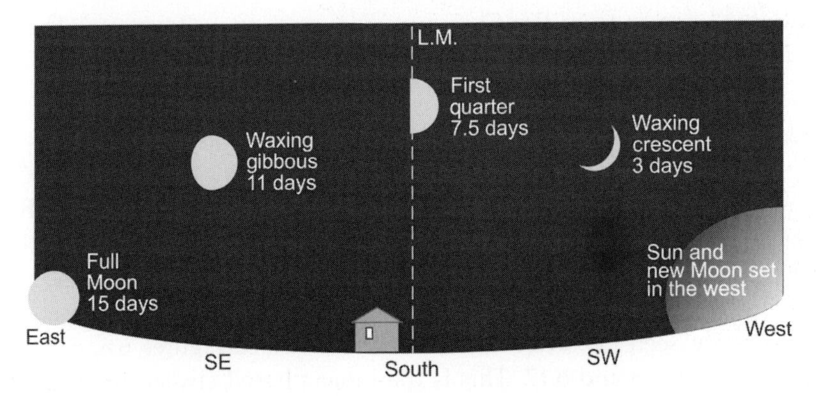

Figure 6.9 For observers in the northern latitudes the waxing lunar phases are lit on the right-hand side of the Moon's face. The apparent path of the Moon lies across the southern sky near the ecliptic line. Notice the direction of the horns of the crescent moon.

Table 6.1 *Lunar phase rise and set times*

Phase	Rises	Transits	Sets
New	Sunrise	Noon	Sunset
Waxing crescent	Morning	Afternoon	Evening
First quarter	Noon	Sunset	Midnight
Waxing gibbous	Afternoon	Evening	Early morning
Full	Sunset	Midnight	Sunrise
Waning gibbous	Evening	Early morning	Morning
Third (last) quarter	Midnight	Sunrise	Noon
Waning crescent	Early morning	Morning	Afternoon

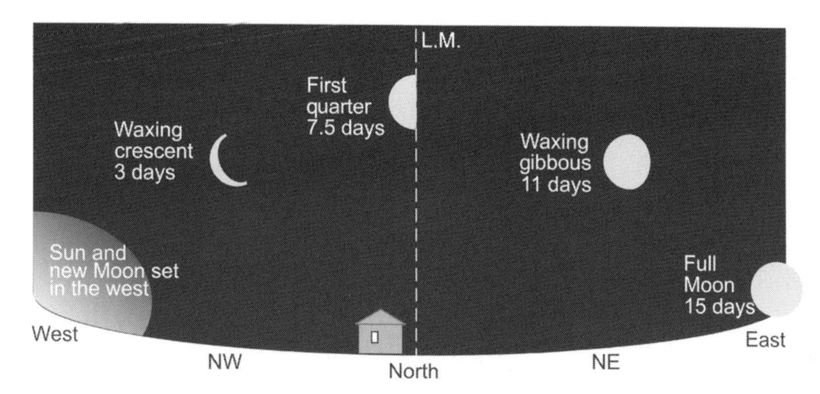

Figure 6.10 For observers in the southern latitudes the waxing lunar phases are lit on the left-hand side of the Moon's face. The apparent path of the Moon lies across the northern sky near the ecliptic line. Notice the direction of the horns of the crescent moon.

Figure 6.11 This photo of the waxing crescent Moon (taken from 43° N latitude) shows a fine example of earthshine. Light reflected from the Earth strikes the Moon, which in turn reflects some of the light back to Earth. This allows us to see a dimly lit night-side of the Moon. (Photo courtesy of Kevin S. Jung.)

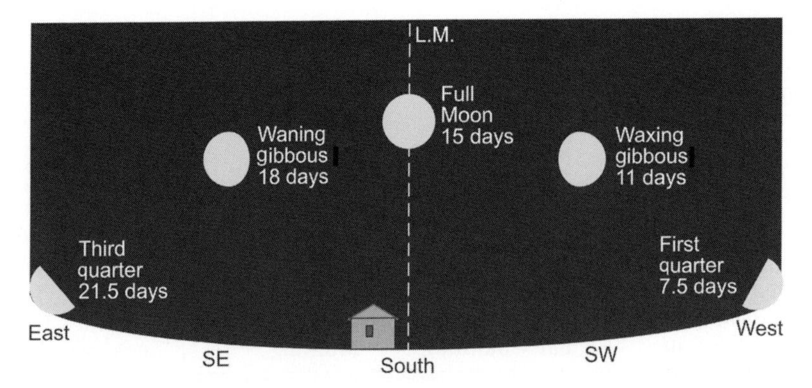

Figure 6.12 These are the sky positions of the lunar phases at midnight for northern latitudes. The Sun is on the antimeridian.

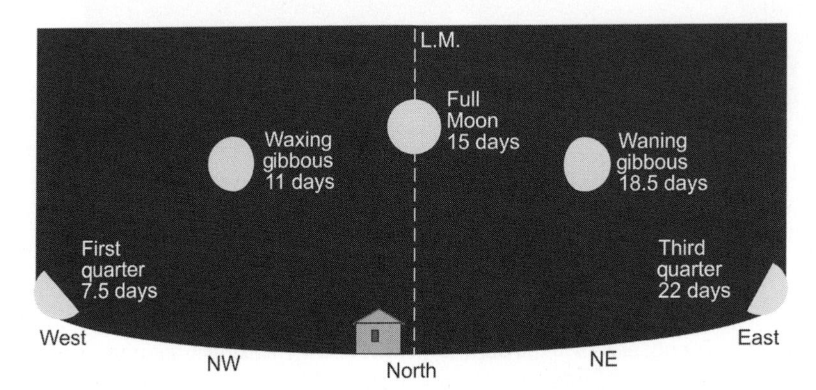

Figure 6.13 These are the sky positions of the lunar phases at midnight for southern latitudes. The Sun is on the antimeridian.

Example: If the date is 20 April, the Sun's celestial longitude is (approximately) 30°. The waxing crescent Moon (Figure 6.11) could be plotted anywhere from 31° to 119°. For general discussions, the waxing crescent is plotted 45° from the Sun. So here, the Moon would be positioned at 75° longitude.

First quarter Moon With the **first quarter Moon**, the west half of the Moon is sunlit. In this phase, the Sun and Moon are at a right angle (90°) to each other in the sky, with the Sun to the west of the Moon. In the northern latitudes, this phase is seen close to the local meridian in the southern sky at sunset. In the southern latitudes, it is seen in the northern sky at sunset. The first quarter Moon rises at noon and sets at midnight. See and compare Figures 6.6, 6.12, and 6.13, and Table 6.1, with Figures 6.16 and 6.17 later.

Note well: this is not called a "half Moon." There is no such phase! The name "first quarter" may come from two different notions: (1) the Moon is one-quarter of the way in its orbit around the Earth; (2) we can see half of the lit half of the

Moon's surface, thus we can see a total of one-quarter of the Moon's surface. This phase lasts only the few hours while the Moon is 90° away from the Sun. The true angle between the Sun and Moon is slightly less than 90° because the Sun is located near the center of our orbit around it. That is, the Sun's light rays are not truly parallel at our distance. The true first quarter phase happens with the right angle located at the Moon's position, not the Earth's. The exact angle at the Earth's position varies slightly because of the Moon's and the Earth–Moon system's slightly elliptical orbits. However, using the approximation that the Sun is 400 times farther than the Moon, the angle at the Earth's position is about 89°.9 (arctangent of 400). On the *Equatorial Chart*, the Moon's longitude is 90° greater than the Sun's.

Example: If the date is 20 February, the Sun's longitude is 330° (in the constellation Aquarius). On this date, a first quarter Moon is located at $330° + 90° = 420° - 360° = 60°$ longitude (in the constellation Taurus).

Waxing gibbous Moon The **waxing gibbous Moon** has more than half (from the western limb) of the Moon sunlit. This phase rises in the east in the afternoon and sets in the west in the early morning. It can easily be seen during the daytime but is best seen in the evening to early morning western sky (Figures 6.6, 6.9, and 6.10). The gibbous phase lasts about seven days, between first quarter and full Moon. "Gibbous" comes from the Latin word *gibbous* – "hump on the back." The hump is on the side of the Moon opposite from the Sun. It is very difficult to tell exactly when the gibbous phase starts and ends. Most people confuse an early gibbous Moon with a quarter Moon or a late gibbous Moon with a full Moon. Doing the calculations to find and thus know the time of the quarter and full phases is the only way to know for sure when the gibbous phase starts and ends. The waxing gibbous phase lies between 91° and 179° away from the Sun (Figure 6.6).

Example: For the date of 11 May, the longitude of the Sun is 30°. The waxing gibbous phase could be plotted anywhere from 121° to 209° in longitude. The waxing gibbous phase can be plotted 135° ahead of the Sun for most purposes in general discussions. For this example then, the waxing gibbous Moon can be plotted at longitude 165°.

Full Moon

The **full Moon** (see Table 6.2 for a list of some common names) has the face of the Moon fully sunlit. The Moon is on the opposite side of the Earth from the Sun so when you are looking at the Moon, the Sun is behind you. Looking at Figure 6.6, the Sun and Moon are 180° apart in both celestial longitude and in the sky. When the Sun is setting in the west, the full Moon is rising in the east. As the full Moon sets in the west, the Sun rises in the east. In its orbit, the Moon is on the opposite side of the Earth from the Sun. The full Moon lasts only a few hours. See and compare figures 6.9, 6.10, 6.14, and 6.15. This phase can only be

Table 6.2 *Some names for the full Moon*

Month	Name[a]
January	Wolf Moon, Winter Moon
February	Snow Moon, Hunger Moon
March	Sap Moon, Crow Moon, Fish Moon
April	Grass Moon, Planter's Moon
May	Milk Moon, Mother's Moon, Flower Moon
June	Rose Moon, Strawberry Moon
July	Thunder Moon, Buck Moon, Summer Moon
August	Grain Moon, Sturgeon Moon
September	Harvest Moon, Fruit Moon
October	Hunter's Moon, Harvest Moon
November	Beaver Moon, Hunter's Moon
December	Cold Moon, Christmas Moon

[a] This list is certainly not complete. There are many other names that can be found with a simple search of websites.

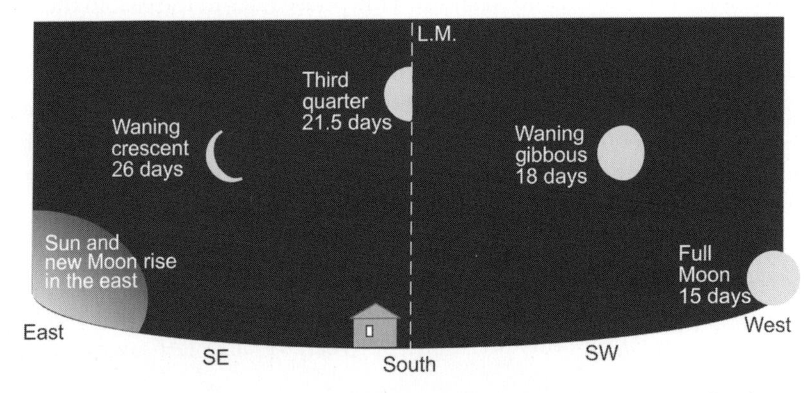

Figure 6.14 These are the sky positions of the lunar phases at sunrise for northern latitudes. The waning phases light the left-hand side of the Moon's face.

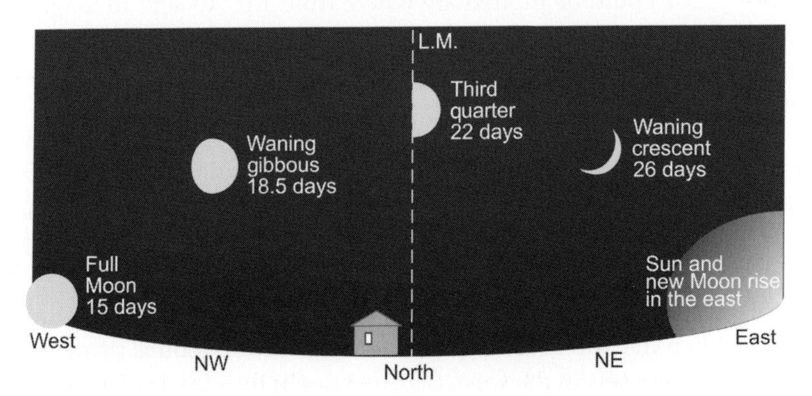

Figure 6.15 These are the sky positions of the lunar phases at sunrise for southern latitudes. The waning phases light the right-hand side of the Moon's face.

seen at night, because we must be on the night-side of the Earth to see the Moon. If we are on the day-side of the Earth, the Earth blocks our view. If the full Moon happens to pass through the Earth's shadow, we see a *lunar eclipse* (Section 7.3). This occurs when the Moon and Sun are at opposite nodes (Figure 7.24). If it is a *total lunar eclipse*, we see a "true" full Moon.

When a full Moon occurs twice in one calendar month, the second full Moon is called a *blue Moon*. The phrase "once in a blue Moon" refers to a rare occurrence, because months with two full Moons generally occur at least 2.5 to 3 years apart. The true origin of the term "blue moon" is hotly debated and is certainly not settled (Hiscock, 1999; Olson and Sinnott, 1999). One proposed origin is that the name comes from a very rare condition in the Earth's atmosphere which filters the Moonlight, giving it a bluish tint. The second full Moon of the month and the (supposed) bluish tint are two unrelated events.

Although blue Moons are rare, 1999 was an unusual year. January had a blue Moon (full Moons on the 1st and 31st), February had no full Moon, and March had another blue Moon (full Moons on the 2nd and 31st). February is the only month short enough to miss a full Moon, which happened in 1961, 1999, and will happen again in 2018. These dates are easily verified with most planetarium programs.

On the *Equatorial Chart* the full Moon is plotted 180° away from the Sun. Simply add 180° to the Sun's longitude, subtracting 360° if the result is greater than 360°. Because this is half a circle, subtracting 180° from the Sun's position works just as well.

Waning phases

The **waning phases** are all the lunar phases where the eastern side of the Moon is lit. "Waning" comes from the Middle English word *wanien* – to lessen. These phases are best seen just before sunrise, and in the morning hours of the day.

Waning gibbous Moon The **waning gibbous Moon** has more than half of the east side of the Moon sunlit. This phase rises in the east before midnight and sets in the west after sunrise, but before noon. It can also be seen in the western sky during the morning hours of daylight. See and compare figures 6.12, 6.13, 6.14, and 6.15. It lasts a little less than seven days – between the full and third quarter Moon phases. The waning gibbous phase is 181° to 269° ahead of the Sun. For general discussions, the waning gibbous can be plotted at 225° ahead of the Sun.

Example: For 10 December, with the Sun at longitude 258°, the waning gibbous Moon could be plotted at $258° + 225° = 483° - 360° = 123°$ celestial longitude.

Third (or last) quarter Moon The **third quarter Moon** (or **last quarter Moon**) has the east half of the Moon sunlit. Like the first quarter phase, the Sun and Moon are again at a right angle to each other, but now the Sun is to the east of the

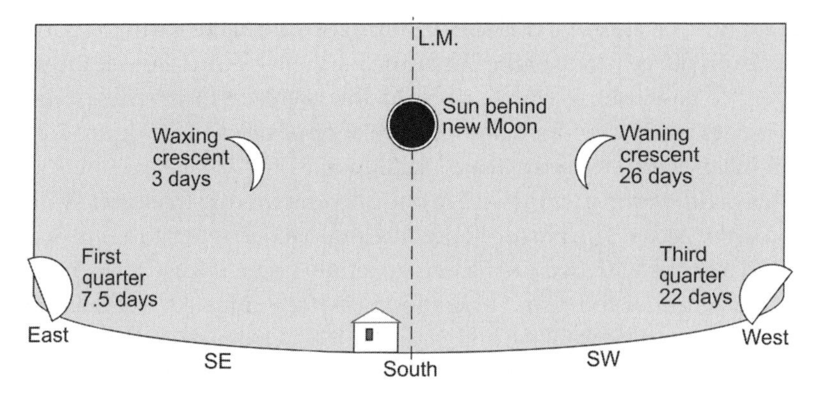

Figure 6.16 These are the sky positions of the lunar phases at high noon for northern latitudes. The waning phases light the left-hand side of the Moon's face, the waxing phases light the right-hand side. The new Moon is shown eclipsing the Sun but this is not the usual case. The full Moon is on the antimeridian.

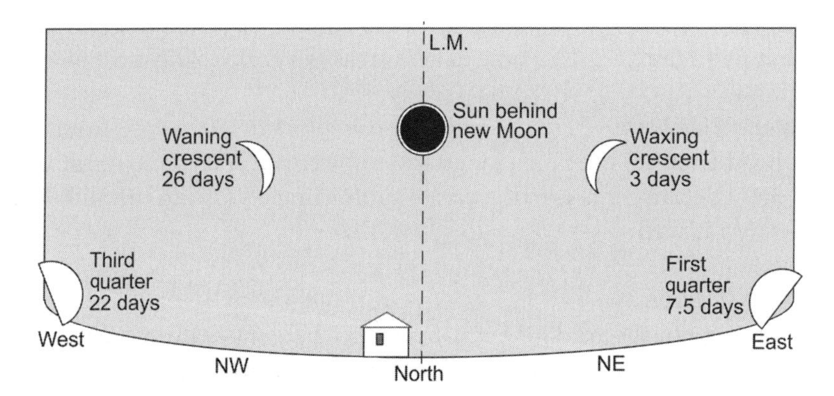

Figure 6.17 These are the sky positions of the lunar phases at high noon for southern latitudes. The waning phases light the right-hand side of the Moon's face, the waxing phases light the left-hand side. The new Moon is shown eclipsing the Sun but this is not the usual case. The full Moon is on the antimeridian.

Moon. In the northern latitudes, this phase is seen close to the local meridian in the southern sky at sunrise. In the southern latitudes, it is seen in the northern sky at sunrise. It rises at midnight and sets at noon. It is probably best observed two hours before sunrise. See and compare Figures 6.12, 6.13, 6.16, and 6.17. The Moon is three quarters of the way in its orbit around the Earth. This phase lasts only a few hours.

The third quarter Moon is 270° ahead (or 90° behind) the Sun. Note that the discussion of the true lunar–solar angle given for first quarter phase also applies to the third quarter phase.

Example: For 1 May with the Sun at longitude 40°, the third quarter Moon is located at $40^\circ + 270^\circ = 310^\circ$ celestial longitude.

Waning crescent Moon The **waning crescent Moon** has less than half of the eastern side of the Moon sunlit. The Moon has a crescent shape again. The "horns" of the crescent point away from the Sun. This phase is best seen in the eastern sky in the morning hours. See Figures 6.14 and 6.15. It lasts about seven days, between the third quarter and new Moon phases. It rises just before sunrise and sets just before sunset. The waning crescent is located anywhere from 271° to 359° ahead of the Sun (or 1° to 89° behind the Sun).

Example: For 1 June, the Sun's longitude is 70°. The waning crescent Moon is then anywhere from 341° to 69° in longitude. For general discussion it can be plotted at 315° ahead (or 45° behind) the Sun. The Moon would then be positioned at 25° longitude.

The phase cycle

The new, first quarter, full, and last quarter phases are collectively called the *quarter phases*. Each of these four phases last about four hours. The two crescent and the two gibbous phases are collectively called the *intermediate phases* or the *inter-phases*, with each lasting about a week. It takes a little less than 30 days (the synodic period) for the Moon to complete one cycle of phases. Therefore, it takes about one week to go from a new Moon to first quarter, and about one week to go from first quarter to full Moon, and so forth. Because the Moon orbits the Earth in a slightly elliptical orbit, the exact duration of the inter-phases depends on the orientation of the Moon's orbit with the position of the new Moon phase. Because this is unique to each orbit, the duration of the inter-phases is unique to each orbit and may vary by nearly a day from orbit to orbit.

Figure 6.6 shows the orbital motion of the Moon is counterclockwise around the Earth, as seen from "above the Solar System" (explained on p. 132). For the Earth–Moon system, everything is counterclockwise. The Earth and Moon both rotate on their axes counterclockwise. The Moon orbits the Earth counterclockwise and the Earth–Moon system orbits the Sun counterclockwise.

Figure 6.6 shows the phases of the Moon when the Sun is to the right on the diagram. Three months later, the Sun would be at the top on the diagram. The arrows showing the incoming sunlight would be pointing downward. The phase cycle (positions) is also rotated by 90° counterclockwise. When the Sun is at the top of the diagram, the new Moon phase is at the top of the (lunar orbit) circle.

Three months later, the Sun is at the left, with the sunlight arrows pointing to the right. The phase cycle is rotated another 90°, so the new Moon is on the left of the diagram. The reader should be able to draw the cycle of phases no matter where the Sun is located relative to the Earth–Moon position. Just remember, everything moves counterclockwise. If you know the order of phases, starting with the new Moon, you can draw this diagram. Do not forget to show the position of the Sun in your diagram! No matter where the Sun is found relative to the Earth–Moon system, a first quarter Moon is always 90° counterclockwise from the new Moon position on the Earth–Sun line (Figure 6.18).

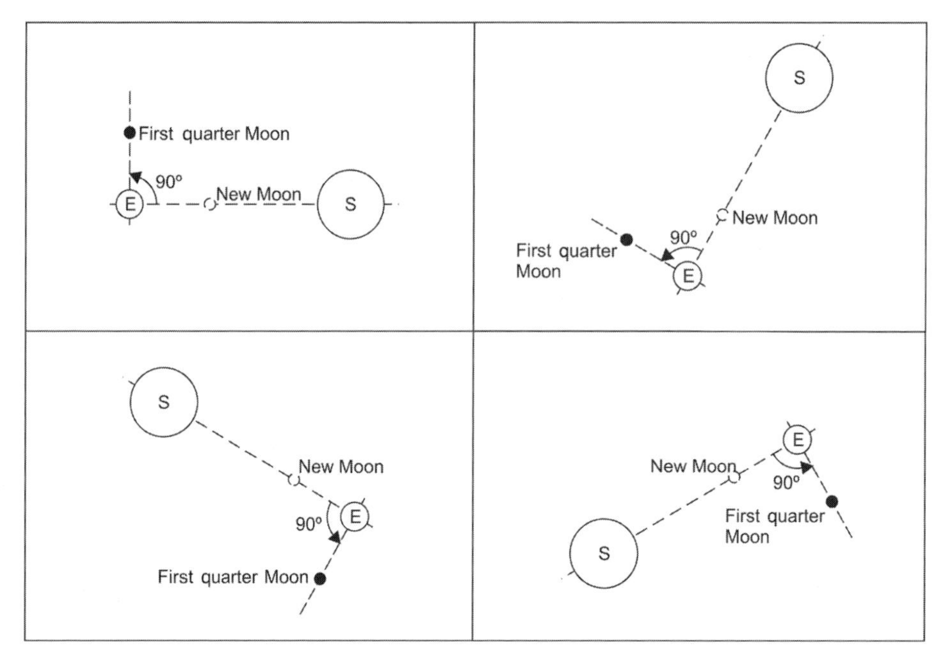

Figure 6.18 In these drawings the large circle is the Sun, the medium circle is the Earth, and the small circle is the Moon. In all four cases the phase position of the Moon is first quarter. No matter where the Earth–Moon system may be found with respect to the Sun, the first quarter Moon position is always 90° from the new Moon position. The new Moon position is always on the line directly between the Earth and the Sun.

6.3 The rising and setting of the Moon

Because the Moon's orbit is close to the ecliptic, its rising and setting points move along the horizon in the same way as the Sun's (Section 5.3, p. 129). The Moon's orbit can carry it as far as $5^\circ.1$ above or below the ecliptic so the variation in the Moon's rising and setting points along the horizon is slightly larger than the Sun's.

As the Moon moves along its orbital path its position changes (on average) by $13^\circ.18$ per day by sidereal reckoning, or $360^\circ \div 29.53$ days $= 12^\circ.18$ per day by solar reckoning. (The difference of about 1° per day is caused by the Earth–Moon system's orbital motion around the Sun.) This causes the Moon to rise and set each day about $12^\circ.18 \div 15^\circ \times 60$ min $= 48.8 \approx 49$ minutes later than the previous day. For instance, if the Moon rises at 3 : 05 p.m. on one day it will rise at 3 : 54 p.m. the next day.

This 49-minute change is an average over the synodic motion of the Moon and observer latitudes. For observers at the equator, deviation (during the synodic period) from this average is the smallest. At greater northern or southern latitudes, the deviation gets larger. North of the Arctic Circle or south of the Antarctic Circle, the Moon does not rise or set at all during some days of the

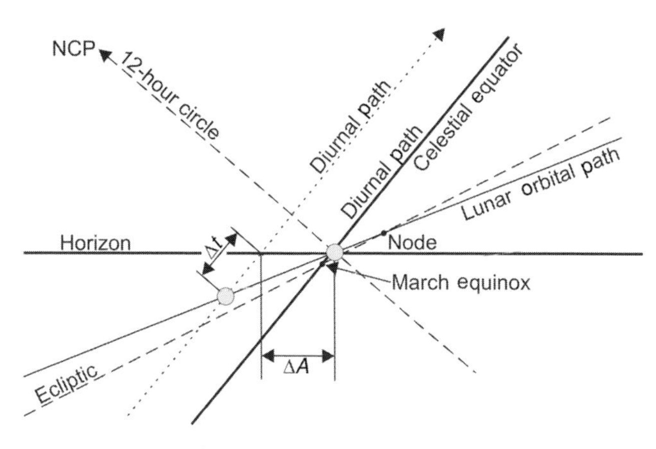

Figure 6.19 In this figure at 40° N, Δt is the change in the moonrise time around the Harvest Moon – the full Moon nearest the date of the September equinox (the first day of fall). The Moon rises with the March (vernal) equinox and the Sun sets with the September (autumnal) equinox – so the date is 21 September. To increase the effect, the Moon's ascending node is shown near the equinox. Compare the Δt and ΔA in this drawing with Figure 6.20.

month. When observing from the Earth's poles the Moon is up for two weeks – for the same reason the Sun is up for six months.

Calculating the azimuth and time of moonrise and moonset is more complicated than the calculations for sunrise and sunset. Sunrise and sunset depend only on the date, but moonrise and moonset depend on the date and on the lunar phase. The inclination of the Moon's orbit to the ecliptic and the regression of the lunar nodes (p. 185) complicate determination of the Moon's declination. The eccentricity of the Moon's orbit combined with perturbations caused by the Sun and the other planets complicate determination of the Moon's celestial longitude. This in turn complicates calculation of the angle between the Sun and the Moon (the lunar phase) and the Moon's hour angle for conversion to horizon coordinates.

We can, however, look at two notable special cases – full Moon rise and full Moon set at the equinoxes – at mid-latitudes. We start by observing moonrise on 21 September at 40° N. The Moon is located at – and rises with – the March (vernal) equinox (Figure 6.19). At moonrise, the December (winter) solstice is transiting the local meridian at $90^\circ - 40^\circ - 23^\circ.5 = 26^\circ.5$ off the southern horizon. The Sun is setting with the September (autumnal) equinox on the western horizon (making the date 21 September). At this latitude the ecliptic line makes an angle of 26°.5 with the horizon and the orbital path of the Moon makes a similar angle with the horizon. Depending on the position of the nodes the lunar path's angle may be as low as 21°.4 (with the Moon's ascending node near the March equinox) at this latitude.

Because of the low angle between the lunar path and the horizon, the Moon moves almost parallel with the horizon. This makes the day-to-day change in

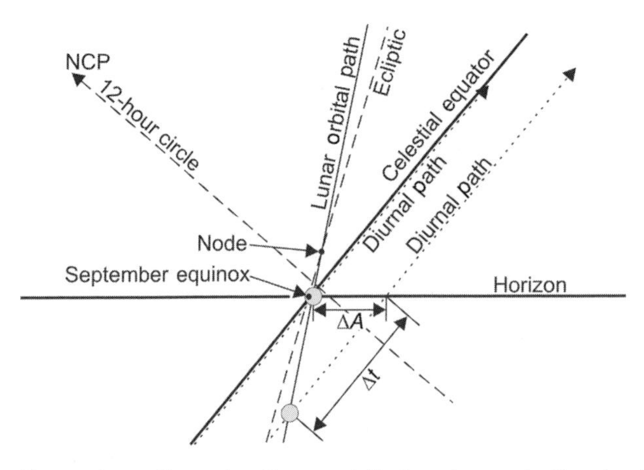

Figure 6.20 Observing from 40° N, the change in the rise time of the Moon is greater than average with the full Moon nearest the date of the March equinox (the first day of spring). The Sun is setting with the March (vernal) equinox so the date is 21 March. Δt is the change in the moonrise time. Compare Δt and ΔA in this drawing with those in Figure 6.19.

the time of moonrise short and the change in moonrise azimuth large. The time difference between moonrise on successive days is about 30 minutes – significantly less than the average. This small change in moonrise time is also known as the *Harvest Moon effect*.

On 21 March at 40° N we find the full Moon located at – and rising with – the September (autumnal) equinox (Figure 6.20). At moonrise, the June (summer) solstice is transiting the local meridian at $90^\circ - 40^\circ + 23^\circ.5 = 73^\circ.5$ off the southern horizon. The Sun is setting with the March (vernal) solstice on the western horizon. The ecliptic line makes an angle of 73°.5 with the horizon and the orbital path of the Moon is making a similar angle with the horizon. Depending on the position of the nodes, the lunar path's angle may be as high as 78°.6 at this latitude.

Because of the high angle between the lunar path and the horizon, the Moon moves almost perpendicular with the horizon. This makes the day-to-day change in the time of moonrise large, as shown by Δt, and the change in moonrise azimuth small, as shown by ΔA. In this scenario the day-to-day change in moonrise time is about 70 minutes – well over the average. Compare Figures 6.19 and 6.20.

This set of scenarios can also be made for the southern hemisphere, except they happen at the opposite dates. The Harvest Moon effect happens on 21 March and the opposite extreme in moonrise time difference happens on 21 September. The drawings equivalent to Figures 6.19 and 6.20 are shown in Figures 6.21 and 6.22. If you look carefully at all four drawings you should notice that the changes in moonrise time are as opposite in the hemispheres as the seasons. On the September equinox the northern hemisphere observes small changes in

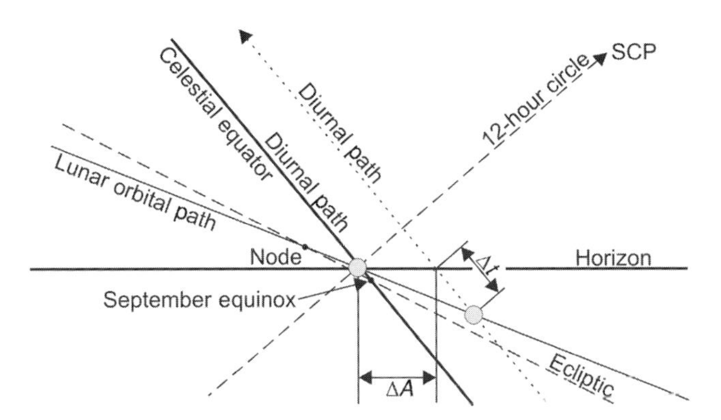

Figure 6.21 In this figure at 40° S, Δt is the change in the moonrise time around the Harvest Moon – the full Moon nearest the date of the March equinox (the first day of fall). The Moon rises with the September equinox and the Sun sets with the March (vernal) equinox – so the date is 21 March. Compare this with Figure 6.22.

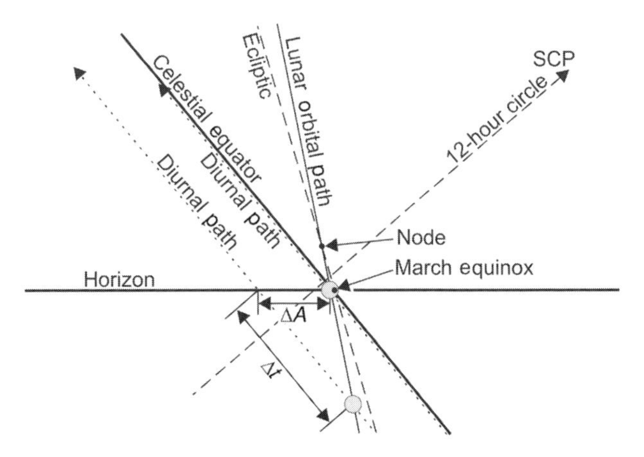

Figure 6.22 Observing from 40° S, the change in the rise time of the Moon is greater than average with the full Moon nearest the date of the March equinox (the first day of spring). The Sun is setting with the September equinox so the date is 21 September. Δt is the change in the moonrise time. Compare this with Figure 6.21.

moonrise time and the southern hemisphere observes large changes in moonrise time.

6.4 Calendars

Why should we care about the phases of the Moon? All seafaring cultures need to know the lunar phases because the harbor tides are directly connected to the phases. All cultures and civilizations now or in the past, have used the lunar phases as the basis for their calendars. The word *month* has as its root, the word *moon*. The Greeks, Romans, Egyptians, Hebrews, Hindus, Chinese, and Muslims

Table 6.3 *The development of the Gregorian calendar*

Roman, 738 BC	King Numa, 713 BC	Julian, 47 BC	Augustan, 8 BC	Gregorian, AD 1582
Martius, 31	Martius, 31	Januaris, 31	Januaris, 31	January, 31
Aprilis, 29	Aprilis, 29	Februaris, 29/30	Februaris, 28/29	February, 28/29
Maius, 31	Maius, 31	Martius, 31	Martius, 31	March, 31
Junius, 30	Junius, 29	Aprilis, 30	Aprilis, 30	April, 30
Quintilius, 31	Quintilis, 31	Maius, 31	Maius, 31	May, 31
Sextilis, 30	Sextilis, 29	Junius, 30	Junius, 30	June, 30
Septembris, 31	Septembris, 29	Julius, 31	Julius, 31	July, 31
Octobris, 30	Octobris, 31	Sectilis, 30	Augustus, 31	August, 31
Novembris, 31	Novembris, 29	Septembris, 31	Septembris, 30	September, 30
Decembris, 29	Decembris, 29	Octobris, 30	Octobris, 31	October, 31
	Januaris, 29	Novembris, 31	Novembris, 30	November, 30
	Februaris, 28	Decembris, 30	Decembris, 31	December, 31
303 days	355 days	365/6 days	365/6 days	365/6 days

used (or use) a calendar based on the lunar cycle. The design of these calendars was the work of politicians and priests.

A short history of calendars

Our modern calendar (Table 6.3) is a modified version of a calendar developed by the Romans. There is some controversy about exactly how the Roman calendar was structured, but there are some obvious remnants of it within our own. The original Roman calendar started the new year in March, probably on the first day of spring, and had only ten months. The seventh, eighth, ninth, and tenth months are now named, September (*septem* – seven), October (*octavus* – eight), November (*novem* – nine), and December (*decem* – ten). Even numbers were considered unlucky, so all (at least at first) the months had an odd number of days. The eleventh (Januaris) and twelfth (Februaris) months were added in 713 BC by King Numa. Confusion reigned over this calendar because it was the subject of political and priestly patronage. If one religious sect was in favor, its religious month was made longer than the others. Favoritism came and went, so the calendar was unstable.

Julius Caesar ended the confusion. He appointed the Greek-Egyptian astronomer, Sosigenes, to make a new calendar that started on the day Roman magistrates took office – 1 January. January is named after the Roman two-faced god Janus. Janus had a face on the front and back of his head, allowing him to see forward and behind. This is perhaps symbolic of our tendency to look forward to the new year and behind at the past. Caesar extended the year 46 BC to 466 days to bring the old and new calendars into step. The first day of spring was 25 March.

This new *Julian Calendar* had twelve months, with each month alternating in length of 30 and 31 days, except for February, which had only 29. This gives a total of 365 days. However, it was known even then that the year is 365.25 days long. To make sure the astronomical (and perhaps astrological?) events, like the first day of spring, occurred on the same day of each year, a leap day was added to the end of February every four years. Thus, in any year divisible by four, February was 30 days long. Julius Caesar was assassinated on 15 March, 44 BC. The calendar he commissioned worked so well and thus was so impressive, that after his death the month in which he was born was renamed in his honor. It is now called *July*.

After the reign of Julius Caesar came his nephew, Augustus Caesar. Augustus was probably the greatest of the Caesars. He lived a long life and the Roman Empire prospered under his rule. Upon his death the month containing the greatest number of days celebrating his achievements as Caesar was renamed in his honor as "Augustus" (now *August*). Then, July was 31 days long and August was only 30 days long. The month honoring Augustus could not be shorter than that for Julius, so August became a 31-day month. This made the year too long. To fix this problem the remaining months (following August) were swapped in length and the extra day was taken from (the already abused) month of February. February became and is now only 28 days long.

Leap years

The tropical (calendar) leap year is not exactly 365.25 days long. It is actually 365.242 199 . . . days long. A small error, but this small error built up over sixteen centuries to about ten days: the first day of spring and thus the religious holidays that are scheduled according to the *lunar* months, like Easter, were falling on the wrong days. In 1582, Pope Gregory XIII announced the creation of a new calendar. He eliminated the extra ten days by making the day after 4 October into 15 October. This brought the first day of spring back to 21 March.

To account for the tiny error in the length of the year, the rule for leap years was modified. For a century year to be a leap year in this new calendar, it must be divisible by 400. This means the year 1900 was *not* a leap year. The year 1900 is divisible by four, but not by 400. The year 2000 *is* a leap year, because 2000 is divisible by 400. This adjustment makes the error in the length of the year about one day in 3300 years. No further adjustment should be necessary until AD 5000. This new calendar is called the *Gregorian Calendar*.

Only those countries with strong ties to the Church made this adjustment. In others, like England and her colonies (USA, etc.), the adjustment was not accepted until 1752. During those years when the calendars may be confused when posting dates, the letters "os" or "ns" were used to designate Julian calendar (old style) or Gregorian calendar (new style). Isaac Newton was born on Christmas day of 1642os, 4 January 1643ns. The Christian World, and most governments and businesses, use this calendar. In those cultures where they wished to avoid

Table 6.4 *The days of the week*

English	German	French	Spanish	Latin	Norse/Saxon	Planet
Sunday	Sonntag	dimanche	domingo	Dies Solis	Sun's day	Sun
Monday	Montag	lundi	lunes	Dies Lunae	Moon's day	Moon
Tuesday	Dienstag	mardi	martes	Dies Martis	Tiw's day	Mars
Wednesday	Mittwoch	mercredi	miércoles	Dies Mercurii	Woden's day	Mercury
Thursday	Donnerstag	jeudi	jueves	Dies Lovis	Thor's day	Jupiter
Friday	Freitag	vendredi	viernes	Dies Veneris	Frigg's day	Venus
Saturday	Sonnabend, Samstag	samedi	sábado	Dies Saturni	Saterne's day	Saturn

connecting the calendar with the birth of Jesus Christ, they use CE (Common Era) in place of AD and BCE (Before the Common Era) in place of BC.

Century years The Hindu–Arabic number system, which includes the concept of place values and the number zero, was not introduced to the Western World until the thirteenth century. Until then the West, which was still using the old Roman numerals, had no symbol for zero. Thus, there is no zero year. (Notice that each month also starts with one, not zero.) The year AD 1 is preceded by the year 1 BC. The first century begins with the year AD 1. The second century begins with the year AD 101, the third with AD 201, and so forth. Because of this, the twenty-first century started on 1 January 2001, not 1 January 2000. The year 2000 was the last year of the twentieth century.

Days of the week

The seven days of the week probably originate from the seven wandering celestial objects or planets: the Sun, the Moon, Mercury, Venus, Mars, Jupiter, and Saturn. In the ancient world, the Sun and Moon were considered planets. This is evident in the English names of the days: *Monday* for the Moon, *Sunday* for the Sun, *Saturday* for Saturn. Other languages (especially the romance languages) show even stronger evidence for this suggestion (Table 6.4). It is also possible that the seven-day week is based on the Jewish *Shabbat* (Sabbath day) or on the Babylonian–Torah/Biblical creation story. There is some disagreement among scholars on this subject.

Lunar calendars

The traditional Jewish calendar is based on the lunar phases. There are 354 days in twelve lunar months – eleven days short of a seasonal year. To have the religious holidays fall on approximately the same day of the year, a thirteenth lunar month (an *intercalary* month, or *leap month*) is inserted every few years.

The Islamic calendar is based strictly on the phases of the Moon. No correction is made to align the Islamic calendar to the seasons. The holy month of Ramadan may come in any season. Islam started with desert peoples of the Middle East, where there was little concern for the seasons of the solar year.

The Chinese calendar is also based on phases of the Moon, and on a cycle of 12 years. The 12-year cycle is linked to the position of Jupiter in the zodiac. The Chinese zodiac is different from the Western (Babylonian) zodiac. Thus, the Chinese calendar has "The Year of the Dog," "The Year of the Dragon," etc.

The Mezzo-American civilizations had calendars based on the solar and lunar cycles, similar to the European calendar. Their observatories in Central America were built to very high precision. Their timings of astronomical events, such as the summer solstice, were nearly perfect. When Hernando Cortez arrived in Mexico in 1518 (before the 1582 Gregorian calendar) he was surprised to see how accurately the Mayans had measured time. The Mayan calendar matched the European (Julian) calendar, except *they* (according to the Europeans) were ten days out of step!

The International Date Line

Imagine taking a journey around the world, heading east from London, England, on 2 October. Rather than taking the 80 days that Phileas Fogg required, we shall take only one day, traveling one time zone per hour. We shall keep a log of the trip, recording our arrival time in a few cities, setting our wristwatches to the correct local time. We leave London at 12 : 00 hours (noon). As we move eastward the time of day becomes later and later. The leg to Moscow requires three hours of travel through three time zones. We record our arrival as 18 : 00 hours on 2 October. The leg to Tokyo traverses six time zones and requires six hours of travel. Because the wristwatch passed through midnight, our recorded arrival in Tokyo is 06 : 00 hours on 3 October.

We then travel to Hawaii through five time zones, requiring five hours of travel. We record our arrival here as 16 : 00 hours, 3 October. The next leg takes us to New York City. This is another five time zones and five more hours. Our wristwatches pass through midnight again and we record an arrival time of 02 : 00 hours on 4 October. Finally, we return to London passing five time zones and requiring five hours of travel. Our travel log shows we arrive in London at 12 : 00 on 4 October. But is this the correct date? We have moved around the world and watched our wristwatches advance through two 24-hour days. However, we neglected to track the calendar as well as we tracked the time. (This is how Phileas won his bet.)

This thought trip points out the need for the International Date Line. The date line is located at 180° (east or west) longitude, passing through the Pacific Ocean. It deviates east of Siberia and west of the Aleutians and again around Fiji and the Samoan Islands, to avoid having different dates with a political boundary. When we passed this line on our imaginary trip we should have dropped 24 hours (one day) from our wristwatches so that we would end our 24-hour journey in London on the proper date – 3 October.

Table 6.5 *The length of the year*

Year	Description	Length[a]
Anomalistic	Perihelion to perihelion	365.259 6
Average Gregorian	Gregorian calendar	365.242 5
Average Julian	Julian calendar	365.250 0
Eclipse	Lunar node to lunar node	346.620 1
Gaussian	Kepler's law for $a = 1$	365.256 9
Sidereal	Fixed star to fixed star	365.256 36
Tropical	Equinox to equinox	365.242 19

[a] One day is 86 400 TAI seconds.

The length of the year

The next aspect of time keeping that must be considered is the length of the year. The length of the year is given as a number of days, where one day is 86 400 TAI seconds long. If we measure the year as the amount of time between successive transits of a star (perhaps Regulus) on the celestial sphere, we get a **sidereal year**. This year is 365.256 36 days long. The **calendar year** is also known as the **tropical year**. It is measured as the amount of time between the Sun's successive transits of the vernal equinox. The tropical year is 365.242 19 days long. The motion of the vernal equinox makes the tropical year nearly 20 minutes shorter than the sidereal year. This shows us that the precessional motion of the Earth is not just a long term effect, but one which must be dealt with on a yearly basis. Although the sidereal year does not directly enter into civil time keeping, it is the length of the year in terms of an observer outside of the Solar System and so it is the basis for calculations involving the motion of the other planets.

Other ways to measure the length of the year are shown in Table 6.5. The *anomalistic year* is the amount of time required for the Earth to return to perihelion – the point in its orbit closest to the Sun. Notice this is not quite the same as the sidereal year. This means the Earth's slightly elliptical orbit is slowly precessing around the Sun. This is a motion seen in the orbits of all the planets. It is predicted and explained by Newton's law of universal gravity. (However, the precession of the orbit of Mercury requires general relativity to be fully explained.) The two average calendar years are fairly obvious. The difference between these two is due to the difference in the way they determine leap years.

The *eclipse year* is the amount of time for the same lunar orbital node (there are two of them, see p. 186) to return to alignment with the Earth–Sun line. The fact that this year is significantly shorter than the tropical year is important for counting the number of possible eclipses during a calendar year. See Chapter 7 for more about eclipses.

In Newton's universal law of gravity, the gravitational constant, $\mathcal{G}$, is known only to four significant digits ($6.674 \pm 0.012 \times 10^{-11}$ N m^2/kg^2). To create ephemerides for the planets, it is necessary to know the constant of gravitation to higher precision. This is possible with careful observation of the planets' motion. The German mathematician Carl Friedrich Gauss used Kepler's third law of planetary motion to determine what is now called the *Gaussian constant of gravitation*. To do this, he used the following values for the Earth: an orbit semi-major axis of one arbitrary unit, an orbital period of 325.256 383 5 days, and a Sun/Earth mass ratio of 1/354 710. This orbital period is now known as one *Gaussian year*. The Gaussian constant of gravitation is currently used as the defining constant of planetary ephemera.

Julian Day calendar

The civil calendar does not work well for astronomers. Often we need to know the amount of time lapsed between two events. Determining the number of days between two dates on the civil calendar is not an easy task. This need brought about the creation of the *Julian Day calendar*. The **Julian Day**, or *Julian Day* (JD) *number* is a continuous count of days starting from 1 January 4713 BC, on the Julian proleptic calendar. The *Julian proleptic calendar* uses the rules of the Julian calendar but is applied to those dates preceding the start of the Julian calendar.

A Julian Day begins at noon rather than at midnight. This happens because when the Julian Day calendar was introduced, the astronomical day began at noon. (It was changed to midnight in 1925.) The integral portion of the Julian Day refers to the instant of Greenwich mean noon, and the decimal portion is a partial of one day. Hours, minutes, and seconds are not used. Zero hours UT is 0.5 of a Julian Day. For new planetary ephemerides, 1 January 2000 = JD 2 451 544.5 TDB. Using day numbers instead of dates makes the calculation between two days a simple matter of subtraction.

The Julian Days become big numbers rather quickly. To make calculations easier the **Modified Julian Day** (MJD) was introduced. The relationship is simple.

$$\mathrm{MJD} = \mathrm{JD} - 2\,400\,000.5.$$

Questions for review and further thought

1. What is the apparent diameter of the Moon?
2. Describe the difference between terra and maria as seen by naked-eye observation.
3. What type of events caused most of the craters on the Moon?
4. Why does the Moon appear to be larger when near the horizon than when high in the sky?
5. Why does the Moon have a red/orange tint when it is near the horizon?
6. Describe the diurnal motion of the Moon.
7. Why do we not see the Moon during the daytime as often as we see it during the night?
8. Why do we always see the same side of the Moon?

9. What is a synchronous orbit? Why did the Moon's orbital motion become synchronous?
10. How much of the Moon's surface can we actually see from Earth?
11. Describe the apparent monthly motion of the Moon.
12. How fast does the Moon move through the constellations? Through what set of constellations does it move?
13. What is the length of the Moon's sidereal period?
14. Define the sidereal period of the Moon.
15. What is the length of the Moon's synodic period?
16. Define the synodic period of the Moon.
17. About how quickly does the Moon move along its path (both sidereal and solar rates)?
18. What is the average time difference between lunar risings on succeeding days?
19. What is a *lunar phase*?
20. Describe the lunar phases. Include the appearance from Earth, rise/set time, duration, the general time and sky location for best observation, and a description of the orbital position for each phase.
21. What is a *true new Moon*?
22. When do we see a true full Moon?
23. What is a blue Moon (in the calendrical sense)? How often do blue moons occur?
24. How long (how many days) is the lunar cycle of phases and how is the lunar cycle of phases related to the orbital period of the Moon?
25. What group of people within most societies or cultures have been most responsible for the creation of calendars?
26. Who is the Julian calendar named for and what changes did he bring to calendars?
27. Describe the procedure for determining if a year is a leap year, under the Gregorian calendar.
28. What do the calendar designations CE and BCE mean and why are they used?
29. Why does the calendar begin with the year one instead of zero?
30. What is the most probable reason for there being seven days in the week?
31. Describe the need for the International Date Line.
32. Describe the need for the Julian Day calendar.

Review problems

1. Determine the sidereal rotation period of the Moon.
2. If the Moon happens to be at its positive maximum inclination from the ecliptic when its right ascension is $6^h\ 0'$, calculate its transit altitude for an observer at latitude 35° N.
3. If the Moon happens to be at a node when its right ascension is $18^h\ 0'$, calculate its transit altitude for an observer at latitude 38°.5 S.

4. Using the figure shown below, draw the appropriate orbital position of the Moon for each of its phases and label each position with the phase name.

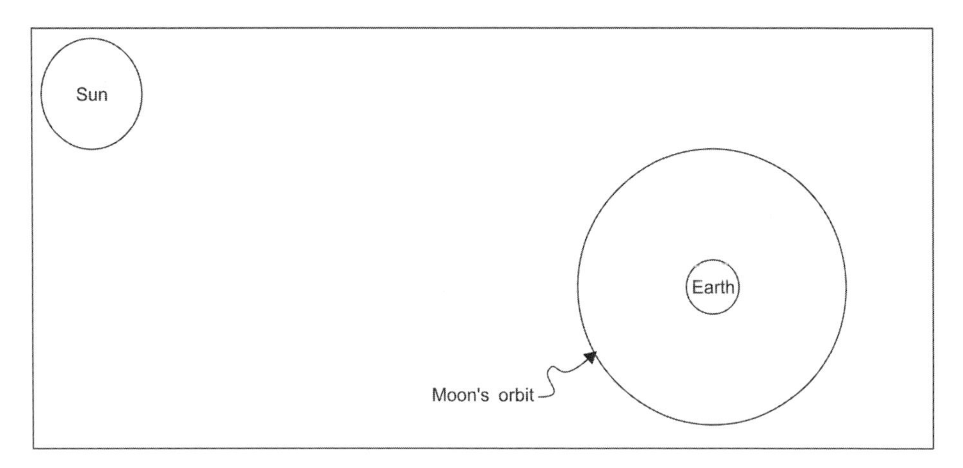

5. Determine which of the century years are leap years on the Gregorian calendar, between its inception and the year AD 4000.

There are more problems dealing with the motion of the Moon in Chapter 8, Observation Project 1.

7 Eclipses

Unlike the phases of the Moon, eclipses are actually caused by shadows. The Earth and Moon glow by reflecting sunlight and they also act like any other opaque object in a path of light – they cast shadows. We generally don't notice this detail of life simply because there is (usually) nothing out there in space on which the shadow may fall. We only notice the existence of a shadow when the shadow causes some other object to be dimly, or abnormally, lit. This is what happens during an eclipse.

From time to time, the Moon may pass through the Earth's shadow and the Moon becomes dimly lit. This is a lunar eclipse. We can then notice the fact that the Earth has a shadow. Also, the Earth may pass through the Moon's shadow, causing a solar eclipse. It is only during these events that we come to understand the shadowy side of the Earth and Moon.

7.1 Shadows

Shadows have two parts: the *penumbra* and the *umbra* (Figure 7.1.) The **penumbra** (from the Latin words *paene* + *umbra*, meaning "almost shade") is the region of partial shadow. The penumbral region is darkened, but not black. The **umbra** (Latin for "shade") is the region of full shadow. This region is in complete darkness.

The umbra is a cone shaped region of total darkness, coming to a point "behind" the object. The penumbra is a cone shaped region of partial darkness spreading outward behind the object until, gradually, the shadow no longer exists because the apparent size of the blocking object is too small to block any appreciable portion of the total amount of light.

The cause of this shadow structure and the shape of the shadow regions is the extant size of the Sun (Figure 7.2). In the penumbral region, only part of the Sun's disk is blocked by a planet or a moon. If the Sun were a true point-sized light

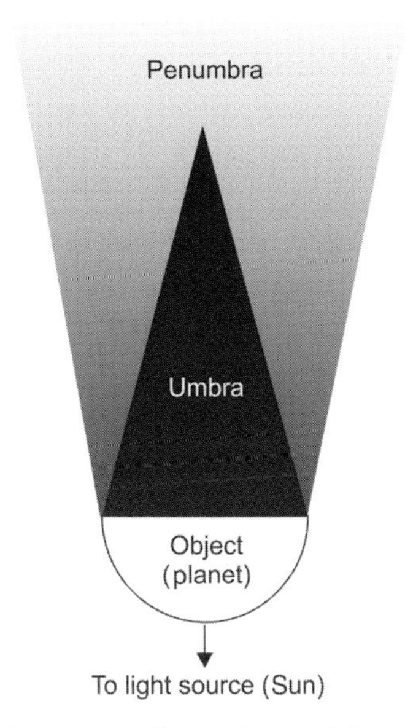

Figure 7.1 The umbra and the penumbra are the two regions of any object's shadow.

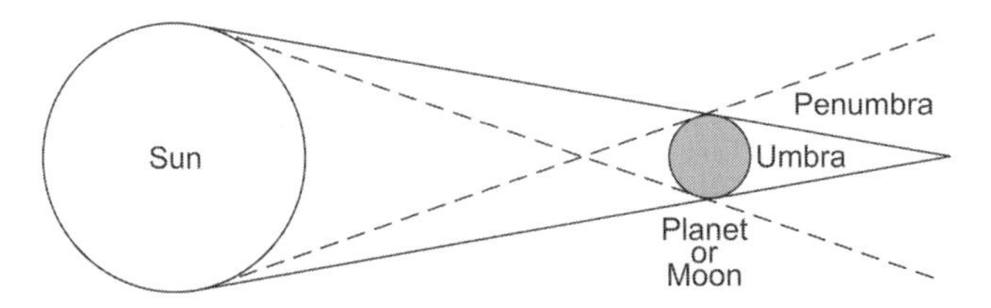

Figure 7.2 The structure of shadows is caused by the extant size of the Sun. If the Sun were point sized the penumbra would not exist – there would only be umbra. Redraw this figure a few times, each time using a smaller solar disk. Watch what happens to the lines outlining the penumbra and umbra regions as the disk gets smaller. In your final redraw, use a point-sized Sun.

source this shadow structure would not exist – the penumbra would become umbra and there would be no penumbra.

7.2 Eclipse types

There are two types of eclipses – *lunar* and *solar*. If you interpret the word **eclipse** to mean "to overshadow," or perhaps "to disappear," it may help you to remember the alignment of the Earth, Moon, and Sun for each type of eclipse. Lunar eclipse means "Moon disappears." The Moon enters the Earth's shadow, so the alignment is Sun–Earth–Moon. Solar eclipse means "Sun disappears." The Earth passes through the Moon's shadow, so the alignment is Sun–Moon–Earth.

These two types of eclipses are each further broken into different kinds. The chart shown in Figure 7.3 shows the relationship between the different types and

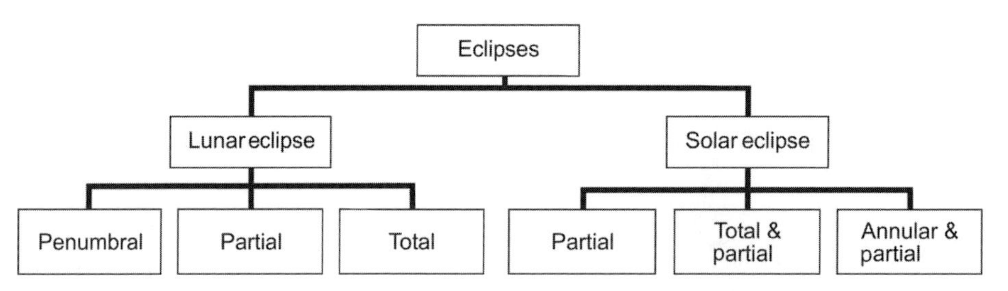

Figure 7.3 This is a flow chart showing the relationship between the different types of eclipses.

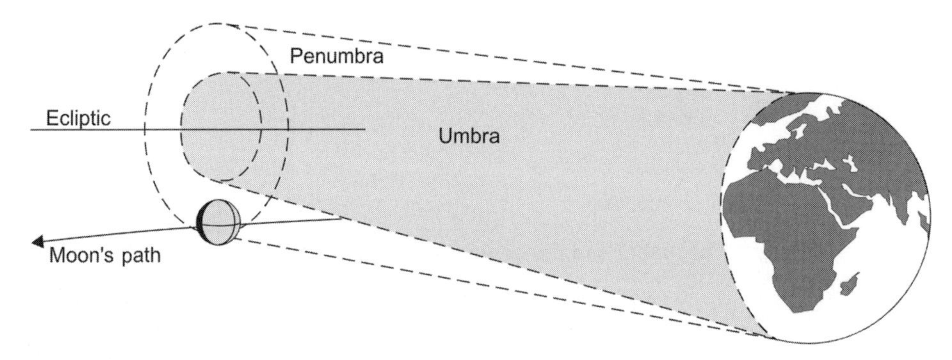

Figure 7.4 During a penumbral lunar eclipse the Moon moves through only the penumbral portion of the Earth's shadow. Any change in the appearance of the Moon is barely noticeable.

kinds of eclipses. The three kinds of solar eclipses – partial, total, and annular – are arranged in an apparently strange manner. As we will see later, total and annular solar eclipses are also seen as partial solar eclipses, depending on the observer's location on Earth.

7.3 Lunar eclipses

During a **lunar eclipse**, the Moon passes through the Earth's shadow. The Moon (Luna) "disappears." Actually, the Moon never *completely* disappears. At best, during a total lunar eclipse, it turns to a very dark red, "coppery" (or "blood red," as the ancients may have described it) color. There are three kinds of lunar eclipses – *penumbral*, *partial*, and *total*. Figures 7.4 through 7.6 should help you understand the difference between them.

> *For any lunar eclipse to occur, the Moon* must *pass into the Earth's shadow. The shadow is, of course, on the other side of the Earth from the Sun. Therefore, a lunar eclipse can occur only during a full Moon phase. Also, because the Earth's shadow is always on the ecliptic (because the Sun is always on the ecliptic), the eclipse must occur on the ecliptic line.*

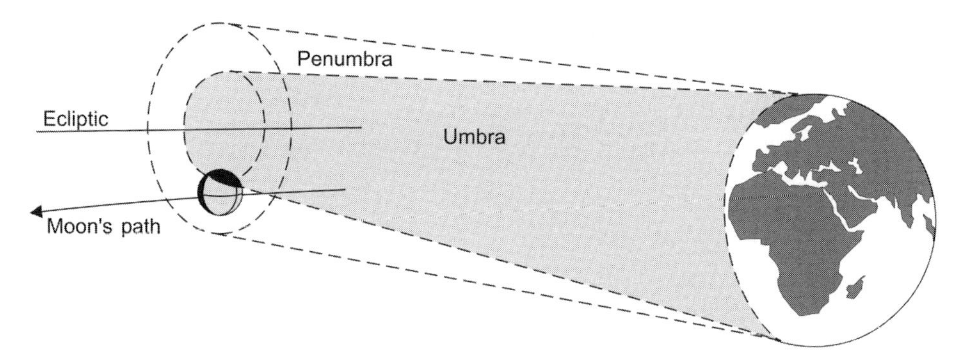

Figure 7.5 During a partial lunar eclipse part of the Moon moves through the umbra portion of the Earth's shadow. One can easily see the Earth's curved shadow move across the Moon's face.

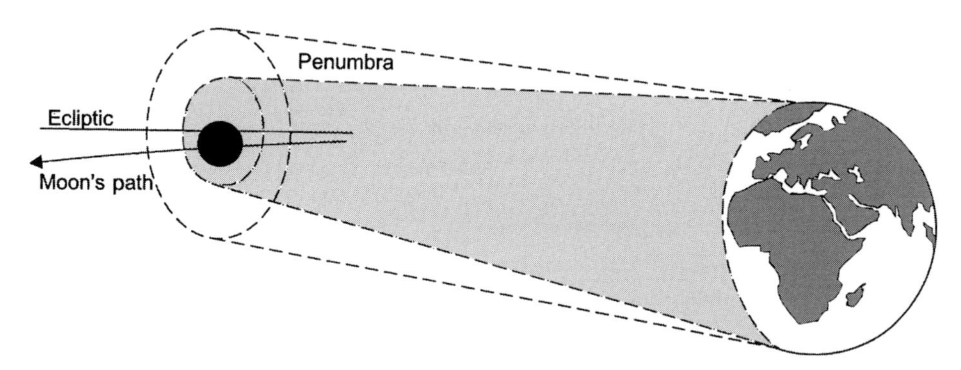

Figure 7.6 During a total lunar eclipse the entire Moon moves through the umbra portion of the Earth's shadow. During totality the Moon turns a coppery red color.

If the Sun is so bright we cannot see the stars behind it, how do we determine its path against the background stars? The fact that lunar eclipses must happen on the ecliptic (the path of the Sun against the background) allows us to map the position of the ecliptic by careful observation of lunar eclipses and thus determine the path of the Sun on the celestial sphere. This is one way we can determine the path. There are other, and better, ways as well.

Penumbral lunar eclipses

During a **penumbral lunar eclipse**, the Moon passes through only the penumbral portion of the Earth's shadow (Figure 7.4). These lunar eclipses are barely noticeable. A penumbral graze may last only a few minutes and cause no difference in the appearance of the Moon. Even a full immersion of the Moon into the Earth's penumbra may go completely unnoticed by an unsuspecting observer. If all or a significant portion of the Moon is immersed in the penumbra, the Moon may lose a partial magnitude in brightness. This would be like having a very thin cloud pass in front of the full Moon. Because there are no drastic

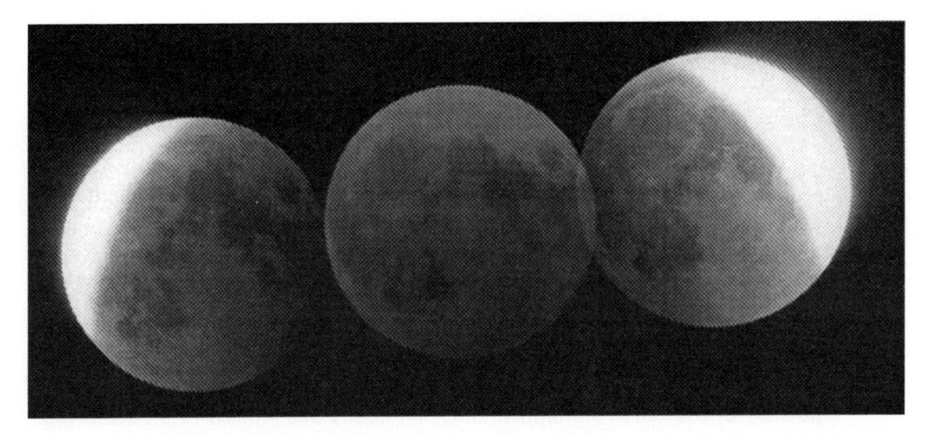

Figure 7.7 The Moon passes through the Earth's umbra showing the curvature of the Earth's shadow on its face while implying the size of the umbra at the distance to the Moon. This is a collage of photos from the total lunar eclipse of 28 October 2004. (Photos courtesy of Kevin S. Jung.)

changes in the appearance of the Moon, this type of eclipse is generally ignored for observation.

Partial lunar eclipses

During a **partial lunar eclipse** (Figure 7.5), part of the Moon passes into the Earth's umbra. The part of the Moon not inside the umbra is in the penumbra and has nearly the normal brightness of the full Moon. The curved shape of the Earth's shadow on the face of the Moon can easily be seen. This curved shadow led the Greeks (of the Pythagorean school) to conclude that the Earth must be spherically shaped, like other celestial objects. The Pythagoreans could understand the phases of the Moon, if they assumed the Moon is spherical. Based on this and other observations, the Greeks were convinced the Earth was round (spherical) long before "Columbus sailed the ocean blue." Columbus knew the World was round. He did not set out to prove this. Rather, he was interested in finding a shorter path to India. His mistake was in his misconception of the *size* of the World, which the Greeks also knew quite well, but was somehow lost to Columbus.

The *magnitude* of a lunar eclipse is an expression of the fraction of the Moon's diameter covered by the umbra, in units of the Moon's diameter. Because its calculation involves the distance between the center of the Moon and the center of the umbra, the eclipse magnitude may be greater than one, but it will never be negative. It therefore also indicates how deeply the Moon is immersed in the umbra.

Total lunar eclipses

During a **total lunar eclipse** (Figures 7.6 and 7.7), the Moon passes completely into the Earth's umbra, turning to a "coppery red color" during totality. It does

Table 7.1 *The Danjon lunar eclipse luminosity scale*

Degree	Description
0	Very dark eclipse, the Moon is hardly visible
1	Dark eclipse, grey-brown color, surface details hardly discernible
2	Dark, red colored eclipse with a dark region at the center of the shadow and brighter edges
3	Medium red eclipse with a border of brighter yellow
4	Very bright, copper colored eclipse with a bluish edge

not completely disappear. The color comes from refractive and scattering effects of the Earth's atmosphere. The sunlight glancing the Earth passes through the Earth's atmosphere. The atmosphere scatters the blue portion of the sunlight and allows the red sunlight through. This is the same atmospheric effect (Rayleigh scattering) that causes the red/orange moonrise, but the effect is being used in a different way. This time the light passing through the atmosphere on its way *to* the Moon is affected, where as with the moonrise it is the light coming *from* the Moon that is affected. The red and yellow portion of the spectrum that does make it through the atmosphere is refracted and dispersed just like a prism breaks sunlight into the visible spectrum. Because of this combination of scattering and refraction, the Earth's umbra is not completely black at the Moon's distance from the Earth. Rather, it is red with slightly yellow edges. During a total lunar eclipse the Moon passes through the red light and this is the reason for its coppery red color. Some light still hits the Moon making it slightly visible, even though it is fully eclipsed.

Exactly how dark the Moon gets during totality is a function of atmospheric conditions. If there are many cloud systems around the limb of the Earth during totality, those clouds will block the sunlight that would normally pass through the atmosphere making the Moon darker. If there has been a major volcanic eruption previous to the eclipse and the particulate material has been well distributed around the stratosphere, the Moon may be a darker red than normal. If the air is fairly clear the Moon will be a brighter red.

André Danjon (1890–1967) established a descriptive luminosity scale for lunar eclipses (Table 7.1). Published in the 1920s, his work and the work of other astronomers demonstrated a link between the luminosity of the Moon during eclipses and solar activity. The luminosity of lunar eclipses increases linearly from one solar minimum to the next, with a sudden drop in luminosity during the minimum between an old and new sunspot cycle. This property of eclipse luminosity allowed astronomers to confirm the times of solar minima in earlier years where sunspot observations are inconsistent.

The important points in time for the eclipse event are shown in Figure 7.8. A complete description of an eclipse event includes the times of these positions, usually given in universal time. The penumbra entry and exit points, P1 and

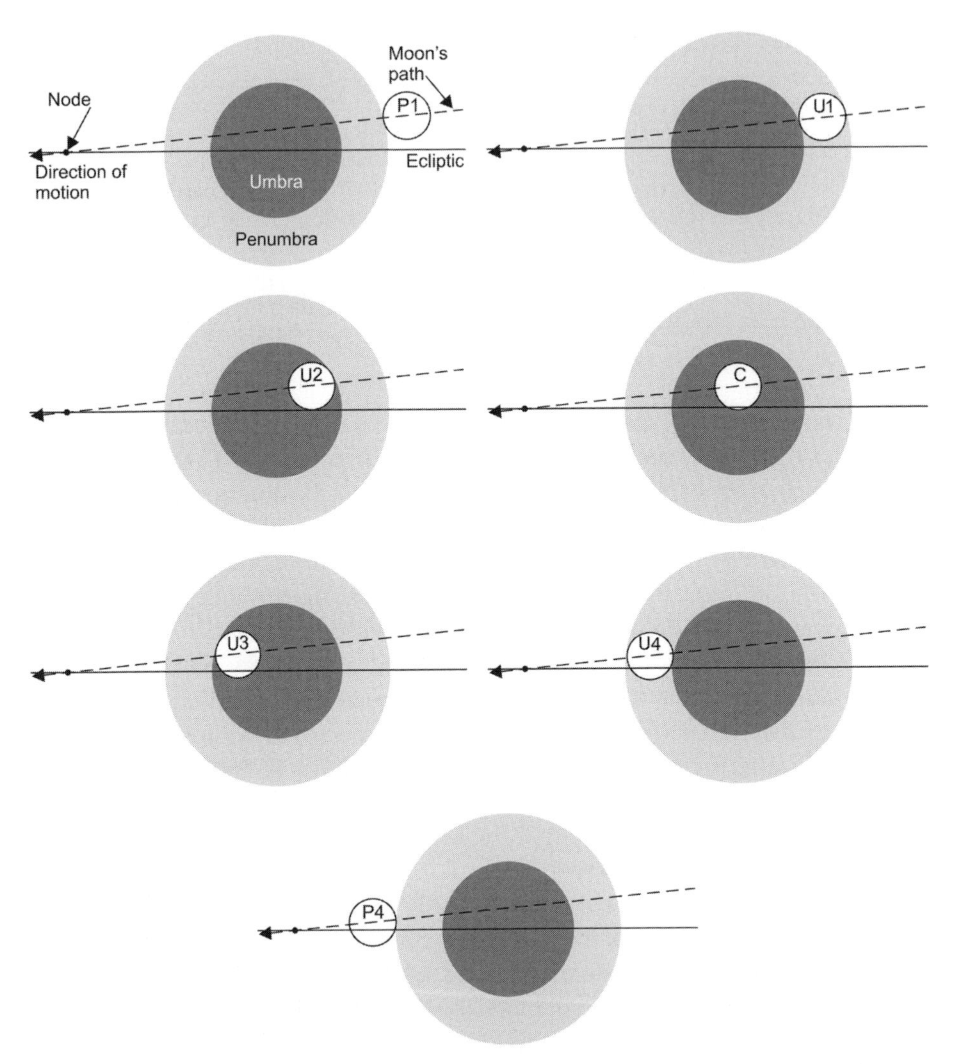

Figure 7.8 As the Moon moves through the Earth's shadow during a total lunar eclipse it passes through stages. An eclipse is described by the timings of these stages in Universal Time. Eclipse duration, from U1 to U4, may be as long as four hours. For the total lunar eclipse of 28 October 2004, the timings (UT) were: P1 = 00 : 05, U1 = 01 : 05, U2 = 02 : 23, C = 03 : 04, U3 = 03 : 44, U4 = 04 : 53, P4 = 06 : 02.

P4, are generally the least interesting. **First contact**, point U1, is the beginning of a total eclipse. **Second contact**, point U2, is the beginning of totality for the eclipse. The *center point*, C, is when the Moon is at its darkest. **Third contact**, point U3, is the end of totality, and **fourth contact**, point U4, is the end of the total eclipse. There are two other penumbral points, P2 and P3. The meaning of P2 and P3 is equivalent to points U2 and U3. However, because the distance between the penumbra edge and umbra edge is one Moon diameter, P2 occurs at the same time as U1 and P3 occurs at the same time as U4. Therefore P2 and P3 are generally ignored.

Duration of lunar eclipses

At the Moon's distance from Earth, the penumbra is about 2°.4 and the umbra is about 1°.4 in diameter. The Moon just fits in the penumbra region, touching the edge of both the umbra and penumbra during the penumbral phase of a total or partial eclipse. A total lunar eclipse may last more than four hours, including the partials during immersion and emersion. The Moon first moves into the Earth's penumbra, taking about 1.5 hours. Its movement through the umbra may take only a few minutes if it is barely within the umbra, or up to 1.5 hours if it passes exactly through the center of the umbra. Then it moves out of the umbra, back into the penumbra (on the other side), and back into sunlight. Therefore, a total lunar eclipse may last from three to four and a half hours, depending simply on how close the Moon passes to the exact center of the Earth's shadow.

7.4 Solar eclipses

We are fortunate enough to live during the era when the apparent size of the Moon and the apparent size of the Sun are about the same. Because of tidal forces between the Earth and Moon, the Moon is slowly receding from the Earth. (Its orbit is getting larger in radius.) Thus, its apparent size is gradually decreasing with time. Eventually, the Moon will be too far from the Earth (its apparent size will be too small) to entirely cover the Sun. Then, total solar eclipses will no longer happen.

A total solar eclipse is the one astronomical event all astronomers, professional and amateur alike, should make the pilgrimage to see. I've used the word "pilgrimage" because you must be in the path of totality to see the eclipse in full glory. Usually this entails a trip to another country, sometimes traveling to the farthest reaches of the World. But the trip is more than worth it (Ehmann, 1999).

Only during the totality portion of a total solar eclipse is it safe to look directly at the Sun. Otherwise,

| *NEVER LOOK DIRECTLY AT THE SUN WITHOUT EYE PROTECTION.*

The best way to observe a solar eclipse is to use the pin-hole camera effect. Poke a small hole through a piece of cardboard, such as an index card. Hold this card in the path of sunlight, and project the image onto a wall, or the ground, or another index card. If you are determined to look directly at the Sun, use a commercially made, specifically designed solar filter to protect your eyes. Inspect the filter for scratches or damage before trusting your eyes to the filter.

Do not use any kind of homemade device. Sunglasses, smoked glass, welder's glass, exposed film, and other sorts of tricks have been claimed to be safe by various sources. They are not safe! While these "filters" may darken the Sun to a comfortable level in visible light, they usually allow non-visible light (infrared

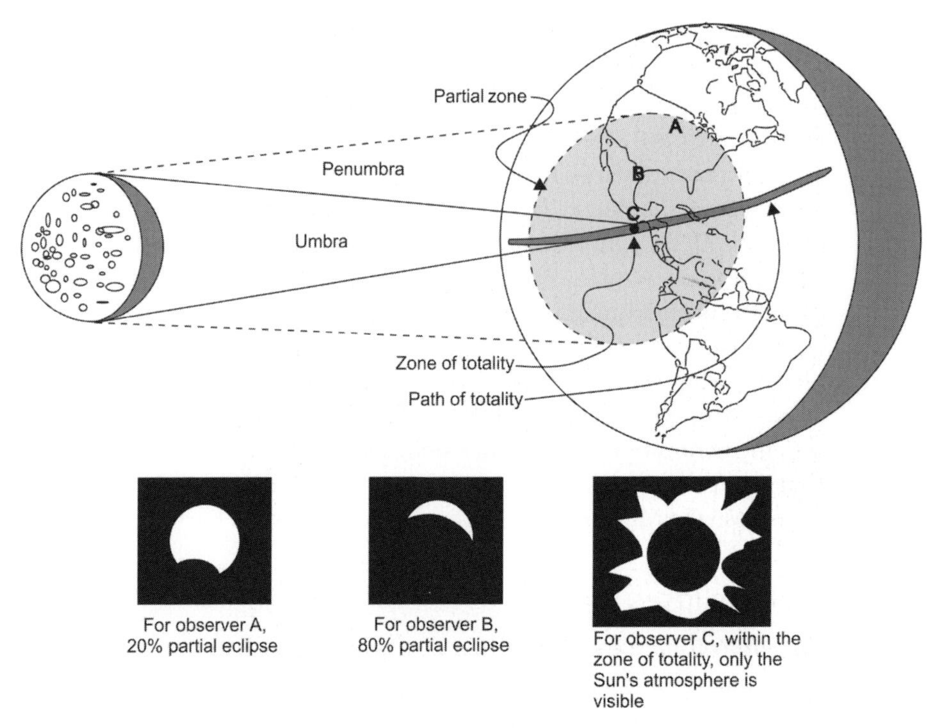

Figure 7.9 When then Moon passes in front of the Sun, which can occur when the Sun and the Moon reach the same node at the same time, we see a solar eclipse. Those who are standing in the path of totality see a total solar eclipse and those in the partial zone (in the Moon's penumbra) see a partial solar eclipse. While the penumbral zone can be over 6000 km (3700 miles) across, some people are so far from the path of totality that they do not see an eclipse at all. This is a result of geocentric parallax.

or ultraviolet) to pass through to your eyes. At the Sun's intensity, these light wavelengths will burn your eyes just as fast, or faster, than visible light. The retina does not have any nerve endings for the sensation of pain. Your eye may be burned well before you even realize it – that is, until you are unable to see.

Think of it this way – if you make yourself blind by improperly watching an eclipse, you will lose the main source of enjoyment of astronomy.

Total solar eclipses

A **solar eclipse** occurs (Figures 7.9, 7.10, and 7.14) when the Moon's shadow passes over the surface of the Earth. This can only occur during a new Moon phase. The observer must be on the Earth's daytime side and within the Moon's umbral shadow as it falls on the Earth. The partial zone covers about one-third of the daylight side. The zone of totality is never more than a 270 km (170 miles) circle of umbral shadow. It sweeps over the Earth's surface at more than 1600 km/h (1000 miles/h) from west to east.

Figure 7.10 The total solar eclipse of 8 June 1937. (Courtesy of the US Naval Observatory library.)

The observer sees either a partial or total solar eclipse depending on where the observer is located relative to the path of totality. In the partial (penumbral) zone, a **partial solar eclipse** is observed. In the zone of totality (the umbra), a total solar eclipse is observed. The path of totality can be plotted for each eclipse (see Figures 7.11 and 7.12, and Figure 7.18 later, for examples). These maps are published by many sources, including the *Astronomical Almanac* of the US Naval Observatory (USNO) and the *Eclipse Ephemeris* from the National Aeronautics and Space Administration (NASA; Espenak, 1987). Some of the more powerful planetarium programs can plot the path of totality for any solar eclipse. The path runs about parallel to the line of the ecliptic plane as it passes through the Earth's surface. However, due to the Earth's rotation, the altitude of the Sun during the eclipse, and other details, the path is generally curved on the Earth's surface.

Total solar eclipses are particularly interesting to astronomers because they allow us to study the thin atmosphere of the Sun – at least for a few minutes. What we refer to as the surface of the Sun is a thin layer of gas called the *photosphere*. This is the gas layer that emits the greatest amount of light. Above the photosphere is the *chromosphere*. The chromosphere emits mostly red light. (*Chromos* comes from the Greek word for *color*.) Above the chromosphere is a hazy white layer called the *corona* (from the Latin word for "crown").

Normally, the upper two layers of the Sun's atmosphere cannot be seen because of the intense radiation of the photosphere. However, during totality they

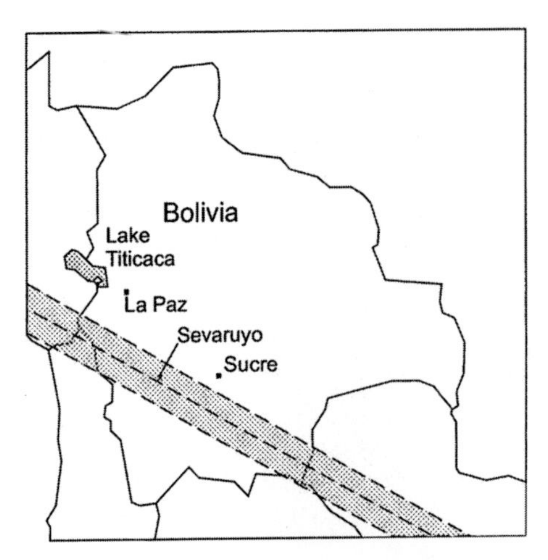

Figure 7.11 The path of totality for the 4 November 1994 eclipse passed through Sevaruyo, Bolivia, at about 09 : 00 local time. This small town was near the center line of the path.

Figure 7.12 The path of totality for the 4 November 1994 total solar eclipse, passing through South America.

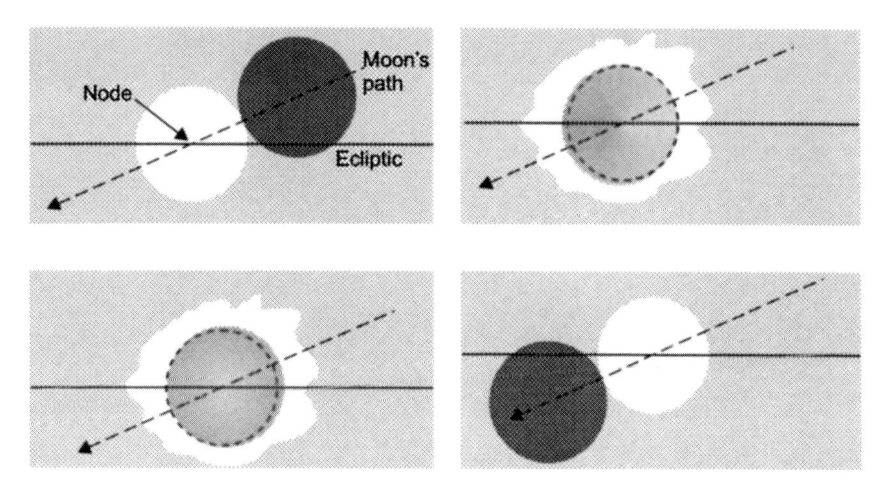

Figure 7.13 The contacts, or stages, of a total solar eclipse. First contact (upper left) occurs when the eastern limb of the Moon touches the western limb of the Sun, signaling the beginning of the eclipse. Second contact (upper right) is the beginning of totality. Third contact (lower left) is the end of totality. Fourth contact (lower right) is the end of the eclipse.

can be observed and photographed with some detail. The chromosphere is so thin that our ability to see it during totality depends on the angular diameter of the Moon during the particular eclipse. The corona on the other hand, is millions of kilometers in height. It is strongly linked to the Sun's (sometimes very) active magnetic field. We can see curvilinear structure in the corona that is linked to the magnetic field lines. Sometimes long streamers of gas can be seen coming out from the corona. All of this depends on the activity level of the magnetic field, which is also linked to surface activity, such as sunspots and active regions.

The eclipse begins with *first contact* (Figure 7.13). Recall that due to the Earth's rotation, the Sun and the Moon move from east to west during the day. However, the Sun and the Moon move along (or for the Moon, nearly along) the ecliptic line from west to east. Therefore first contact occurs on the western limb (disk edge) of the Sun. It takes about an hour for the Moon to move in front of the Sun. At the onset of totality, "shadow bands" move across the landscape. The probable cause of these bands is atmospheric turbulence, making the narrow, crescent Sun "twinkle," thus producing alternating bands of light and dark. However, this effect is still only poorly understood. Just before totality, we see *Bailey's Beads*. This effect is named for Francis Bailey (1774–1844), one of the first European astronomers to see a total solar eclipse. He was also an amateur scientist, principally responsible for establishing the field of solar physics. The beads are the last bits of sunlight peeking through the uneven surface features (mountains and valleys) on the lunar limb.

Second contact is the beginning of totality. During totality you notice a rapid drop in temperature accompanied by an *eclipse wind*. It gets as dark as late

Figure 7.14 The diamond ring effect, photographed at Sevaruyo, Bolivia, on 4 November 1994. (Photo courtesy of Herbert DeVries.)

twilight. The chromosphere and corona are visible (Figure 7.10) and sometimes major solar prominences or flares can be seen. The brighter stars can be seen – even those near the Sun. Often, Venus is visible, appearing as a bright star. After these few minutes the Moon moves from in front of the Sun. At *third contact*, just as totality ends, we get to see the spectacular *diamond ring* (Figure 7.14). This happens for the same reason as Bailey's Beads – a small piece of Sun peeks through a larger valley at the limb of the Moon. The emerging partial takes roughly another hour, until the eclipse is over at *fourth contact*.

Total eclipses have been recorded since the beginning of writing. They appear in the mythology and history of most cultures. One story from Greek history is about a battle between the Lydians and the Medes in 585 BC. In the height of the battle, a total solar eclipse occurred, with the zone of totality passing over the battlefield. The warriors were so awestruck by this event, they laid down their weapons and ended the battle – and their war (Krupp, 1999). There is a claim that the ancient Greek scientist Thales of Miletus predicted this particular eclipse by using the previous eclipse observations of the Babylonians. This claim has been disputed and is not settled.

Modern *coronagraphs* can create artificial eclipses. These instruments are used to study the solar atmosphere. This important field of study allows us to make connections between events on the Sun and events in the Earth's atmosphere and changes in climate.

Table 7.2 *Apparent size of the Sun and Moon, distance between Earth and Moon, and the Moon's shadow length*

Datum	Maximum	Average	Minimum
Angular diameter of the Moon	33′ 30″	31′ 05″.16	29′ 22″
Angular diameter of the Sun	32′ 36″	31′ 59″.26	31′ 32″
Lunar distance from Earth (center to center)	406 700 km, 252 710 miles	384 500 km, 238 900 miles	356 400 km, 221 460 miles
Length of lunar shadow (umbra)	381 000 km, 236 700 miles	372 000 km, 231 100 miles	365 000 km, 226 800 miles

Figure 7.15 Two photos of the full Moon showing its apparent size at perigee (closest to Earth, left) and apogee (farthest from Earth, right) points in its elliptical orbit. When a solar eclipse occurs with the Moon near apogee, it is not large enough to cover the disk of the Sun and an annular eclipse occurs.

Annular solar eclipses

The Moon is in an elliptical path around the Earth, causing its apparent size to vary slightly (Figure 7.15). The Earth is in an elliptical path around the Sun, causing the Sun's apparent size to vary slightly. During an **annular solar eclipse** not all of the Sun is covered by the Moon. A small annulus (Latin for "ring") of the Sun remains around the Moon as it passes in front of the Sun.

This happens because of the Moon's elliptical orbit, changing the apparent size of the Moon as seen from Earth (Table 7.2). The Moon's shadow has a variation of 16 000 km (9900 miles) in length due to the Earth–Moon system's elliptical orbit around the Sun. (This can be calculated by using the size of the Sun and the Moon and the distance between them.) Because the Moon's shadow is shorter than its average distance from the Earth, an annular eclipse is more likely than a total. Only when the Moon is closer to the Earth do we see a total

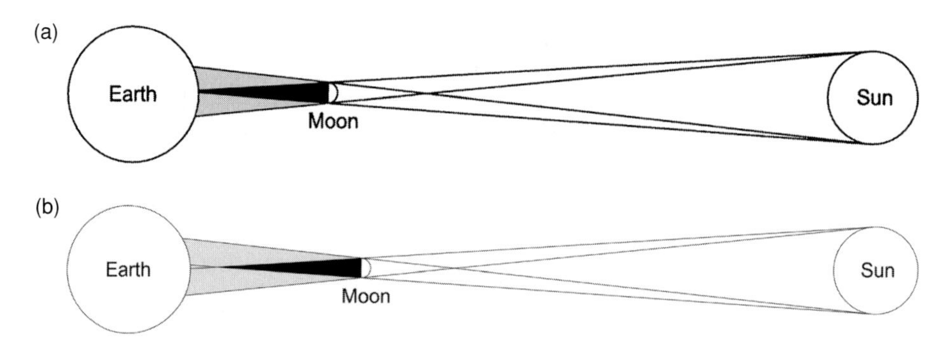

Figure 7.16 The difference between a total solar eclipse and an annular solar eclipse is caused by the elliptical orbits of the Moon and Earth. If the eclipse occurs while the Moon is in the farther region of its orbit (around apogee) the Moon's umbra is not long enough to reach the surface of the Earth. (a) During a total solar eclipse, the Moon's shadow can reach the Earth's surface. (b) During an annular eclipse, the Moon is too far from the Earth and its shadow cannot reach the Earth's surface.

solar eclipse (Figure 7.16). This can be verified by comparing the average angular diameters of the Sun and Moon. Because the Moon's diameter is slightly less, annular eclipses are more likely than total. Under certain circumstances the curvature of the Earth contributes to a particular eclipse event having both annular and total portions to its path. These are called *annular-total eclipses* or *hybrid eclipses*.

The appearance of an annular eclipse is not as exciting as a total eclipse. Some darkening and cooling occur but neither are as drastic as during a total eclipse. An annular eclipse cannot be directly viewed any more than a normal, un-eclipsed Sun. You must use eye protection during the entire annular solar eclipse. The sequence of events for an annular eclipse is similar to those of a total eclipse and are shown in Figure 7.17. Just like a total solar eclipse, you must be in the path of annularity. This path is never wider than 313 km (194 miles).

Duration of solar eclipses

During a total or annular eclipse, it takes about an hour for the submerging partial, during which the Moon is slowly covering the Sun, and about another hour for the emerging partial. While the elliptical orbit of the Moon changes its apparent size as seen from Earth, annual variation of the Earth–Sun distance also changes the apparent size of the Sun. When the Sun is farthest away and the Moon is closest, we may have a very long (7.5 minutes) total solar eclipse like the one that occurred in July of 1991. The longest period of totality possible is 7 minutes and 40 seconds. When the Sun and Moon are not at extremes in their distances from Earth, we have the more typical three to four minute total solar eclipse. The shortest duration solar eclipse is a grazing partial lasting only a few minutes. You would hardly see it. The longest is an annular eclipse, lasting

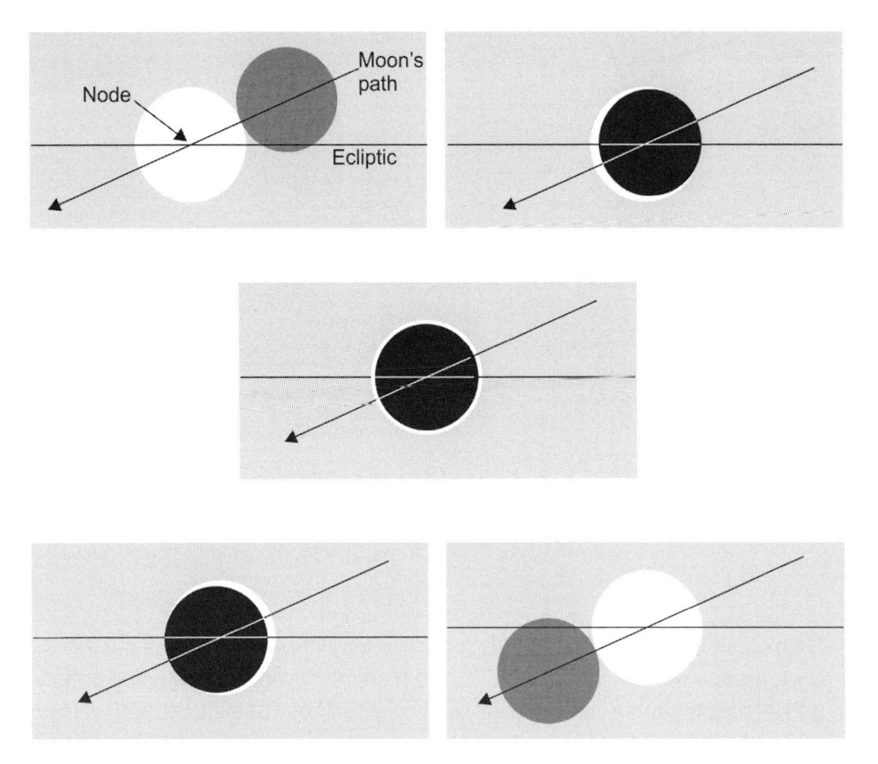

Figure 7.17 The contacts, or stages, of an annular solar eclipse. First contact (upper left) is the same as for a total eclipse. Second contact (upper right) is the beginning of annularity. Center contact (center) is maximum annularity where the ring is symmetrical. Third contact (lower left) is the end of annularity and fourth contact (lower right) is the end of the eclipse.

a total of about two hours, including the time taken for the Moon to cover and uncover the Sun.

Solar eclipses also have a magnitude rating. The *eclipse magnitude* is defined as the fraction of the Sun's diameter covered by the Moon at maximum eclipse. *Maximum eclipse* occurs when the axis of the Moon's shadow comes closest to the center of the Earth. For partial eclipses, the magnitude is always less than one. For annular eclipses the magnitude is also less than one, but the value is calculated using the ratio of the apparent diameters of the Sun and Moon. For total solar eclipses, the magnitude is also based on the apparent diameters and is always equal to or greater than one.

7.5 The frequency of eclipses: eclipse seasons

As we discussed on p. 132, the orbital path of the Earth, as it goes around the Sun, lies in a flat plane called the ecliptic plane. For the same reasons, the Moon's orbital path, as it goes around the Earth, also lies in a flat plane. The Moon's orbital plane is not the same plane as the ecliptic plane. It is inclined from the

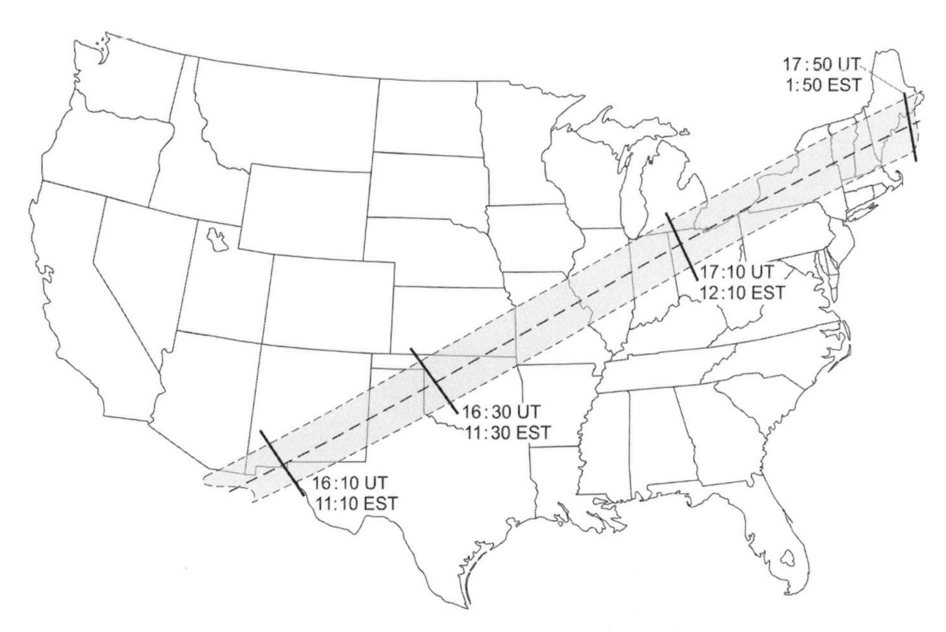

Figure 7.18 The path of annularity for the 10 May 1994 solar eclipse passed through the USA. Anyone standing in this path saw the annular eclipse and anyone outside it, in the path of the penumbra, saw a partial solar eclipse. Those outside the path of the penumbra saw no eclipse.

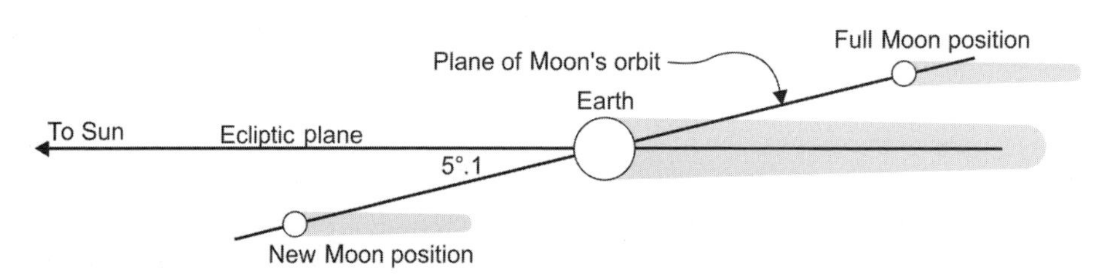

Figure 7.19 The Moon's orbital plane is inclined by 5°.1 to the ecliptic plane. At the Moon's distance from the Earth, this small angle is enough to allow the Moon to miss the Earth's shadow. This is a "side view" of a non-eclipse season arrangement as shown in Figure 7.22 at position A or C.

ecliptic plane by 5°.1. If it were the same as the ecliptic plane, the Moon would move exactly along the ecliptic line with the Sun and we would have lunar and solar eclipses every full and new Moon phase. The Moon must stay in its orbital plane and because it is inclined to the ecliptic, we always find the Moon within 5°.1 of the ecliptic line. Look back to Figures 6.3 and 6.4.

Five degrees may seem like a small angle but at an average distance of 384 400 km (240 000 miles) from the Earth, five degrees means the Moon can be as much as 34 200 km (21 200 miles) above or below the ecliptic plane (Figures 7.19 and 7.20). The Moon is only 3480 km (2200 miles) in diameter. The ecliptic line represents the position of the Sun and the position of the center of

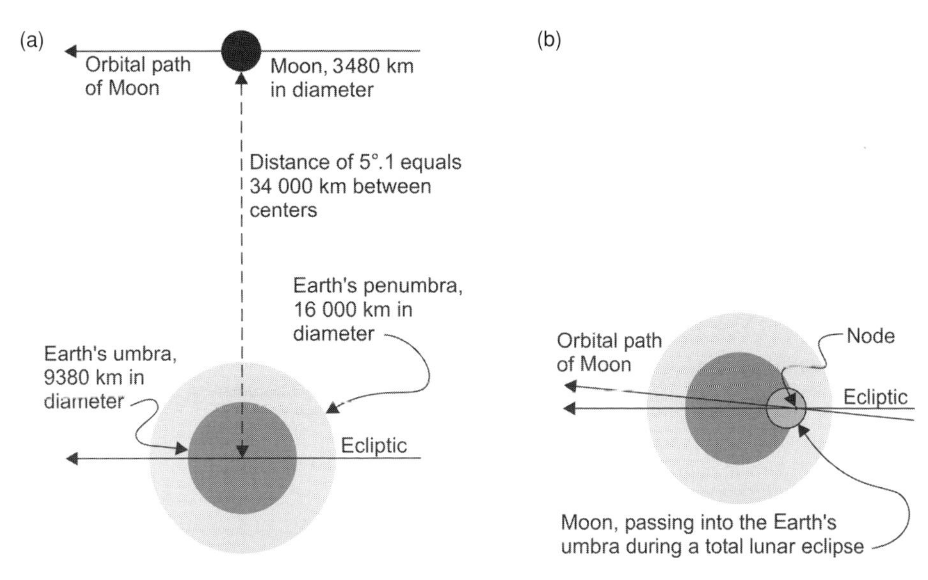

Figure 7.20 The Moon crosses the ecliptic – at a node – twice each month. A lunar eclipse occurs if the Moon and the Earth's shadow are near the same node at the same time. (a) When it is not an eclipse season, the Moon can miss the Earth's umbra by as much as 16 000 km (drawn to scale). (b) A lunar eclipse may occur only during an eclipse season, when the Moon and the Earth's shadow cross a node simultaneously (drawn to scale).

the Earth's shadow, which is always on the opposite side of the celestial sphere from the Sun. At the orbital distance of the Moon, the Earth's umbra is 9380 km (5800 miles) and its penumbra is 16 000 km (9920 miles) in diameter. (These numbers are averages.) The Moon could be as high as twice the height of the Earth's penumbra, above the ecliptic. These numbers tell us it is easy for the Moon to miss the Earth's shadow.

So, if the Moon's orbital plane is inclined by 5°.09, why does it enter the Earth's shadow at all? The answer lies in the motion of the Earth–Moon system around the Sun. Recall from p. 143, the Moon must pass through the ecliptic plane at least twice each month, or at least 24 times per year. The two points where it crosses the ecliptic line (the orbital nodes) are on opposite sides of the celestial sphere. A straight line drawn between the nodes represents the intersection of the Moon's orbital plane with the ecliptic plane. This line is called the **line of nodes** (Figure 7.21).

> *An* **eclipse season** *occurs when the line of nodes is nearly aligned with the Earth–Sun line, or when the Sun is near a lunar node. This happens at least twice each year, a little less than six months apart, or once at each end of the line of nodes (Figure 7.22, see also Figures 7.23 and 7.24).*

The Moon's orbital plane precesses or "wobbles" about the ecliptic poles with a period of 223 lunar phase cycles (or synodic periods). This precession causes the line of nodes to rotate (or precess), with the nodes moving along the ecliptic

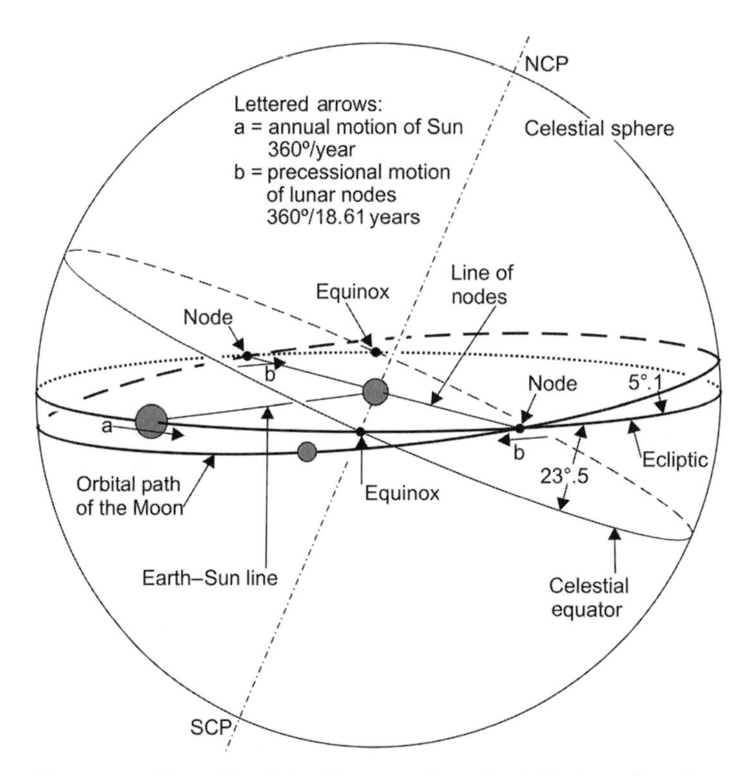

Figure 7.21 The orbit of the Moon on the celestial sphere lies close to the ecliptic. The nodes are the points where the Moon's orbit crosses the ecliptic, one on each side of the celestial sphere. The line of nodes is drawn between these two points, passing through the Earth. As the Sun moves along the ecliptic and it passes through a lunar node, the Earth–Sun line then matches the line of nodes, creating an eclipse season. If the Moon is at a node at the same time we see either a lunar eclipse or a solar eclipse.

from east to west, opposite to that of the Sun (Figure 7.21). The rate of precession is about 19°.3 per year, so compared with the motion of the Sun during a one-year period, the line of nodes moves slowly around the ecliptic. This motion of the nodes causes the possibility of three eclipse seasons in a calendar year. Because of this relative motion, the Sun moves through the two nodes (returns to the first node) in 346.6 days. This length of time is called an **eclipse year**. Because the eclipse year is shorter than the tropical year, it is possible to have three eclipse seasons in one calendar year.

The saros

The Moon's orbit around the Earth is slightly elliptical, with a perigee (closest approach) of 356 400 km (221 460 miles) and apogee (farthest point) of 406 700 km (252 710 miles). The amount of time between the Moon's successive passes of the perigee is called the **anomalistic month** and it is equal to 27.55 days.

In Table 7.3 we can see a coincidence of three periods of time that otherwise have nothing to do with each other. The period of 223 lunar cycles is called

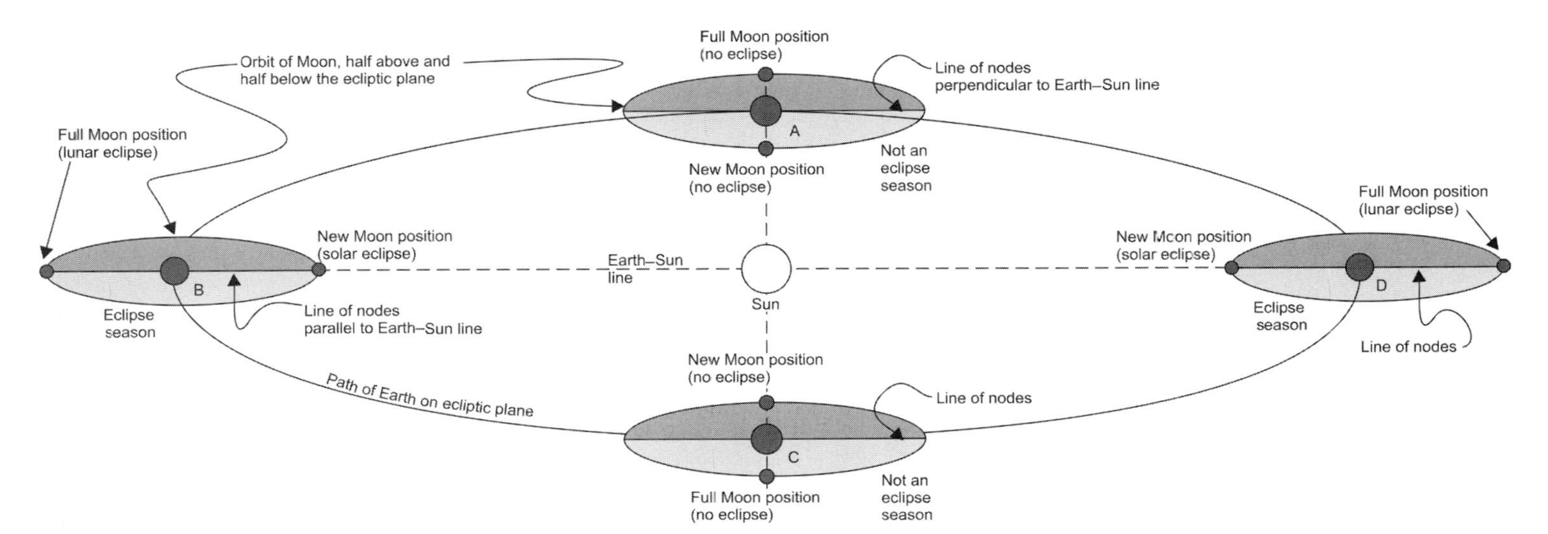

Figure 7.22 An eclipse season occurs when the Moon's line of nodes points toward the Sun. This occurs twice each year, as shown here at positions B and D. Compare this drawing at A with Figure 7.20a – the full Moon is above the ecliptic plane and the Earth's shadow. At position C, which also is not an eclipse season, the full Moon is below the ecliptic plane and the Earth's shadow. When the Earth–Moon system has moved along its solar orbital path to a position in an eclipse season (positions B and D), the situation is like that shown in Figure 7.20b – a lunar eclipse will occur. The discussion is the same for the new Moon positions and solar eclipses. Eclipses can only happen during an eclipse season. Compare this figure with Figures 7.23 and 7.24.

Table 7.3 *The saros: a coincidence of periods*

Period name	Period length (days)	Multiple	Total period (days)
Synodic period	29.5306	223	6585.32
Anomalistic month	27.5545	239	6585.53
Eclipse year	346.620	19	6585.78

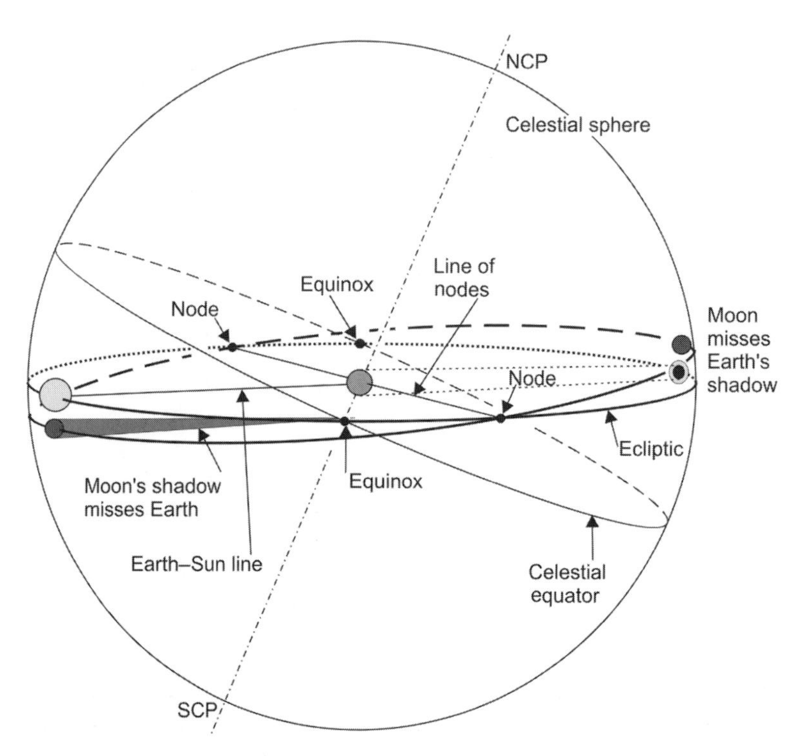

Figure 7.23 When the Earth–Sun line is not near the line of nodes – the Sun is not near a node – there can be no eclipses. The Moon's shadow passes below (or above if on the other side of the sphere) the Earth and the Moon passes above (or below) the Earth's shadow. The circles representing the Sun, Moon, Earth, and Earth's shadow are not drawn to any scale. Compare this figure with Figures 7.22 and 7.24.

the **saros**. It is 0.46 days short of 19 eclipse years. The precessional motion of the lunar nodes causes the *saros cycle of eclipses,* believed to be discovered by the ancient Chaldeans. ("Saros" is a Chaldean word.)

The **saros cycle** *is a near repetition in the pattern of eclipses and contributes to our ability to predict the occurrence of eclipses. If a total eclipse occurs at a given Earth location, then* 6585.32 *days later there will be another total (or near total, because of the slight difference in the anomalistic period) eclipse at almost the same latitude, but slightly north or south of the original, depending on the particular saros.*

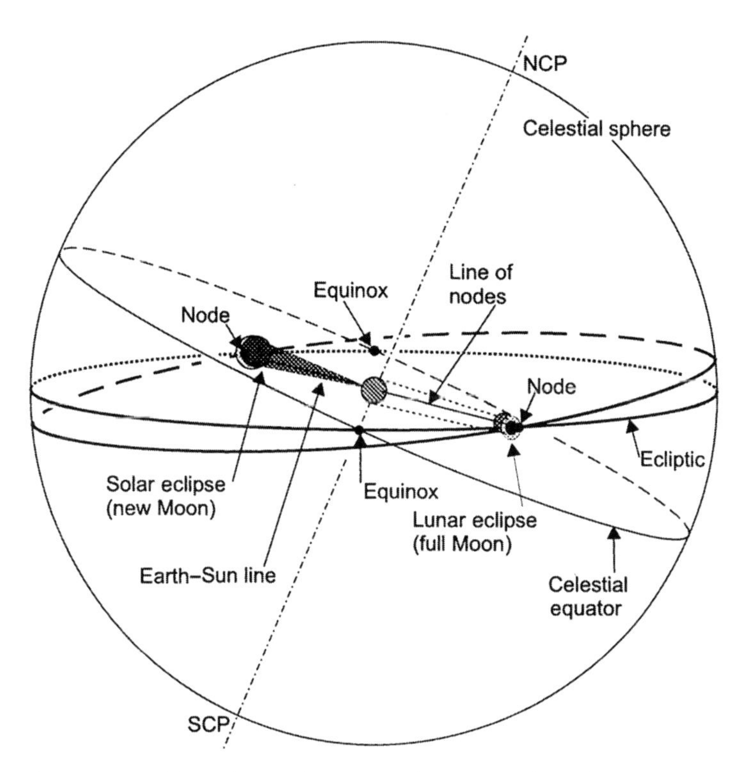

Figure 7.24 When the Earth–Sun line is near the line of nodes – the Sun is near a node – then eclipses are "in season." There will be a solar eclipse and possibly a lunar eclipse. The circles representing the Sun, Moon, Earth, and Earth's shadow are not drawn to any scale. Compare this figure with Figures 7.22 and 7.23.

Due to the timing between the saros cycle and the rotation of the Earth the next eclipse will be 120° (one-third of a day) farther west in longitude. After three saros cycles, the longitude of the eclipse is nearly repeated. The saros cycle is generally associated with solar eclipses but it can be applied to lunar eclipses as well. For lunar eclipses, there is no north–south displacement to the eclipse because lunar eclipses are visible to an entire hemisphere of the Earth.

Required conditions for a lunar eclipse

For a lunar eclipse to occur, the Moon must pass through the Earth's shadow. Because the Earth's shadow is on the opposite side of the Earth from the Sun, the Moon must be in its full phase. Thus, lunar eclipses can occur only during a full Moon phase. To have at least a partial lunar eclipse, the center of the Moon must come to within a degree of the ecliptic (Figure 7.25). This condition of proximity can occur within the **lunar ecliptic limit**, which is on average about 11° along the ecliptic.

A variation between 9°.5 and 12°.3 occurs because of the variation in the apparent size of both the Moon and the Earth's shadow. The *minor ecliptic limit*

Table 7.4 *The number of solar and lunar eclipses per year*

	Two eclipse season years		Three eclipse season years	
	Minimum	Maximum	Minimum	Maximum
Lunar	0	2	0	3
Solar	2	4	3	5

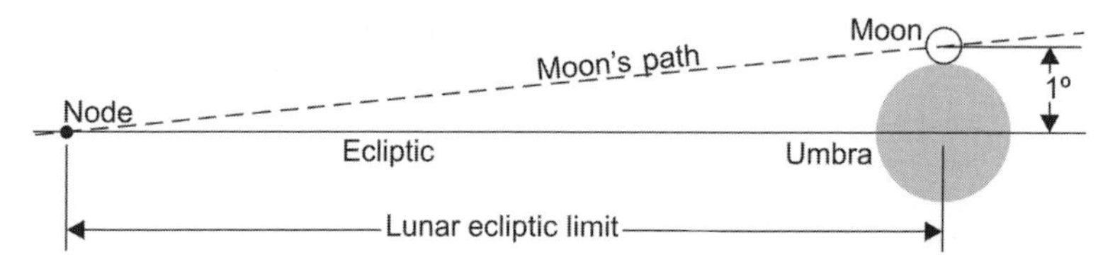

Figure 7.25 The lunar ecliptic limit determines the farthest distance from the node for which a lunar eclipse may occur. The major lunar ecliptic limit is twice this distance, because the eclipse may occur on either side of the node.

of 9°.5 occurs when the Moon is at apogee and the Earth is at perihelion. The *major ecliptic limit* of 12°.3 occurs when the Moon is at perigee and the Earth is at aphelion.

Because a partial lunar eclipse could occur with the Moon at the top (as shown in the figure) or bottom of the shadow (on the other side of the node), the angle for a possible lunar eclipse is at least 19° and at most 24°.6 along the ecliptic. This is called the **major lunar eclipse period**. Because the Earth's shadow is related to the Sun, it must move along the ecliptic at the same rate as the Sun – about one degree per day. Thus there is, at best, a 25-day period during an eclipse season (while the shadow is close enough to the node) when a lunar eclipse may occur. Because the synodic period of the Moon is 29.5 days, it is possible for the full Moon to miss an eclipse season entirely.

For either of the two nodes, a lunar eclipse will occur in three out of four eclipse seasons. There can be only one lunar eclipse per eclipse season. In some years the Moon will miss both eclipse seasons and in other years it will hit both. To observe the lunar eclipse, the observer must be on the night side of the Earth, for which there is a 50 : 50 chance. Thus, about one-half of all lunar eclipses are visible from a particular location on Earth.

In those (calendar) years where there are three eclipse seasons, it may be possible to have a lunar eclipse during all three seasons. This happened in 1982 for only the third time during the twentieth century. Thus, having three lunar eclipses in one calendar year is very rare. Typically, there is one or two per year (Table 7.4).

This discussion has ignored the possibility of penumbral lunar eclipses, which because of the larger sized penumbral region, have a higher probability

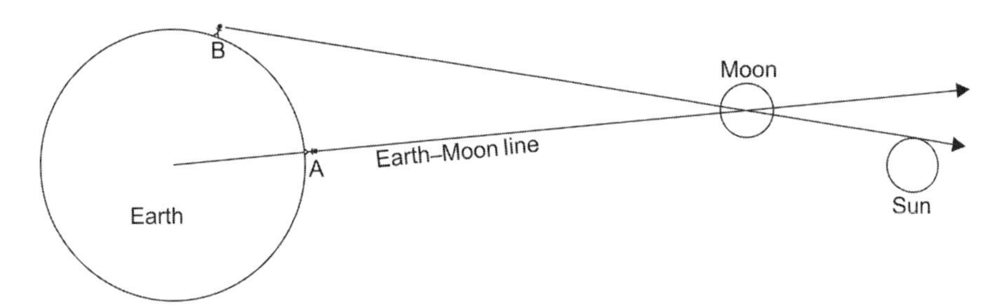

Figure 7.26 Geocentric parallax is the result of an observer being located on the surface of the Earth instead of at the center. It creates an apparent misalignment of the Moon and the Sun which may lead to a solar eclipse for observer B, but not for observer A, who is closer to the true Earth–Moon center line.

of occurrence. The reason for ignoring them is that they are generally not worth observing. Appendix B, p. 258 contains a list of dates for some partial and total lunar eclipses.

> *For a lunar eclipse to occur, the full Moon must pass through the Earth's shadow. This condition of proximity can occur within the major lunar ecliptic limit, which is on average about 22° along the ecliptic. There is, at best, a 25-day period during an eclipse season in which a lunar eclipse can occur. Because the synodic period of the Moon is 29.5 days, it is possible for the full Moon to miss an eclipse season entirely.*

Required conditions for a solar eclipse

Determining the requirements for a solar eclipse is a little more difficult. The *Earth–Moon line* passes through the center of the Moon and the center of the Earth. Only those observers on the Earth–Moon line will see the center of the Moon lying on the Moon's true celestial sphere orbital path. Because the Earth is an object of some extent and the Moon is near the Earth, observers at different latitudes see the Moon in slightly different positions on the celestial sphere (against the background stars), either slightly above or below its true orbital path. This apparent displacement of the Moon is the result of *geocentric parallax*. **Geocentric parallax** is the apparent displacement of an object due to the observer being on the surface of the Earth rather than at its center (Figure 7.26). Geocentric parallax is not important for lunar eclipses because the position of the Moon and the position of the Earth's shadow are affected in the same way.

To determine the **solar ecliptic limit** we must allow for the geocentric parallax of the Moon. Geocentric parallax gives us the arrangement shown in Figure 7.27. Because of the parallax effect, the solar ecliptic limit is 50% longer than the lunar ecliptic limit. The *minor solar ecliptic limit* is 15°.4 and the *major solar ecliptic limit* is 18°.5 (Figure 7.27). The Sun moves along the ecliptic about 1° per day, thus providing a **minor solar eclipse period** of 31 days and a **major solar eclipse period** of 37 days. Because both are longer than the synodic period of the Moon, a new Moon phase and a solar eclipse must occur during every eclipse season.

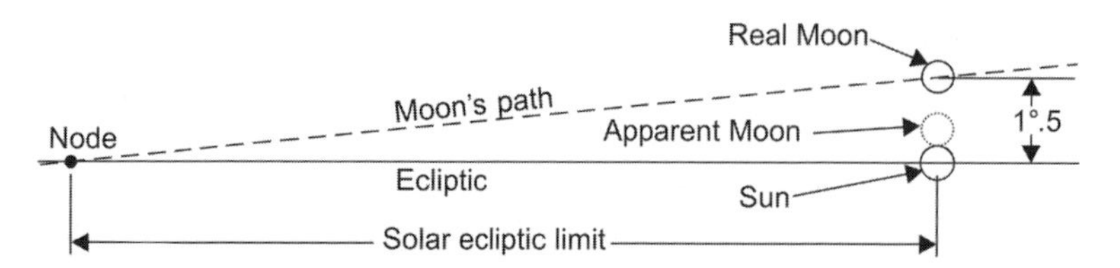

Figure 7.27 The solar ecliptic limit determines the farthest distance from the node for which a solar eclipse may occur. The limit is increased over its true length by geocentric parallax. The major solar ecliptic limit is twice this distance, because the eclipse may occur on either side of the node.

This means there *must* be at least two solar eclipses per year. It is also possible for a solar eclipse to occur at the beginning of an eclipse season and again at the end. If this happens, there are at least three solar eclipses during the year (Table 7.4).

If a double solar eclipse happens during an eclipse season then the Earth's shadow must be near a node and the full Moon that occurs between the two new Moons of the two solar eclipses must pass through the Earth's shadow. In other words, when we have a double solar eclipse, a lunar eclipse is unavoidable.

Annually, the maximum number of solar eclipses (of all types) is four. The minimum is two, with the average a little more than two. Total solar eclipses occur at the rate of about 70 per century. The remaining eclipses are usually annular solar eclipses or from time-to-time, simply a partial solar eclipse. (The Moon's umbra passes above or below the Earth, so the partial eclipse is observable at one of the poles.) Because of the saros cycle and geocentric parallax, a *total* solar eclipse is seen at any specific place on Earth about once every 300 years. This is about one in every 210 total eclipses. However, this is a simple statistical average and is not true for every location (Meeus, 2000). There is a list of solar eclipses in Appendix B, p. 257. By percentage, the frequency of solar eclipses is: total, 26.9%; annular, 33.2%; annular-total, 4.8%; and partial, 35.2%.

For a solar eclipse to occur, the Earth must pass through the new Moon's shadow. This condition can occur within the solar ecliptic limit, which is at least 31° along the ecliptic. This creates a 31-day period during an eclipse season in which a solar eclipse can occur. Because the synodic period of the Moon is 29.5 days, it is impossible for the new Moon to miss an eclipse season.

Questions for review and further thought

1. What causes eclipses?
2. What are the two parts of a shadow? Describe them.
3. Why does a shadow have two parts?
4. What object disappears during a lunar eclipse?
5. What object disappears during a solar eclipse?
6. Describe the arrangement of the Sun, Moon, and Earth during a lunar eclipse.

7. Describe penumbral, partial, and total lunar eclipses. Include a description of the path of the Moon and the Moon's appearance during the eclipse.
8. Explain why the Moon turns red during a total lunar eclipse.
9. List and describe the stages of a lunar eclipse.
10. There will come a time when total solar eclipses will no longer happen. Why?
11. When is it safe to look directly at the Sun?
12. Why is it not safe to use a "homemade" solar filter?
13. Describe the relationship between an observer's location in the shadow path and the observed solar eclipse.
14. Why are astronomers so interested in total solar eclipses?
15. What do we mean by the "path of totality?" Describe it.
16. Describe the diamond ring. How is it created?
17. Why would a solar eclipse be annular instead of total?
18. What is the longest duration (of totality) for a total solar eclipse?
19. How long is totality for a typical total solar eclipse?
20. What is a saros and what is the saros cycle?
21. How close to the ecliptic must the Moon come to have at least a partial lunar eclipse?
22. During the average calendar year, how many lunar eclipses occur? (Minimum, maximum, and typical.)
23. Why must a solar eclipse occur every eclipse season?
24. How many solar eclipses are there in a calendar year? (Minimum, average, and maximum.)

Review problems

1. Verify the relationships between the angular diameter and distance diameter (kilometers) of the Earth's umbra and penumbra at the Moon's average distance.
2. What is the maximum and minimum length of the major lunar eclipse period?
3. Given a lunar eclipse season of 24.5 days, calculate the probability of a lunar eclipse. Hint: approach this problem as a simple ratio.

8
Observation projects

Now that you've read all about the sky it's time to go out and observe. Here is a set of observation projects to get you started on learning the properties of the celestial sphere. When you've completed these projects, you may want to see the book *Practical Astronomy: A User-Friendly Handbook for Sky Watchers* (Mills, 1994). It contains many practical projects for building instruments and making observations of the night sky.

8.1 Constellation charts

The objective of this project is to become more familiar with the charts included with this book. You should also learn more about the relationship of the motions of the Sun and Moon.

Part I Celestial coordinates

The *Equatorial Chart* shows the stars located within 60° declination of the *celestial equator* (the line dividing upper and lower halves of the chart). Coordinates above and below the celestial equator are called *declination* (Dec). Declination is measured in degrees, with positive values above the celestial equator and negative values below. Coordinates from left (east) to right (west) are called *right ascension* (RA). Right ascension is measured in hours, minutes, and seconds, starting from the center of the chart, i.e. from the *vernal equinox*. Notice that right ascension increases to the east (left) because objects rise in the east as the Earth turns on its axis and time passes.

The first part of this exercise is to locate a few objects on the chart in order to become familiar with equatorial coordinates. The declination angles are marked along all five vertical lines on the chart. The right ascension hours are shown at the top and bottom of the chart and along the equator. The first row of the table below is given as an example. Some of the stars are labeled by their Greek letter designations (e.g. β Leonis). In these cases, use the Bayer designation (see p. 45) to label the star.

Object name	Object type	Right ascension	Declination (°)
Leo	Constellation	$10^h\ 30'$	+20
	Constellation	$2^h\ 15'$	+25
	Constellation	$18^h\ 30'$	−30
	Star	$23^h\ 0'$	−30
	Star	$6^h\ 45'$	−17
β Boötis	Star		
γ Trianguli	Star		
		$3^h\ 45'$	+25
Delphinus	Constellation		+15
Arcturus		$14^h\ 10'$	+20
Canis Minor			
	Star	$6^h\ 0'$	+45
	Star cluster	$8^h\ 40'$	
Vernal Equinox	Sky point		
Summer Solstice	Sky point		

The line across the center of the map is the *celestial equator*. Find the constellation Orion along the celestial equator. Find the proper names of the four bright stars whose Bayer designation is given below.

Bayer des.	Proper name	Bayer des.	Proper name
α Orionis		β Orionis	
γ Orionis		κ Orionis	

Due to the Earth's motion around the Sun, different stars and constellations are visible in the evening sky during the year. Notice the dates along the *bottom* of the *Equatorial Chart*. These dates indicate when the stars above are best viewed at midnight. Based on these dates, what is the best month of the year to view the following constellations?

Object	Declination (°)	Month
Orion	0	
The Great Square	+20	
Scorpius	−30	

Using these same dates, name four constellations that should be visible in the sky tonight. Choose constellations with brighter stars, making them easier to locate.

Answer:

The hours of right ascension are based on the hourly motion of the stars across the night sky. For example, a star with a right ascension of 8 hours rises three hours after a star whose right ascension is 5 hours. At any given time, we have a 12-hour window on the sky. Answer these questions based on the right ascension position of these constellations along the celestial equator.

What constellation rises about four hours after Orion?
Answer:

What constellation is rising in the east, as the constellation Ophiuchus is setting in the west?
Answer:

Part II Apparent magnitude

The description of the brightness of a star, as we see it from Earth, is called its *apparent magnitude*. A magnitude number is assigned to each star with the *dimmer stars having higher numbers*. A key to the brightness dots, used in the charts, is given at the upper right corners of the chart. Based on this key, what is the approximate apparent magnitude of the following stars?

Star	Approximate position	Apparent magnitude
Sirius	$7^h, -15°$	
Altair	$20^h, +10°$	
α Cancri	$9^h, +10°$	
Canopus	$6^h, -53°$	

Give the apparent magnitude of each of these stars in the constellation Leo the lion. Some of Leo's stars are not in the stick figure pattern. In some cases, it may be necessary to estimate the magnitude. Use half magnitudes if you cannot decide between two whole magnitudes from the chart scale.

alpha, α	Epsilon, ϵ	Lambda, λ
Beta, β	Zeta, ζ	Mu, μ
Gamma, γ	Eta, η	Omicron, o
Delta, δ	Theta, θ	Rho, ρ

The number of stars visible in the sky depends both on the star magnitudes and the sky conditions, particularly the level of light pollution. For instance, more stars are visible in the center of the Pacific Ocean than in the center of New York City. Using the magnitude information on the chart for each star in the constellation, count the number of visible stars in the constellations Orion and Aquarius under the three conditions listed in the table below. The "lowest magnitude" column states the dimmest visible star magnitude under the particular observing condition.

Seeing condition	Lowest magnitude	Orion	Aquarius
City	2		
Country field	4		
Perfect	6		

Part III The motion of the Sun

The curved line that snakes across the *Equatorial Chart* marks the apparent path of the Sun as it moves among the stars during the year. This path is the *ecliptic line* and it crosses the celestial equator at the *equinox points*. The dates given along this line indicate when the Sun occupies that position. The degree numbers just above the ecliptic line give the Sun's celestial longitude in degrees.

Determine the Sun's right ascension and celestial longitude on the dates listed in the table below. Be careful not to confuse celestial longitude degrees with declination degrees.

Date	Sun's RA (h, min)	Sun's celestial longitude (°)
Vernal equinox		
Autumnal equinox		
21 January		
7 September		

The constellations that the Sun passes through are known as the signs of the zodiac. In which zodiac sign is the Sun located on these days?

Date	Constellation	Date	Constellation
Nov. 18		April 12	

According to astrology, what is your zodiac sign? (See Table 4.1 on p. 86 for zodiac sign dates.)

Answer:

Does this agree with the actual position of the Sun on your birthday? Explain the cause of any difference in these dates. (Hint: It is due to one of the Earth's motions studied in Chapter 4.)
Answer:

What constellation does the Sun occupy on December 12?
Answer:

As the Sun moves among the stars, it changes in declination. This causes it to appear higher or lower in the sky at different times of the year. Our seasons are related to these changes in position. What is the declination of the Sun on these dates?

June solstice	December solstice

What is the latitude of your location?
Answer:

What is the altitude of the celestial pole at your latitude?
Answer:

What is the transit altitude of the celestial equator at your latitude?
Answer:

What is the *high noon altitude* of the Sun at your latitude when it is passing through the solstices?

June solstice	December solstice

These dates are known as the solstices because the Sun stays at the same noon altitude for several days. Compare the change in altitude of the Sun around the June solstice with that around the vernal equinox by completing this chart. After calculating the noon altitude for a pair of dates, calculate the change in altitude between the dates. The change in altitude determines whether the Sun has passed through a solstice.

Date	Declination (°)	Noon altitude (°)	Change in altitude (°)	Solstice? (Y/N)
March 5				
April 5				
June 5				
July 5				

Part IV Culminations

We now look at the culmination altitudes of some objects. These altitudes depend on your location. The symbol for latitude is λ (the Greek letter lambda).

What is the latitude of your location?

$\lambda =$

Which of the celestial poles is *not* visible from your location?
Answer:

Calculate the value,

$Q = 90^\circ - \lambda =$

All sky objects located within the number of degrees equal to your latitude of the celestial pole are never visible at your latitude. In the northern hemisphere, all objects with declination less than $-Q$ can never be seen from your latitude. In the southern hemisphere, all objects with declination greater than $+Q$ can never be seen from your latitude. Can the objects in this table ever be seen at your latitude? (For a constellation, use the declination of its center point.)

Object	Object type	RA	Dec. (°)	Yes, No, Partly
Columba	Constellation	6^h		
α Centauri	Star	15^h		
Ursa Major	Constellation	12^h		
Vela	Constellation	9^h		

Given that the altitude of the celestial equator equals 90°minus the location latitude, what prominent line on the chart represents the horizon to an observer standing on the North Pole?
Answer:

Would all of the constellation Orion (RA = 5^h), ever be visible while standing at the North Pole? Explain your answer. (Hint: How do the stars move at the North Pole?)
Answer:

How would the view of Orion differ between observers standing on the Earth's North and South Poles?
Answer:

In either hemisphere, all objects with declination between $-Q$ and $+Q$ rise in the east and set in the west. Their upper culmination (transit altitude) depends on their declination. What is the upper culmination of these objects for your latitude?

Object	RA	Dec. (°)	Upper culmination
Antares	16^h		
Sirius	7^h		
Vega	$18^h.5$		
Sun (21 March)			
Sun (21 June)			

Part V The path of the Moon

The Moon's orbital path lies very near the ecliptic line. For the purposes of this discussion, we'll consider the Moon's path to be the same as the ecliptic line. While the Sun requires a year to complete one trip along the ecliptic, the Moon requires only about one month, or 27.3 days. As the Moon moves along the ecliptic, it goes through a sequence of phases. The lunar phase depends on its location relative to the Sun. The figure below shows the Sun at its position for

21 March (vernal equinox) and the position of the Moon for each of its possible phases on that date.

During the year, the Sun moves to the east (left), and the position of each of the lunar phases shifts to the east with the Sun. In the following two charts, the celestial longitude of the Sun and each Moon phase is given in degrees. The celestial longitude of the Moon phase is found by adding the lunar phase angle to the celestial longitude of the Sun. The lunar phase angle can be found on Figure 6.6. For the intermediate phases, use the halfway phase angle. That is, for the waxing crescent phase use 45°, which is halfway between 1° and 89°.

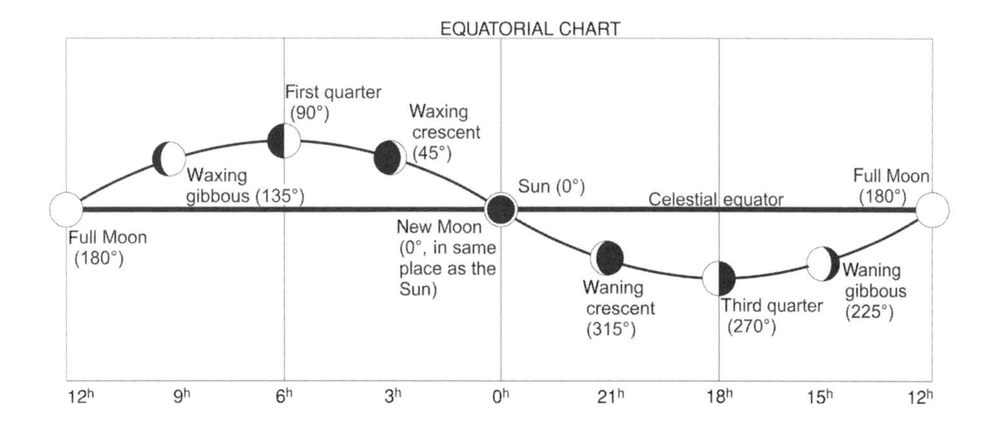

As the Sun moves along the ecliptic line, the phases of the Moon shift in a proportional amount.

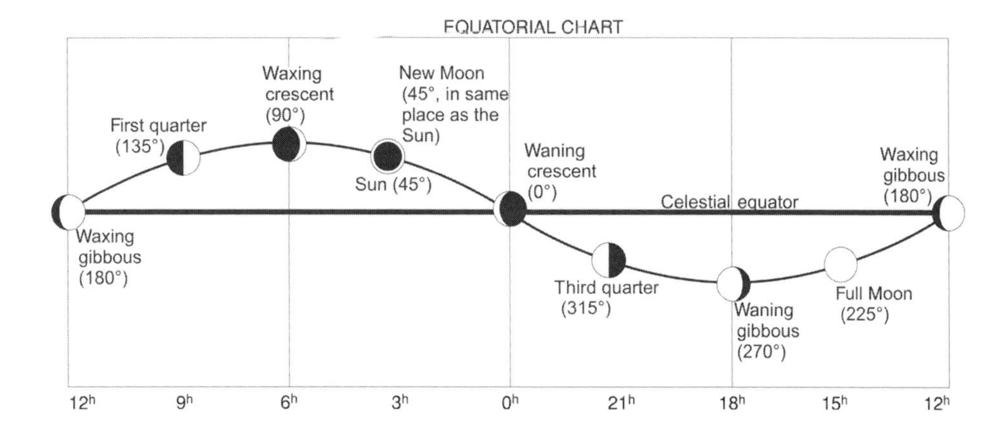

Fill in the drawing on p. 202 by shading in the unlit portion of the moon for each phase, and label the phase with its name and its celestial longitude. Your

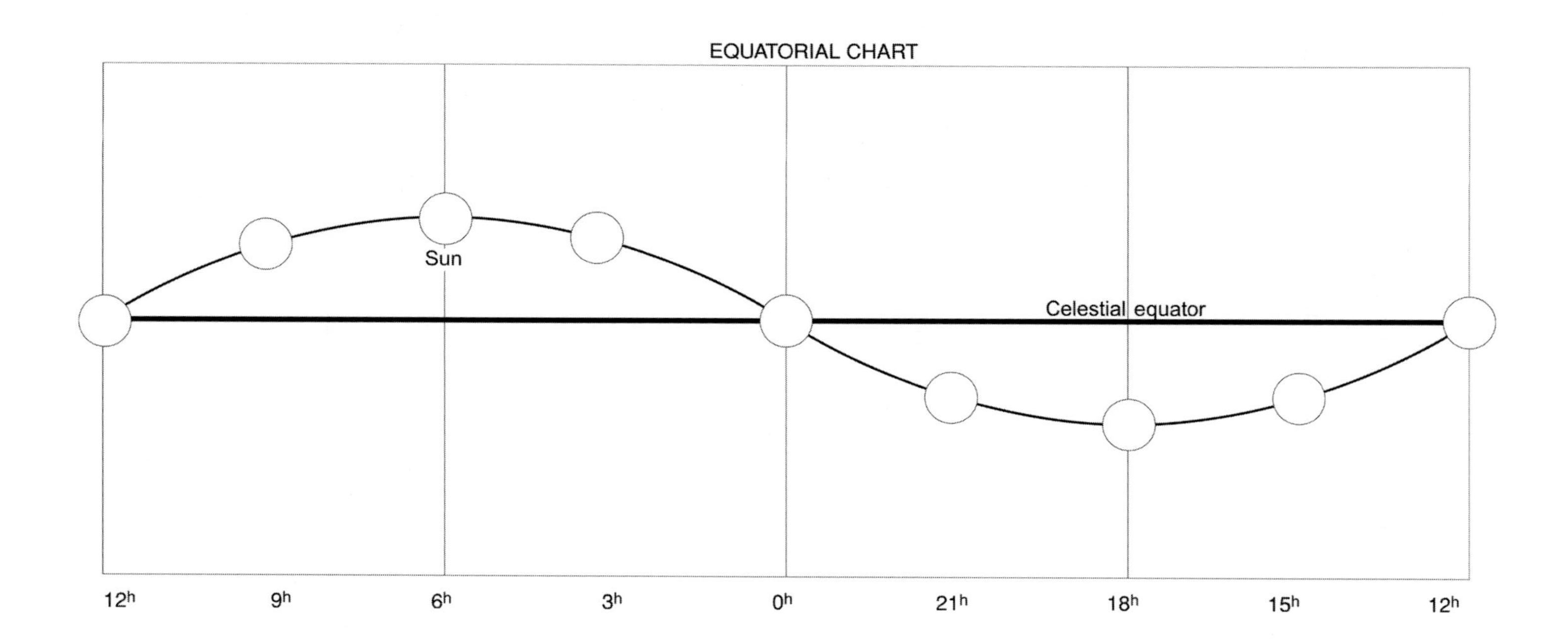
EQUATORIAL CHART
Sun
Celestial equator
12h
9h
6h
3h
0h
21h
18h
15h
12h

completed drawing should look similar to the two drawings above, noting the change in the position of the Sun. In the table that follows, state the appropriate position of the Moon during each phase, and name the zodiac constellation at the Moon's location. Using the figure you just completed as a guide, fill in the table below.

Moon phase	Sun RA (h)	Moon RA (h)	Zodiac constellation
New	6		
First quarter	6		
Full	6		
Waning gibbous	6		
Third quarter	6		
Waning crescent	6		

Use the same reasoning to complete this next table. The Sun will be in different locations. This time, use the celestial longitude, marked off in degrees along the ecliptic line. The first line is shown as an example. Use Figure 6.6, on p. 146, as a guide. Angles measured counterclockwise (from the Sun, new Moon position) on this figure are equivalent to celestial longitude angles measured toward the left (from the Sun or new Moon position), along the ecliptic line of the chart.

Date	Sun long. (°)	Moon long. (°)	Phase
March 21	0	180	Full
June 21			Full
		237	New
	310	130	
Sept 21			First quarter
May 10		20	
	142	232	

In what constellation is the full moon located on these dates?

February 15	October 22

On what day of the year is the full moon seen highest in the midnight sky?
Answer:

If a total lunar eclipse happens when the Moon is at 250°, what is the date?
Answer:

8.2 A simple quadrant

Here is a description of how to make the simple quadrant shown in Figure 2.19. You will need this quadrant (or one like it) to complete some of the observation projects in this chapter.

Parts list and tools

First you need to gather the materials.

1. One plastic protractor of radius 7.5 to 10 cm (3 to 4 inches). If you can, get one with a small hole at the center point, it will save work later.
2. One ruler (metric or inches) or simply a straight piece of wood of about the same size as a ruler.
3. Two small machine screws with nuts, long enough to pass through the protractor and ruler together.
4. One piece of string, about 30 cm or 12 inches long.
5. One small weight, such as a large flat washer or large machine nut.

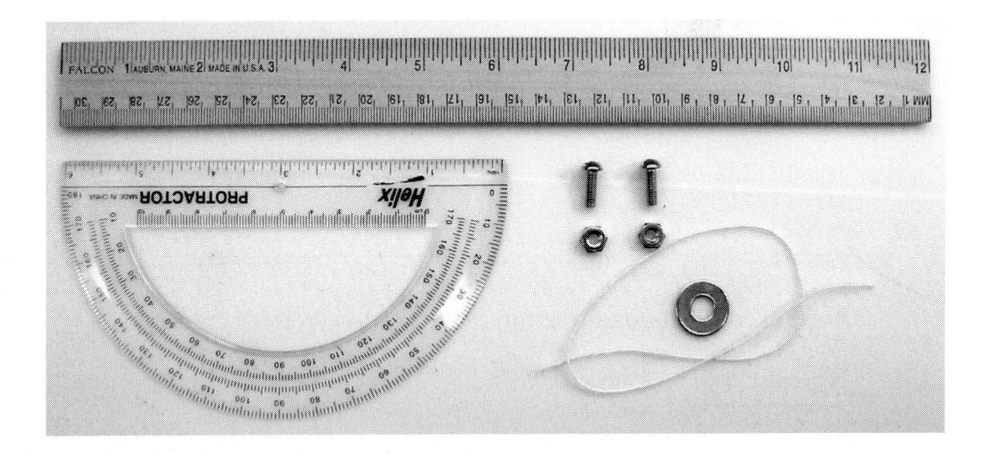

For tools you will need a drill, a small bit just large enough to let the string pass through, and another bit for holes for the two machine screws.

Assembly procedure

1. Carefully align the center line of the protractor (center point of semi-circle to 90° mark on the scale) along the bottom edge of the ruler. See the figure below. Holding it there, drill two small holes for the machine screws.

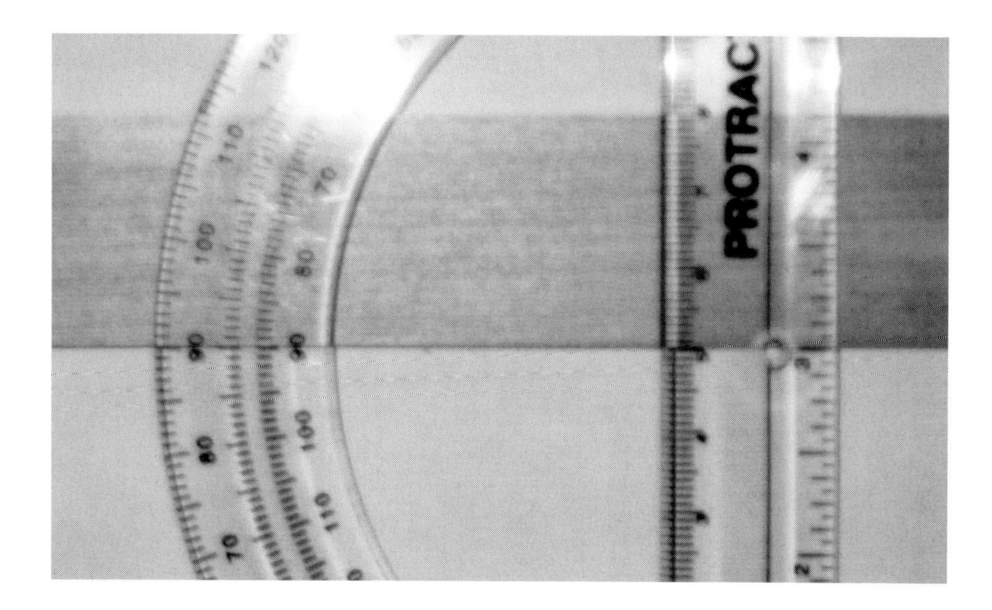

2. Use one machine screw and nut through the hole farthest from the center of the protractor to hold the protractor and ruler together.
3. If the protractor does not already have one, drill a small hole through the protractor at its center point, for the string. Make sure the ruler does not get in the way of the string's passage through the hole in the protractor.
4. Pass the second screw through the hole closest to the protractor center, and start the nut on the screw. Pass the string through the center point of the protractor and secure it to the second screw by wrapping it around the screw, once. See the figure below.

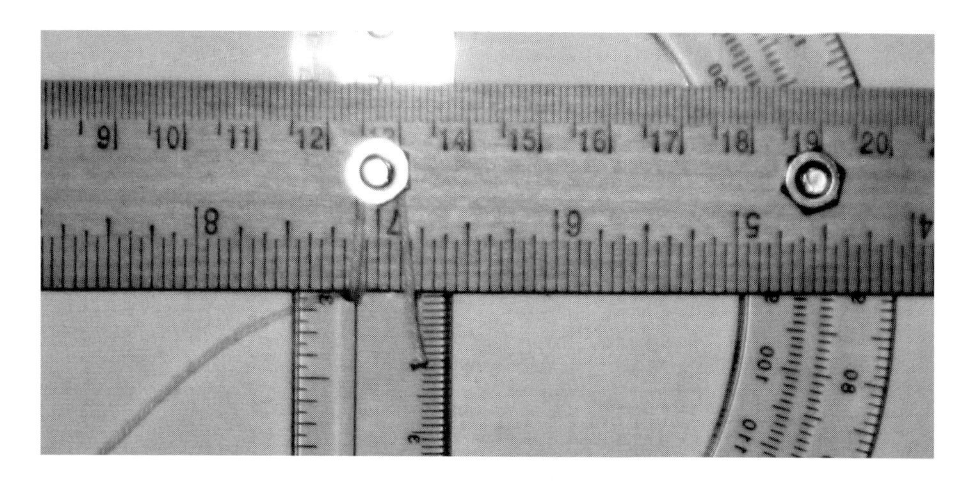

5. Tighten the nuts, making sure not to crack the plastic protractor. Finger-tight should be enough.

6. Tie the small weight to the free end of the string. This completes assembly. See the figure below.

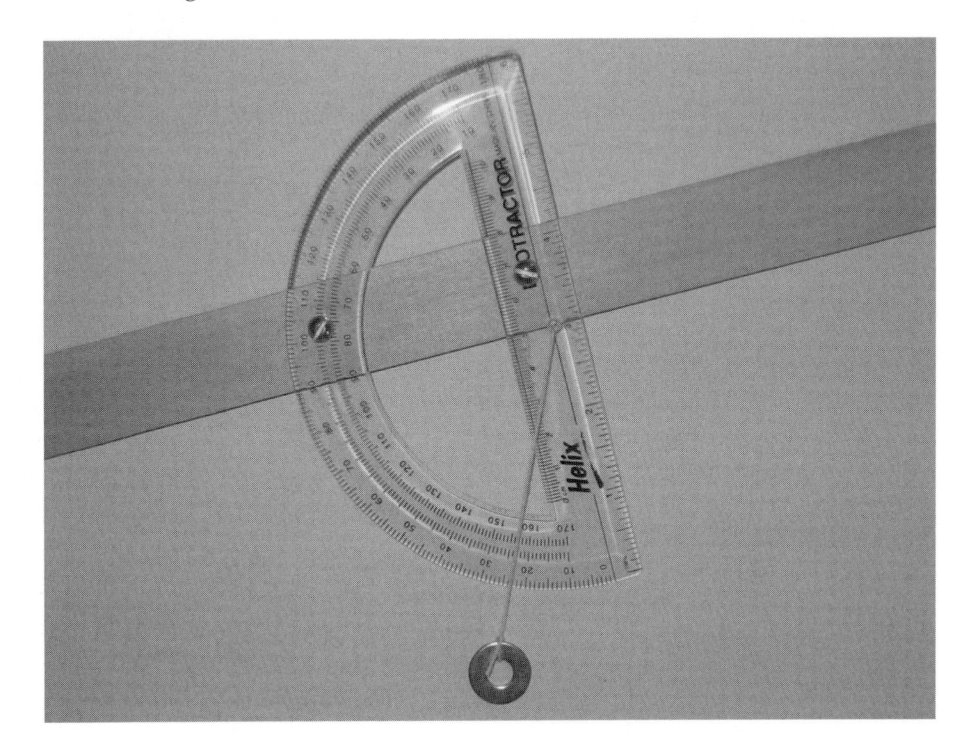

Use

To use the quadrant, hold the end of the ruler up close to your eye (within a few centimeters at most). The curve of the protractor should be toward your face. Sight to object of interest along the top edge of the ruler. Wait for the string/weight to stop moving, then pinch the string onto the protractor. Holding the string in place, bring the quadrant down so that you can read the position of the string on the scale. This is the altitude of the object.

8.3 Culminations

Objectives

1. To practice the calculation of upper and lower culminations.
2. To check those calculations against measurement for a few bright stars.

Procedure

You will need to choose at least three bright stars, visible from your latitude. You can use stars that you already know, or use the list of bright stars in Appendix B, p. 254 and the star charts for this task. In any case, you must be able to locate the stars you choose in the night sky to make observations. The chosen stars should have nearly the same right ascension so they all can be observed in one short observation session.

Choose one star that is circumpolar so that it has a measurable lower culmination. The other two stars should be stars that rise and set, with one of them needing the correction for the greater than 90° altitude result.

You will also need to know when the chosen stars are on your local meridian. This can be found by using a planetarium program, or by making a set of observations around the time when the star is close to your local meridian. In the second case, make the best guess you can for that time of night.

Your latitude, $\lambda =$

Altitude of celestial equator, $90^\circ - \lambda =$

See Section 4.2 or Appendix A, p. 231, for the correct procedure for calculating the upper and lower culminations for your stars. Use the table on the next page (or one like it) to record your data and calculations. Under "Star," put the name of the star you are observing. Under "Dec.," record the declination of the star. Then calculate the star's upper and lower culminations. If the lower culmination is below your horizon, you can leave the entry blank, or record the value if you wish. (Obviously, you are not going to observe it.) Under "Transit time," record the (approximate) local solar time at which the star will transit the meridian for the culmination.

Star	Dec. (°)	Upper culmination	Transit time	Lower culmination	Transit time

Go out and make measurements of the culmination altitudes for your chosen stars. Compare your measurements with your calculations. Use the next table to record your observations. This table is built so you can make a set of six observations around the transit time and then take the best measurement for comparing to the calculated result. For the upper culmination, the best measurement is the highest recorded altitude. For the lower culmination, the best measurement will be the lowest recorded altitude.

Star	Culmination	Measurements
	Upper	
	Lower	
	Upper	
	Lower	
	Upper	
	Lower	

Now use the last table to compare your measured values with your calculated values. Compute the percent difference with the following formula,

$$\% \text{ Difference} = \frac{2(\text{Calculated} - \text{Measured})}{\text{Calculated} + \text{Measured}} \times 100\%,$$

where Calculated is the calculated value and Measured is the measured value for the culmination.

Star	Culmination	Calculated	Measured	% difference
	Upper			
	Lower			
	Upper			
	Lower			
	Upper			
	Lower			

You've done pretty well if your percent difference is less than 5%, even better if it's less than 2%. Can you think of ways to reduce the possible errors in your measurements?

Answer:

8.4 Sunrise and sunset

During the year, the position of sunrise and sunset changes along the horizon. This is due to the Sun's changing declination during the year, which, in turn, is due to the obliquity of the ecliptic.

Project objectives

1. To observe the change in the Sun's horizon location at either sunrise or sunset, or possibly both.
2. To find the point on the horizon directly west (sunset), or east (sunrise), for your location.

Procedure

1. Decide whether you will observe sunrises or sunsets or both.
2. Choose an observing spot. A window will do, if it provides a wide view of your horizon. Distant trees or buildings are helpful, but nearby objects may cause problems. All observations must be made from the same location. If you can mark your observation location, perhaps with a narrow pole or stick, you can use this pole to make your observations more accurate.

3. Make a sketch of your horizon, using prominent landmark objects (trees, poles, houses, barns, etc.) visible along your horizon on a standard size piece of paper. There is a sample sketch shown below.

4. Observe a series of sunrises or sunsets. Carefully mark the Sun's location among your horizon objects. Mark its position on your sketch with an arrow, and label it with the date. See the sample sketch shown above.
5. Repeat step 4, at least five more times. The observations should be made at least ten days apart. One observation must be made within five days of the fall or spring equinox.
6. Based on your observations and your knowledge of the diurnal path of the Sun on the equinox days, mark on your sketch, the direction of true east or west with a big E or W. See the sample sketch.

This project could be extended to a full six-months duration if you would like to determine the location of sunrises or sunsets on the solstice days as well as the equinox days. There have been some people who have built their own rendition of Stonehenge (although usually with tall wooden poles).

8.5 Solar altitude

Objectives

1. To use the changes in length of a shadow created by a short stick and the Sun, to measure changes in the Sun's altitude at the same time (by a solar clock) over a period of several months.

Procedure

1. Select a viewing location where the Sun is visible around noon. This may be inside, looking out through a window, or somewhere outside. In any case, there must be no obstructing objects creating their own shadows that might interfere with the observations of the stick's shadow during the series of observations. If such a place cannot be found, it is perfectly acceptable to use different locations around the area.

2. Select an object to cast a shadow. A piece of tape on the window, a clothes-line pole, a soda bottle, or any narrow object that will stand vertically. This same object must be used for all the observations. Enter the height, h, of the object in the space below. Generally, an object of 25 to 50 cm (10 to 20 inches) works best. If you would really like to do this project with a touch of accuracy (and class), you could build a simple set-up like the one shown in the photo below.

3. On a sunny day, at about noon, measure the length of your object's shadow, L. *Make sure your shadow stick is vertical and the ground (or other surface) on which you are measuring the shadow is horizontal (level).* Use a carpenter's level to check both.
4. You must use the same time of day for each observation. Do not deviate more than 10 minutes before or after the noon-time observation period. If you cannot make the observation within this time constraint, wait another day.
5. Enter the date of the observation and the length of the shadow in the table below. Repeat this step and step 3 for at least five other sunny days. The observations should be at least ten days apart.
6. Your observations are based on the wall clock, which is subject to politics. If your observation series passes through the time shift that occurs with switching between standard and daylight savings modes, you must adjust your observation time. Adjust your observation time in the same direction as the time change. When daylight savings is started (spring) the clock is shifted one hour ahead. Therefore your observations should occur at 13:00 hours rather than 12:00 hours. When daylight savings is stopped (fall), your observations should shift to 11:00 hours rather than 12:00 hours.

   An alternative to all this trouble is to use a wall clock only for the purposes of this project. That is, when the time shift occurs, leave your "special clock" as it is. This completely avoids the time shift problem.
7. To calculate the altitude of the Sun from your data, you must use a little trigonometry. The figure below shows the concepts involved. The altitude of the Sun is given by

$$A_{\mathrm{Sol}} = \theta = \arctan \frac{h}{L},$$

where L is the length of shadow and h is the height of the object making the shadow.

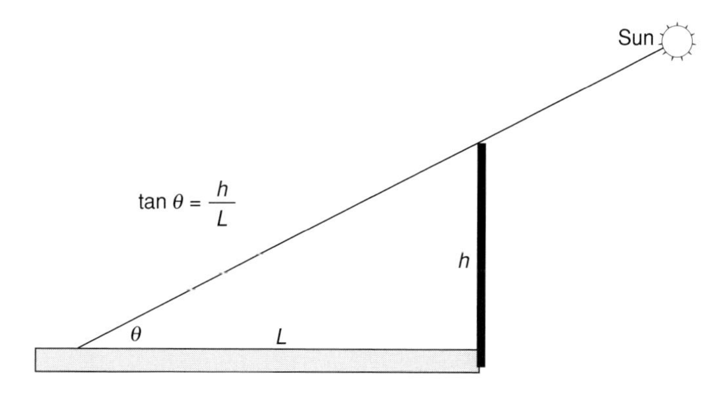

8. Based on your readings of Section 5.2, write a brief explanation for why the shadow length is changing. What connections are there between these observations and the seasons?
   Answer:

Height of object, h =

Observation time: S.T./D.S.T. (S.T., Standard Time; D.S.T., Daylight Savings Time)

Date	Shadow length	Solar altitude (°)

8.6 The duration of daylight

This project looks at the duration of daylight during some portion of the year. You may find it more interesting to do this project for the period between the solstices. Also, it could be done around each solstice, in order to verify the dates of the longest and shortest days of the year. These dates will not actually be on the solstice dates because of atmospheric refraction.

Purpose

1. To measure and note the change in the length of daylight during some portion of the year.

Procedure

1. In the table given below, record the date and the times of sunrise and sunset. You can obtain the times of sunrise or sunset off of any weather report from television or the newspaper.
2. Make six dates of observation. Each observation date must be separated by at least ten days from the previous date.
3. Calculate the number of minutes of sunlight for each day. You will need to convert hours to minutes.
4. After all observations and records are complete, calculate the total change in the length of the day, from your first observation to the last.
5. Based on your readings of Chapter 5, create a brief explanation of why the length of daylight changes. Does the amount and direction of change fall within your expectations?
   Answer:

	Rise time	Set time		Rise time	Set time
Date	Minutes of daylight		Date	Minutes of daylight	

Total change in duration of daylight =

8.7 High noon

Objectives

1. To measure the altitude of the Sun between 11 : 30 and 14 : 30 and determine the time it reaches maximum altitude during that day.

Procedure

1. Select a location where the Sun is visible for at least three hours, around noon.
2. Select an object to cast a shadow. A piece of tape on the window, a clothes-line pole, or any narrow object that will stand vertically. Enter the height,

h, of the object in the space below. An object of 25 to 35 centimeters (10 to 20 inches) is generally best. Make sure the object is vertical and the ground (or other surface) is horizontal. Use a carpenter's level to check both. If one or the other is not truly vertical or horizontal, make sure it is at least reasonably close.

3. Choose a clear sunny day, and enter the date in the space provided below. At 11:30 hours, measure the length of your object's shadow, L. Enter the shadow length in the chart beside 11:30.
4. Repeat step 3 every 20 minutes, within two minutes, until 14:30 hours.
5. Circle the time when the Sun reached its maximum altitude. This is the time nearest high noon at your location.
6. Give a brief explanation for why the shadow changes length. Discuss how close to 12:00 noon your time of high noon came. Why is there any difference?
Answer:

Height of object, h =

Observation date:

Time	Shadow length	Time	Shadow length
11:30		13:10	
11:50		13:30	
12:10		13:50	
12:30		14:10	
12:50		14:30	

8.8 Quarter phase Moon

Objectives

1. To gain some familiarity with lunar phases.
2. To take a detailed look at, and describe, the face of the Moon.

Procedure

During either the first quarter or third quarter lunar phase, make a drawing of the surface of the Moon. At the quarter Moon phases, the angle of the sunlight as it strikes the lunar surface is best for highlighting the surface features. Draw all the mountains, craters, and maria you can see in the appropriate hemisphere below. Be sure to state the lunar phase you are drawing. If you use a pair of binoculars, you might label those surface features you find interesting.

Phase:

Observation date and time:

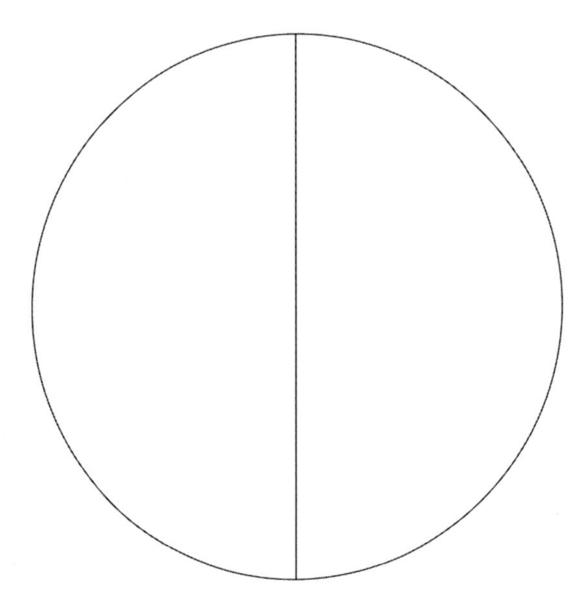

Observation notes:

8.9 Phases of the Moon

Objectives

1. To observe each of the lunar phases.
2. To become familiar with the horizon position of the phases of the Moon.
3. To analyze and understand the relative positions of the Sun, Moon, and Earth for each phase.

Procedure

Observe the phase and location of the Moon within the following restrictions:

1. Each Moon phase is allowed only once. On the following data charts, each phase has already been labeled so only one observation per phase is allowed. If you make your own phase data chart you could observe a particular phase more than once, making note of its horizon position with each observation. This may be particularly enlightening with observations of the intermediate phases.

2. At least one observation (other than the new phase) must be made while the Sun is visible. Most phases are visible during the daylight hours. More than a few times I have met people who did not realize this fact.
3. At least one month must pass during these observations. Simply a fact of nature, but noted to emphasize that it will take you at least a month of observations to complete this project.

Use the correct section of the chart to record each Moon phase observation. Each chart is pre-labeled with the Moon phase. Don't accidentally mix them up.

1. Record the date and time of each phase observation. The observations do not have to be, and will probably not be, in calendar order. Weather conditions may delay observations of the quarter phases.
2. The complete phase name (e.g. waxing crescent) is given. In the circle, sketch the phase as observed, by shading in the *unlit* portion of the Moon. Thus, the paper white portion of the circle is the lit part of the Moon's face. You do not need to include any surface detail in the drawing. We are interested only in the apparent shape of the Moon.
3. Record the azimuth as a compass direction (SE, SW, etc.). This need only be approximated. Using the quadrant built in Section 8.2, measure and record the Moon's altitude. You should notice that one of the observations is of the new Moon phase, which cannot be observed. To get the altitude and azimuth of the new Moon, measure the altitude and azimuth of the Sun. Of course, we should not violate the warning given on p. 175 about looking at the Sun. We can, however, make a quick estimate of the Sun's position and this estimate is good enough for this project.
4. For the "orbital location sketch," draw a small circle to indicate the position of the Moon in its orbit, relative to the Earth and Sun, during this phase. Use Figure 6.6 to help you here. While you may challenge yourself to determine the exact phase angle for the Moon during this particular observation, it is only necessary to understand the general position of the Moon for this phase to understand the creation of lunar phases.
5. The first chart is filled in as an example of the observational data format. The phases can be observed and recorded in whatever order weather and observation opportunity will allow.

Date	Phase name	Direction	Orbital location sketch
Time	Sketch	Altitude	
Oct 15	Waxing crescent	SW	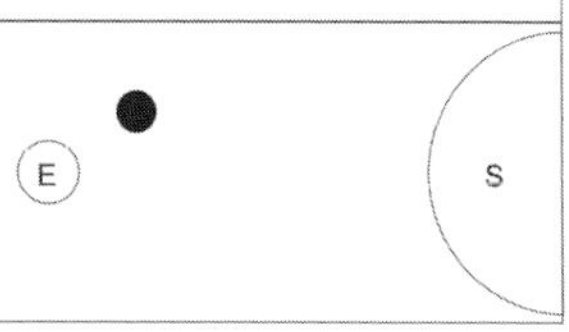 E S
10:00 P.M.	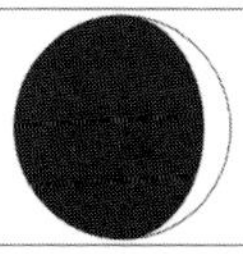	30°	

Date / Time	Phase name / Sketch	Direction / Altitude	Orbital location sketch
	New Moon		E S
	Waxing crescent		E S
	First quarter		E S
	Waxing gibbous		E S
	Full Moon		E S
	Waning gibbous		E S
	Third quarter		E S
	Waning crescent		E S

8.10 Observations log

There are many well-written and superbly illustrated books about finding the stars and constellations, most of which you will find listed in the bibliography. I would recommend any constellation book or charts with illustrations or maps created by Wil Tirion (Tirion, 1991; Moore and Tirion, 1997; Tirion and Sinnott, 1998; Ridpath and Tirion, 2003; Heifitz and Tirion, 2004). There are also a few planetarium programs written for both the PC and MacIntosh computer systems (running under various operating systems). These programs can produce customized sky maps for any location (latitude and longitude) and for any date and time. These programs, such as "The Sky" and "Starry Nights," can be seen at their respective websites (see the bibliography). The definitive list of available planetarium software can be found at http://seds.lpl.arizona.edu/billa/astrosoftware.html.

Because so much information about star and constellation location is already available and customizable to the observer, it seems quite superfluous to create yet another set of printed charts here. Printed charts are always limited to particular latitudes and epochs. Printing enough charts in this book to be useful to everyone would be a waste of paper. Therefore, what I recommend here is that you, as an observer, start a log of your sky observations (if you have not already). The figure below shows an example page from such a log. This page is also available as a PDF file from this book's associated website. The following table gives a brief description of the information that should be recorded in the log page entry.

Object: Name ____________________ Type __________

Observed: Date ______________ Time __________

Location: Lat.________ Long.________ Elevation: __________

Altitude: __________ Azimuth: __________

Seeing conditions: ____________________

Equipment: ____________________

Description: ____________________

Data entry	Purpose
Upper (image) box	Make a sketch of the object. If you are looking at a star, make a sketch of its constellation, pointing out its location within the constellation
Object: name	Enter the name of the object, such as "Betelgeuse" or "Delta Tauri"
Object: type	"Star," "Constellation," etc.
Observed: date, time	The date and time of the observation
Location: lat., long., elevation	If you have access to a GPS or you have such information about your favorite observing spot, this can be filled in fairly accurately. Otherwise, you may just put an approximation or a city or town name here
Altitude, azimuth	The horizon coordinates of the object. You can measure the altitude with the quadrant built in Section 8.2. The azimuth can be measured or estimated with a compass direction
Seeing conditions	Seeing conditions are a judgment call. They are affected by light pollution, air temperature, humidity, turbulence, cloud haze, and other effects. Seeing is described by sky glow, lunar phase, and transparency (see Section 3.6)
Equipment	List any pieces of equipment used for the observation. Sometimes this is just your eyes
Description	Record anything you may notice about the object here

The location of the object (horizon coordinates) can be determined by using the quadrant described in Figure 2.19 or Section 8.2. Mills' book contains projects for building measurement instruments (Mills, 1994). Simple and inexpensive coordinate measuring instruments can also be purchased from hobby stores or the "stores" of organizations such as the Astronomical Society of the Pacific.

The most important aspect of observing the night sky is to dress for the occasion. If you live in a cold (or cool) climate, dress in layers of clothing. This way, if you get warm you can remove a layer or two, and if you get cold you can put on a layer or two. Hands and feet always get cold first – cover them well. If you live in an area with mosquitoes or flies, remember to take a can of repellent spray. Also remember your sky maps, observations log, and, to see these things, a low power, red lens flashlight. The red color leaves your night-vision unaffected, but only to a certain level of brightness, which is the reason for wanting the flashlight to be of low power. Keep the light level as low as possible.

Now, go out and enjoy the night sky!

Appendix A
Topics from math and physics

This appendix provides readers with a slightly more mathematical look at topics either glossed over or simply mentioned in the text. While knowledge of these topics is not critical for understanding the properties of the celestial sphere, such knowledge will certainly deepen your understanding.

Lure math

This section contains the calculations for some of the claims made in Chapter 1.

Grains of sand

Using the typical estimates for galaxies (100 billion) and star count per galaxy (200 billion), we find there are 20 000 billion-billion (2×10^{22}) stars in the Universe. To estimate the number of grains of sand, we assume a beach to be 100 meters wide and the sand to be (on average) 10 meters deep. The world has roughly 1 000 000 kilometers (1 000 000 000 meters) of beaches, giving a volume of sand of 1×10^{12} cubic meters. If we now assume 1.75×10^{8} grains of sand per cubic meter (each grain is 0.25 mm^3, with 70% packing), we have 1.75×10^{20} grains of sand. The grains of sand are fewer in number than stars by a factor of more than 100. Therefore it would take all the beaches of more than 100 Earths to have the same number of grains of sand as there are stars in the Universe. Thanks go to my GRCC Physical Sciences department colleague, geologist Elaine Kampmueller, for information about sand and sandy beaches in doing this calculation.

Traveling to Alpha Centauri

Travel time to the nearest star is calculated in the following manner:

$$\frac{24 \times 10^{12}\ \text{miles}}{7 \times 10^{5}\ \text{miles/h}} = 342\ 857\ 143\ \text{hours}$$

$$342\ 857\ 143\ \text{h} \left(\frac{1\ \text{day}}{24\ \text{h}}\right)\left(\frac{1\ \text{yr}}{365.25\ \text{day}}\right) = 39\ 000\ \text{yr}.$$

Using 20 years for a human generation,

$$\frac{39\,000 \text{ yr}}{20 \text{ yr/generation}} = 1950 \text{ generations.}$$

We could easily put about a 10% error on the estimates.

Numbers and scientific notation

There is no occupation or hobby in which a person does not have to do some form of calculation, whether simple addition or complex formula evaluation. The better we understand the basic concepts of numbers – the language we use to interpret nature and to do calculations – the better we understand nature. Astronomers must talk about objects of extreme size, distances of extreme measure, and counts of extreme number. How do we express the numbers dealing with such large distances and counts?

Mathematicians have a classification system for numbers: natural, whole, integers, rational, real, and complex. The *natural numbers* are those with which we count. These are also known as the *cardinal numbers*, or the *positive integers*. They include all positive numbers, starting at one. This class also includes the *ordinal numbers*, by which we order things. In English, these are "first," "second," "third," etc.

The *whole numbers* are the natural numbers plus the number zero. The *integers* are all the whole numbers with the addition of all negative values of whole numbers. Zero is neither negative nor positive. The phrase "non-negative integers" means all positive integers plus zero. Whereas the phrase "positive integers" means strictly the positive integers, not including zero.

Rational numbers are those that can be expressed as the ratio or quotient of two integers. For example $3/2$ is a rational number. The integers are also rational numbers because they can be expressed as the quotient of that integer over one. For example $5/1 = 5$ tells us that five is both an integer and a rational number. An *irrational number* is then simply a number that cannot be expressed as the ratio of two integers. The prime example of such a number is π. When a rational number is written in decimal form there is either a termination of digits, or there is an infinite number of digits with a repeating pattern to the right of the decimal point (for example, $1/4 = 0.25$ and $1/3 = 0.333\,33\ldots$). Irrational numbers have an infinite number of digits with no repeating pattern. For example, $\sqrt{2} = 1.414\,2136\ldots$

Real numbers are all the rational and irrational numbers. *Complex numbers* are a combination of real and *imaginary numbers* such as, $3 + 6\imath$. The $\imath$ symbol represents the square root of negative one, which does not exist ($\sqrt{-1} = ?$). Imaginary numbers were introduced with situations like the solutions to the algebraic equation $x^2 + 1 = 0$. While it is extremely rare that one might find a use for imaginary numbers in amateur astronomy, I chose to include them simply to make your background in numbers more complete. You just never know where they might turn up.

Power of ten notation

When we speak of astronomical sizes and distances, or of the particles which make all matter, we need a system to write very large and very small numbers in a compact form. This system is called *scientific notation*. Scientific notation uses the concepts of power of ten notation. Thus, we must first understand power of ten notation.

Power of ten notation uses the base of our numbering system, 10, combined with an exponent to create a **power** that then represents a multiple of ten. An **exponent** is a number indicating the number of times the base is to be self-multiplied. The **base** is the number being self-multiplied. The base of a power is not necessarily the same as the base of a numbering system, but in this case it happens to be. This notation is used for very large and very small numbers that are multiples of ten.

> *A power is simply a shorthand for multiplication in the same way that multiplication is a shorthand for addition.*

Example: In this example, 10 is the base, 2 is the exponent, and 10^2 represents the power of ten

$$10^2 = 10 \times 10 = 100.$$

Example: In this example, 4 is the exponent and 10^4 represents the power of ten.

$$10^4 = 10 \times 10 \times 10 \times 10 = 10\,000.$$

Thus, 100 and 10 000 are both powers of ten that can be represented with this shorthand notation for powers. Sometimes the words *power* and *exponent* are used interchangeably.

When the number is completely written out, such as "10 000," it is said to be in *decimal form*. You should be able to translate between the power of ten and the decimal form of a number. (Converting between these forms of numbers is not something that is done every time you go out to look at the sky. However, those who know how to do this will find it easier to read any astronomical or scientific literature.) Notice, in the two examples above, the exponent in the power of ten form is equal to the number of zeros in the decimal form.

There are three cases of exponents we must understand: positive, zero, and negative. First, powers of ten with positive exponents represent numbers greater than one. Second, because any number (except zero) divided by itself equals one, any number (except zero) raised to the zero power also equals one. Thus, we

Table A.1 *Powers of ten and their decimal equivalent*

Power of ten	Decimal form
10^9	1 000 000 000
10^6	1 000 000
10^3	1 000
10^2	100
10^1	10
10^0	1
10^{-1}	0.1
10^{-2}	0.01
10^{-3}	0.001
10^{-6}	0.000 001
10^{-9}	0.000 000 001

can write

$$1 = \frac{10^2}{10^2} = 10^{2-2} = 10^0 = 1.$$

The exception for a base of zero is a result of the fact that division by zero is not defined. (Arithmetic with exponents is discussed below.) Third, powers of ten with negative exponents represent numbers less than one but greater than zero. As examples,

$$10^{-1} = \frac{1}{10} = 0.1,$$

or

$$10^{-3} = \frac{1}{10^3} = \frac{1}{10 \times 10 \times 10} = \frac{1}{1000} = 0.001.$$

With negative exponents, the exponent gives the number of zeros in the number in the denominator of a fraction. Table A.1 shows the conversions between the most commonly used powers of ten and their decimal form.

Powers of ten with positive exponents represent numbers greater than one. Ten raised to the zero power equals one. Powers of ten with negative exponents represent numbers less than one but greater than zero.

To represent negative numbers in powers of ten, the negative sign is placed in front of the base, not in the exponent. So,

$$-10^3 = -1000 \quad \text{and} \quad -10^{-3} = -0.001.$$

As an example of a large number, the total number of subatomic particles (protons, neutrons, and electrons) in the visible Universe is estimated as 10^{80} (Sagan, 1980, p. 219). This is a one, with eighty zeros behind it. The number, 10^{100} is known as a *googol*. (The name, "googol" was invented by a young nephew

10^{100} = 10 000 000 000 000 000 000 000 000 000 000 000 000 000 000 000 000 000
000 000 000 000 000 000 000 000 000 000 000 000 000 000 000 000 000

Figure A.1 This is a googol written in decimal form in nine point printing. A googolplex would have this number of zeros behind its one. At this size print, about 10 000 zeros can be put on a standard (8.5 × 11 inch) page. A googolplex would then require only 10^{96} pages to print – ten-quadrillion times more than the number of protons, neutrons, and electrons in the visible Universe. Yet, the googolplex is nowhere near infinity.

of the American mathematician, Edward Kasner.) This is a number with one-hundred zeros (Figure A.1). A *googolplex* is ten raised to the power of a googol. For this number, there are more zeros behind the one than there are available subatomic particles in the Universe on which to write them (if we could). However, using the power of exponential notation, a googolplex can easily be written as

$$10^{10^{100}}.$$

This number, the googolplex, as large as it is, comes nowhere near infinity. In fact, it is as far from infinity as is the number one. No matter how large a number may be, infinity is larger by at least one. As difficult as this concept is to imagine, we have an easy way to write infinity, as the symbol

$$\infty.$$

To write infinity is one thing, to understand its meaning is quite another.

> *The concept of infinity is one of the most remarkable accomplishments of mathematics. Infinity is the ultimate value of the very large and it is a concept with which we must contend when we look into the Universe.*

In many cosmological models the Universe is described as infinitely large. That is, space-time has no edge or boundaries. This is a statement as bewildering as infinity itself. Although this book does not deal with cosmology, the concept of infinity is one that should be understood by all those who partake – at any level – in the sciences.

Scientific notation

Scientific notation uses powers of ten, with a number multiplying the power of ten called the **mantissa**. The mantissa is always between one and ten, but never equal to ten:

$$1 \leq \text{mantissa} < 10.$$

The exponent on the power of ten can be positive, zero, or negative.

Example: In this example, the 2.354 is the mantissa and the 10^7 is the power of ten. The combination is scientific notation:

$$2.354 \times 10^7.$$

Table A.2 *Scientific notation and equivalent decimal notation form*

Scientific notation[a]	Decimal notation	Significant digits
5.28×10^{2}	528	3
2.9×10^{3}	2900	2
4.28×10^{1}	42.8	3
5.7×10^{8}	570 000 000	2
4.5×10^{-1}	0.45	2
3.87×10^{-3}	0.003 87	3
2.06×10^{-8}	0.000 000 0206	3
5.60×10^{12}	5 600 000 000 000	3
2.5×10^{0}	2.5	2
9.8365×10^{4}	98 365	5
3.0×10^{7}	30 000 000	3

[a] In some countries a comma is used in place of the decimal point shown here. I have used my ingrained American notation habits for these examples.

To convert between scientific notation and decimal notation, simply move the decimal point. When the exponent is positive, move the decimal point to the right by the number of places equal to the exponent. When the exponent is negative, move the decimal point to the left by the number of places equal to the exponent.

Example: If the mantissa does not have enough places to move the decimal point, simply add zeros as place holders so the decimal point can be moved far enough, such as in this case.

$$4.73 \times 10^{5} = 473\,000.$$

Example: Here is a number in scientific notation with a negative power of ten, converted to decimal form:

$$4.73 \times 10^{-5} = 0.000\,0473.$$

Table A.2 shows some additional sample conversions between scientific notation and decimal form.

Arithmetic with exponents

Arithmetic with powers or exponents has a set of rules. The simplest rules deal with multiplication and division.

Multiplication and division

To multiply two powers, they must have the same base. Given the same base, multiplying the two powers is done by adding their exponents.

Example: Multiplying two numbers,

$$100 \times 1000 = 100\,000 \qquad \text{or} \qquad 10^2 \times 10^3 = 10^{2+3} = 10^5.$$

In general notation,

$$10^a \times 10^b = 10^a \cdot 10^b = 10^a 10^b = 10^{a+b}, \tag{A.1}$$

where various forms of indicating multiplication is shown.

Using the definition that powers with a negative exponent mean that the power is in the denominator of a fraction, we can extend this rule for multiplication to a rule for division. Thus,

$$\frac{10^a}{10^b} = 10^a 10^{-b} = 10^{a-b}.$$

Example: Dividing two numbers,

$$\frac{100\,000}{100} = 1000 \qquad \text{or} \qquad \frac{10^5}{10^2} = 10^{5-2} = 10^3.$$

In both multiplication and division, if the power has a mantissa (such as with scientific notation), the mantissas are handled in the usual way as decimal numbers.

If the numbers are in scientific notation and the result of multiplying their mantissas is greater than ten, then the resulting mantissa's decimal must be shifted to bring it into proper range, with the power of ten also shifted appropriately.

Example: With two numbers in scientific notation,

$$\left(2.3 \times 10^3\right)\left(4.70 \times 10^2\right) = 10.8 \times 10^5 = 1.08 \times 10^6,$$

where the last step shifted the decimal point and the power of ten so the mantissa would meet its requirement for scientific notation.

If the bases are not the same, then either one base/power must be converted to the other base/power or both powers must be converted to decimal form before the arithmetic function can be performed. In this rare situation, it is usually easiest to convert both numbers to decimal form and then do the indicated operation.

Addition and subtraction

For addition and subtraction – specifically with scientific notation – the exponents of the powers must be the same before the addition or subtraction can be performed.

This may require changing one of the numbers into a form that is not considered scientific notation.

Example: To do the following addition problem,

$$2.30 \times 10^2 + 4.80 \times 10^3$$

must be changed to

$$2.30 \times 10^2 + 48.0 \times 10^2 \qquad \text{or} \qquad 0.230 \times 10^3 + 4.80 \times 10^3$$

to get

$$50.3 \times 10^2 = 5.03 \times 10^3.$$

This procedure must also be used for subtraction. For an example, simply take the above addition problem and change the plus sign to a minus sign and rework the problem.

Powers of powers

One more property of powers of ten can be shown. What does it mean to raise a power to a power? It is simply the extension of the power notation. That is,

$$(x^a)^b = x^{a \cdot b}.$$

Example: In other words,

$$(10^2)^3 = 10^2 \times 10^2 \times 10^2 = 10^{2+2+2} = 10^{2\times 3} = 10^6.$$

These arithmetic concepts are the foundation of the idea of logarithms, which are discussed on p. 227.

Fractional exponents

Occasionally, you may run into fractional, or decimal, exponents in formulas or relations. A first example of this would be the relationship between the luminosity and mass of a star. That is, the luminosity of a star is proportional to its mass raised to the three and one-half power. What does such a fractional exponent mean?

> *When an exponent is fractional, the numerator represents the power and the denominator represents the root. Power and root are opposite operations, like addition is opposite to subtraction and multiplication is opposite to division.*

The exponent of 3.5 can also be written as 7/2. This means the base is raised to the seventh power and the square root is taken from this result.

Example:

$$4^{3.5} = 4^{7/2} = \sqrt{4^7} = \sqrt{16\,384} = 128,$$

or

$$4^{3.5} = 4^{7/2} = \left(\sqrt{4}\right)^7 = 2^7 = 128,$$

where we see two ways of evaluating the same expression. Logarithms are a form of fractional exponent.

Table A.3 *The most common metric system prefixes*

Prefix (symbol)	Power of ten
Pico (p)	10^{-12}
Nano (n)	10^{-9}
Micro (μ)	10^{-6}
Milli (m)	10^{-3}
Centi (c)	10^{-2}
Deci (d)	10^{-1}
Deca (da)	10^{1}
Hecto (h)	10^{2}
Kilo (k)	10^{3}
Mega (M)	10^{6}
Giga (G)	10^{9}

Table A.4 *Converting measurements with a metric prefix to decimal form*

Given value	Convert prefix	Decimal form
3.4 kilometers	3.4×10^{3} meters	3400 meters
2.4 megaparsecs	2.4×10^{6} parsecs	2 400 000 parsecs
50.1 nanoseconds	50.1×10^{-9} seconds	0.000 000 050 1 seconds
53.6 millimeters	53.6×10^{-3} meters	0.053 6 meters
17.8 centimeters	17.8×10^{-2} meters	0.178 meters
37.2 kilolight years	37.2×10^{3} light years	37 200 light years

Metric system prefixes

The metric system is based on powers of ten. Changing between different sized measurement units is as easy as shifting the decimal place and changing the measurement unit's prefix. The unit's prefix is a word or symbol representing a power of ten by which the measurement's mantissa is multiplied. A short list of metric prefixes and their corresponding power of ten is given in Table A.3. Readers should memorize this table, if they have not already. A set of examples for converting measurements using these prefixes to decimal form is given in Table A.4.

While unit prefix conversions are not needed with every page or every observation, an amateur astronomer should certainly feel comfortable with making such conversions.

Logarithms

Logarithms are used in the magnitude systems for stars and other objects. Although many people tremble when the word logarithm is used, they are really quite simple. There are two primary bases used for logarithms, natural and common. The natural logarithms are based on the number e, or 2.7182... The common logarithms are based on the number 10. The rules and properties of logarithms are independent of the base used.

> *Logarithms are just exponents and can be seen as simply an extension of the concepts of power of ten notation.*

This extension uses fractional exponents, as discussed on p. 226.

Definition

We can define logarithms in the following way: For any number written in the form,

$$x = 10^y,$$

the logarithm of x is y. That is,

$$\log x = y. \tag{A.2}$$

In other words, if the number x, can be written as a power of ten the (common) logarithm of x is simply the exponent of the power of ten that produces x.

Example:

$$100 = 10^2,$$

so $x = 100$ and $y = 2$, and therefore

$$\log 100 = 2.$$

That is, when ten is raised to the second power, we get 100. So the log of 100 is 2. Also, $\log 1000 = \log 10^3 = 3$, $\log 10\,000 = \log 10^4 = 4$, and so forth.

This works well for those numbers that are integer powers of ten, but what about numbers such as five? To what power would ten need to be raised to get five? Using our understanding of fractional exponents as described on p. 226, with a little trial and error we can find quite quickly that

$$5 \approx 10^{7/10} = 10^{0.7},$$

or

$$\log 5 = 0.7.$$

Evaluating this expression gives,

$$10^{7/10} = \sqrt[10]{10^7} = \sqrt[10]{10\,000\,000} = 5.0118\ldots,$$

or again

$$10^{7/10} = \left(\sqrt[10]{10}\right)^7 = (1.2589\ldots)^7 = 5.0118\ldots$$

By adjusting the values of the numerator and denominator of the fractional exponent, a more exact value for the log of five can be found. A calculator gives,

$$\log 5 = 0.678\,97\ldots$$

Because ten raised to the zero power is one, the logarithm of one is zero. There is no way to write zero as a power of ten. That is, there is no exponent we can apply to the base of ten (or any other base) that gives the result of zero and therefore the logarithm of zero does not exist.

Try taking the log of zero on your calculator and it will display its favorite error message.

For common logarithms, all numbers between one and ten have logarithm values between zero and one. All numbers between 10 and 100 have logarithm values between 1 and 2. Numbers between 100 and 1000 have logs between 2 and 3. Similar numbers have similar logarithms. For example,

$$\begin{aligned} \log 5 &= 0.678\,97 \\ \log 50 &= 1.698\,97 \\ \log 500 &= 2.698\,97 \\ \log 5000 &= 3.698\,97 \end{aligned}$$

Notice that the number of zeros behind the five simply determines the integer part of the logarithm value. The reason for this is made clearer below.

Properties of logarithms

There are a few rules or properties of logarithms. The primary advantage of logarithms is that they allow us to change multiplication/division problems into addition/subtraction problems. Look carefully at the following expressions:

$$10^5 = 100\,000 = 100 \times 1000 = 10^2 \times 10^3 = 10^{2+3}.$$

The center expression is a multiplication problem (100×1000) and the last expression is an addition problem (10^{2+3}). Taking the logarithm (on both sides) of the last two expressions and using the definition of logarithms, Eqn (A.2), leads us to the following:

$$\begin{aligned} \log\left(10^2 \times 10^3\right) &= \log\left(10^{2+3}\right) \\ \log\left(10^2 \times 10^3\right) &= 2 + 3 \\ \log\left(10^2 \times 10^3\right) &= \log 10^2 + \log 10^3. \end{aligned}$$

Or, in general,

$$\log(A \times B) = \log A + \log B. \qquad \text{(A.3)}$$

Two other logarithmic properties are,

$$\log\left(\frac{A}{B}\right) = \log A - \log B$$

and

$$\log A^n = n \log A.$$

Proof of these properties is left as an exercise for the reader.

Using Eqn (A.2) and the logarithm property (A.3) we can write,

$$\begin{aligned}\log 5000 &= \log\left(5 \times 10^3\right) = \log 5 + \log 10^3 \\ &= \log 5 + 3 = 0.698\,97 + 3 \\ &= 3.698\,97,\end{aligned}$$

as was given above. These steps clearly demonstrate why the integer portion of a logarithm is related to the power of ten range of the number.

Apparent magnitude

The mathematical statement of the relationship between apparent magnitude and apparent brightness is

$$m_1 - m_2 = -2.50 \log \frac{f_1}{f_2}, \tag{A.4}$$

or

$$\frac{f_1}{f_2} = 2.512^{-(m_1 - m_2)}, \tag{A.5}$$

where f is the energy flux of the two stars and m is the apparent magnitude of the two stars. Translating between (A.4) and (A.5) is left as an exercise for the reader.

Equation (A.5) *shows us that a difference in magnitude of one* ($-(m_1 - m_2) = 1$) *is equal to a ratio in brightness of* 2.512 *(or approximately* 2.5*):*

$$\frac{f_1}{f_2} = 2.512^1.$$

Example: We can use Sirius, $m = -1.42$ as m_1, and Polaris, $m = +1.97$ as m_2, for an example calculation:

$$\begin{aligned}2.512^{-(-1.42-1.97)} &= 2.512^{-(-3.39)} \\ &= 2.512^{3.39} \\ &= 22.7.\end{aligned}$$

This shows us that Sirius has 22.7 times the light energy flux at the Earth than does Polaris. The answer has no units of measure because it is a ratio of energy flux measurements.

Example: As an interesting, but somewhat useless example, let us use the extremes of the apparent magnitude scale to gain understanding of its advantages. The Sun's magnitude is $m = -26$ (rounding off to make the arithmetic easier) and the minimum magnitude (rating) of the Hubble Space Telescope is $m = +29$. The difference between these two is 55 magnitudes. What is the ratio in brightness? The answer is given by (A.5):

$$2.512^{55} = 100^{55/5} = 100^{11} = \left(10^2\right)^{11} = 10^{22}.$$

Ten raised to the twenty-second power is a one with 22 zeros behind it.

$$10^{22} = 10\,000\,000\,000\,000\,000\,000\,000 \qquad 6\,000\,000\,000. \tag{A.6}$$

This number is named, "ten-sextillion." In other words, the Sun is ten-sextillion times brighter than the faintest object observable by the Hubble Telescope. It is far easier to talk about a difference of 55 magnitudes than to talk about a ratio of ten-sextillion. The second number shown in (A.6) is the approximate human population of the Earth.

This should also give you some idea of why we do not look at the Sun or the brighter planets with the Hubble Telescope. Its optical components and cameras simply cannot tolerate the intense light from these objects. It was built to view faint, distant objects.

Calculating the culminations of an object

This section treats the transit altitude (culminations) calculations in a more formal manner. Given the declination of an object, we can calculate its altitude as it crosses the local meridian. We have already discussed how to figure out the transit altitude of the celestial equator on the local meridian (p. 28). The formula is

$$A_{CE} = 90^\circ - \lambda,$$

where A_{CE} is the transit altitude of the celestial equator and λ is the observer's latitude. In the northern hemisphere, A_{CE} is measured off the southern horizon and in the southern hemisphere, A_{CE} is measured off the northern horizon.

Once A_{CE} is known, finding the altitude of the object as it crosses the local meridian – called the object's *culminations* or *transit altitudes* – is simply an addition or subtraction problem. For the formulas below, A_{UC} is the upper culmination altitude and A_{LC} is the lower culmination altitude of the object, each measured off the appropriate horizon, and δ is the object's declination. The symbols "(s)" and "(n)," designate an altitude as being measured off the southern or the northern horizon, respectively.

Northern latitudes

In general, the upper culmination is given by

$$A_{UC} = A_{CE} + \delta \text{ (s)}. \tag{A.7}$$

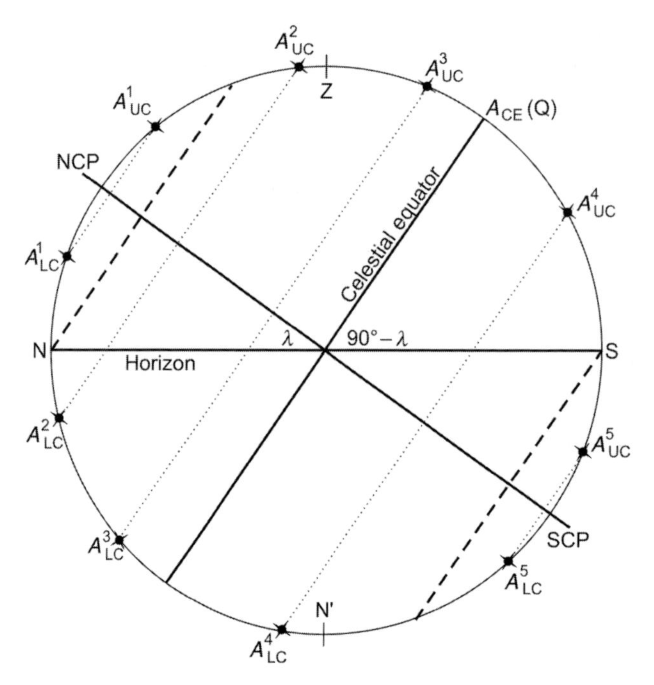

Figure A.2 Stellar culminations for the northern hemisphere. Here we are looking at the upper and lower culmination (transit altitude) of five example stars, numbered 1 through 5. The symbols used are: Z = zenith, N = north, S = south, N′ = nadir.

If the result is greater than $+90^\circ$ then subtract the result from 180° and use this as the altitude off the northern horizon. The correction is

$$A_{\mathrm{UC}} \leftarrow (180^\circ - A_{\mathrm{UC}}) \text{ (n)}. \tag{A.8}$$

The lower culmination is given by

$$A_{\mathrm{LC}} = \delta - A_{\mathrm{CE}} \text{ (n)}. \tag{A.9}$$

As you can see, this measurement is made off the northern horizon. If this result is less than -90° we must make a slightly more complicated correction to make the measurement off the southern horizon. The correction is

$$A_{\mathrm{LC}} \leftarrow -(A_{\mathrm{LC}} + 180^\circ) \text{ (s)}, \tag{A.10}$$

where the original A_{LC} (the one inside the parentheses) has a negative value. If the final result is negative, the object is below the horizon. If the lower culmination is positive, the object is circumpolar at the given latitude. If the upper culmination is negative, the object is never visible at the given latitude.

In the following examples given for latitude 35° N, the altitude of the celestial equator on the local meridian is $90^\circ - 35^\circ = 55^\circ$ off the southern horizon. Figure A.2 shows the altitudes of the upper and lower culminations of five example stars, numbered 1 to 5 with superscripts. In the examples, the star numbers are supplemented with the name of an actual star randomly chosen for the example.

Example: As an example for star 1, shown in Figure A.2, we use Kochab, the second brightest star in Ursa Minor. The dotted line in the figure tells us this should be a circumpolar star. Kochab's declination is approximately $\delta = +75°$. Its upper culmination (A.7) is

$$A_{\mathrm{UC}}^{1(\mathrm{Kochab})} = A_{\mathrm{CE}} + \delta = 55° + 75° = 130°\ (\mathrm{s}).$$

Because this result is greater than 90°, it must be subtracted from 180° and the observation made from the northern horizon (A.8):

$$A_{\mathrm{UC}}^{1(\mathrm{Kochab})} = 180° - 130° = 50°\ (\mathrm{n}).$$

Kochab's lower culmination (A.9) is

$$A_{\mathrm{LC}}^{1(\mathrm{Kochab})} = \delta - A_{\mathrm{CE}} = 75° - 55° = 20°\ (\mathrm{n}).$$

Because this result is positive, Kochab is indeed a circumpolar star at this latitude.

Example: For star 2 we use the bright star Capella, α Aurigae, $\delta = +46°$. Its upper culmination (A.7) is

$$A_{\mathrm{LC}}^{2(\mathrm{Capella})} = A_{\mathrm{CE}} + \delta = 55° + 46° = 101°\ (\mathrm{s}).$$

Again, because the first result is greater than 90°, it must be subtracted from 180° and the observation made from the northern horizon (A.8):

$$A_{\mathrm{LC}}^{2(\mathrm{Capella})} = 180° - 101° = 79°\ (\mathrm{n}).$$

Capella's lower culmination (A.9) is

$$A_{\mathrm{LC}}^{2(\mathrm{Capella})} = \delta - A_{\mathrm{CE}} = 46° - 55° = -9°\ (\mathrm{n}).$$

Because this result is negative, Capella is not a circumpolar star at this latitude.

Example: For star 3 we choose Altair, in Aquila. It has a declination of $\delta = +9°$. Altair's upper culmination (A.7) is

$$A_{\mathrm{UC}}^{3(\mathrm{Altair})} = A_{\mathrm{CE}} + \delta = 55° + 9° = 46°\ (\mathrm{s}).$$

Because this result is not greater than 90°, the measurement off the southern horizon is correct. We now calculate its lower culmination (A.9):

$$A_{\mathrm{LC}}^{3(\mathrm{Altair})} = \delta - A_{\mathrm{CE}} = 9° - 55° = -46°\ (\mathrm{n}).$$

Altair is not circumpolar.

Example: We can use Sirius for star number 4. Sirius has a declination coordinate of $\delta = -17°$. Its upper culmination (A.7) is

$$A_{\mathrm{UC}}^{4(\mathrm{Sirius})} = A_{\mathrm{CE}} + \delta = 55° + (-17°) = 38°\ (\mathrm{s}).$$

Notice its negative declination created a subtraction problem. In the formulas given in this section, the addition signs are algebraic, meaning they could actually stand for subtraction when negative numbers are involved. The lower culmination of Sirius (A.9) is

$$A_{LC}^{4(\text{Sirius})} = \delta - A_{CE} = -17^\circ - 55^\circ = -72^\circ \text{ (n)}.$$

Sirius is not visible during its lower culmination.

Example: As the final example for star 5, we use α Centauri, at $\delta = -62^\circ$. The upper culmination (A.7) is

$$A_{UC}^{5(\alpha\text{ Cen})} = A_{CE} + \delta = 55^\circ + (-65^\circ) = -7^\circ \text{ (s)}.$$

The negative result tells us the upper culmination is below the horizon and therefore α Centauri is never visible at the latitude of 35° N. Although it is obvious that Sirius will not be visible, to complete the example, Sirius' lower culmination (A.9) is

$$A_{LC}^{5(\alpha\text{ Cen})} = \delta - A_{CE} = -65^\circ - 55^\circ = -117^\circ \text{ (n)}.$$

Because the result is less than -90°, the lower culmination is on the other side of the nadir. To get the correct altitude, we must use the negative of adding 180° to the first result. Thus, the lower culmination (A.10) is

$$A_{LC}^{5(\alpha\text{ Cen})} = -(-117^\circ + 180^\circ) = -63^\circ \text{ (s)}.$$

At lower culmination, α Centauri is 63° below the southern horizon.

At the Earth's equator

The situation at the Earth's equator becomes a strange looking special case because the celestial equator passes through the zenith (Figure 4.14). It is also quite easier. The upper culmination becomes

$$A_{UC} = 90^\circ - |\delta|\,. \tag{A.11}$$

Ignore the positive or negative sign on the declination while calculating and always subtract the value from the altitude of the celestial equator, which is 90° at the Earth's equator. The sign is then used to decide from which horizon the altitude is measured. For positive declinations, the upper culmination altitude is measured from the northern horizon. For negative declinations it is measured from the southern horizon. At the Earth's equator, all upper culminations are positive.

The lower culmination is given by

$$A_{LC} = -90^\circ + |\delta|\,. \tag{A.12}$$

This result will always be negative, as it must be because nothing is circumpolar at the equator. All lower culminations must be negative. For positive declinations, the lower culmination altitude is measured from the northern horizon.

For negative declinations it is measured from the southern horizon. We site only two examples here.

Example: For Altair, $\delta = +9^\circ$. The upper culmination (A.11) is

$$A_{\mathrm{UC}} = 90^\circ - |+9^\circ| = 81^\circ \text{ (n)}.$$

Because Altair's declination is positive, this altitude is measured off the northern horizon. The lower culmination (A.12) is

$$A_{\mathrm{LC}} = -90^\circ + |+9^\circ| = -81^\circ \text{ (n)}.$$

Again, because the declination is positive, the altitude is measured off the northern horizon.

Example: For Sirius we have $\delta = -17^\circ$. The upper culmination (A.11) is

$$A_{\mathrm{UC}} = 90^\circ - |-17^\circ| = 73^\circ \text{ (s)}.$$

Because Sirius' declination is negative, this altitude is measured off the southern horizon. The lower culmination (A.12) is

$$A_{\mathrm{LC}} = -90^\circ + |-17^\circ| = -73^\circ \text{ (s)}.$$

Again, because the declination is negative, the altitude is measured off the southern horizon.

Southern latitudes

In general, the upper culmination is given by

$$A_{\mathrm{UC}} = A_{\mathrm{CE}} - \delta \text{ (n)}. \tag{A.13}$$

If the result is greater than $+90^\circ$ then subtract the result from 180° and use this as the altitude off the southern horizon. The correction is

$$A_{\mathrm{UC}} \leftarrow (180^\circ - A_{\mathrm{UC}}) \text{ (s)}. \tag{A.14}$$

The lower culmination is given by

$$A_{\mathrm{LC}} = -(\delta + A_{\mathrm{CE}}) \text{ (s)}. \tag{A.15}$$

As you can see, this measurement is made off the southern horizon. If this result is less than -90° we must make a slightly more complicated correction to make the measurement off the northern horizon. The correction is

$$A_{\mathrm{LC}} \leftarrow -(A_{\mathrm{LC}} + 180^\circ) \text{ (n)}, \tag{A.16}$$

where the original A_{LC} (the one inside the parentheses) is a negative value. If in either case the result is negative, the object is below the horizon. If the lower culmination is positive, the object is circumpolar at the given latitude. If the upper culmination is negative, the object is never visible at the given latitude.

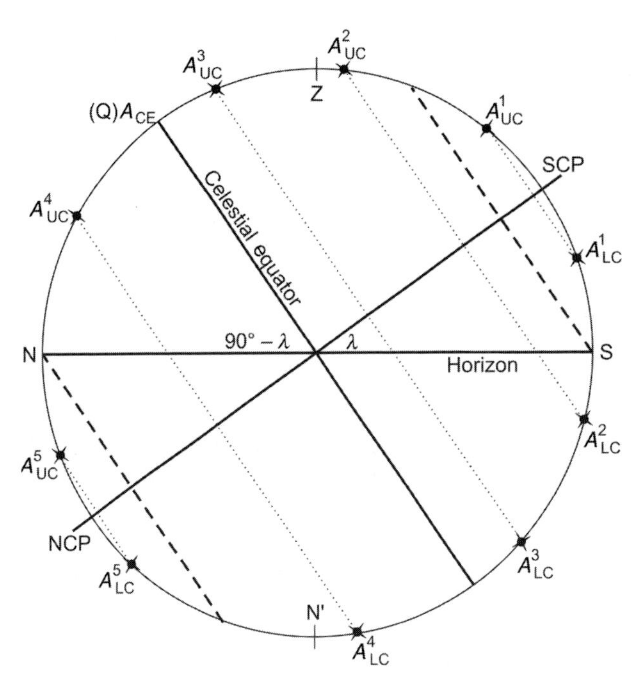

Figure A.3 The star culminations for the examples set at 35° S latitude.

In the following examples given for latitude 35° S, the altitude of the celestial equator on the local meridian is $90° - 35° = 55°$ off the northern horizon. Figure A.3 shows the altitudes of the upper and lower culminations of five example stars numbered 1 to 5 with superscripts. In the examples, the star numbers are supplemented with the names of actual stars randomly chosen as examples.

Example: As an example for star 1, shown in Figure A.3, we use β Hydri, at $\delta = -77°$. The upper culmination (A.13) is

$$A_{\mathrm{UC}}^{1(\beta\ \mathrm{Hyi})} = 55° - (-77°) = 132°\ (\mathrm{n}).$$

Notice the star's negative declination combined with the minus sign in the formula to create an addition problem. In the formulas given in this section, the addition signs are algebraic, meaning they could actually stand for subtraction when negative numbers are involved. Because the first result was greater than 90° we needed to apply the correction and switch measurement horizons. The upper culmination (A.14) is actually

$$180° - 132° = 48°\ (\mathrm{s}).$$

The lower culmination (A.15) is

$$A_{\mathrm{LC}}^{1(\beta\ \mathrm{Hyi})} = -(-77° + 55°) = 22°\ (\mathrm{s}).$$

Because this result is greater than 0°, β Hydri is circumpolar at 35° S latitude.

Example: For star 2 we use α Phoenicis, with a declination coordinate of $\delta = -42^\circ$. Its upper culmination (A.13) is

$$A_{\text{UC}}^{2(\alpha \text{ Phe})} = 55^\circ - (-42^\circ) = 97^\circ \text{ (n)}.$$

Again we must use the correction because this result is greater than 90°. The upper culmination (A.14) is

$$180^\circ - 97^\circ = 83^\circ \text{ (s)}.$$

The lower culmination (A.15) is

$$A_{\text{LC}}^{2(\alpha \text{ Phe})} = -(-42^\circ + 55^\circ) = -13^\circ \text{ (s)}.$$

Because the result is negative, the lower culmination of α Phoenicis is below the horizon. This star is not circumpolar at 35° S. It must rise in the east and set in the west.

Example: For star 3 we choose Rigel, in Orion. It has a declination $\delta = -9^\circ$. Rigel's upper culmination (A.13) is

$$A_{\text{UC}}^{3(\text{Rigel})} = 55^\circ - (-9^\circ) = 64^\circ \text{ (n)}.$$

Because this result is not greater than 90°, the measurement off the northern horizon is correct. We now calculate its lower culmination (A.15):

$$A_{\text{LC}}^{3(\text{Rigel})} = -(-9^\circ + 55^\circ) = -46^\circ \text{ (s)}.$$

Rigel is also not circumpolar.

Example: As star number 4, we use the bright star Pollux, in Gemini, $\delta = +28^\circ$. Its upper culmination (A.13) is

$$A_{\text{UC}}^{4(\text{Pollux})} = 55^\circ - (+28^\circ) = 27^\circ \text{ (n)}.$$

Pollux's lower culmination (A.15) is

$$A_{\text{LC}}^{4(\text{Pollux})} = -(+28^\circ + 55^\circ) = -83^\circ \text{ (s)}.$$

Pollux is not a circumpolar star at this latitude.

Example: As the final example, star 5, we use Kochab. The dotted line in Figure A.3 tells us this should not be a visible star. Kochab's declination is approximately $\delta = +75^\circ$. Its upper culmination (A.13) is

$$A_{\text{UC}}^{5(\text{Kochab})} = 55^\circ - (+75^\circ) = -20^\circ \text{ (n)}.$$

The negative result confirms the non-visibility of Kochab at this latitude. The lower culmination (A.15) is

$$A_{\text{LC}}^{5(\text{Kochab})} = -(+75^\circ + 55^\circ) = -130^\circ \text{ (s)}.$$

The result is less than -90° and thus the correction must be applied. The lower culmination (A.16) becomes

$$A_{\mathrm{LC}}^{5(\mathrm{Kochab})} = -(-130^\circ + 180^\circ) = -50^\circ \text{ (n)}.$$

The lower culmination of Kochab is 50° below the northern horizon.

At the poles

When standing on the Earth's poles, the concepts of upper and lower culmination are no longer valid. The stars follow almucantars around the sky. At the North Pole, the altitude of a star is simply its declination. At the South Pole, the altitude of a star is the negative of its declination.

An exercise

Now that you have seen examples worked at 35° N and 35° S latitudes, work the same examples at your latitude. Make sure you have reasonable answers by drawing a picture like the ones given for the examples. Use a protractor to make your picture accurate. See also Observation Project 8.3.

Force and linear momentum

We have all heard of **force**. Physicists have a strict definition of force, but for our current purposes we can describe force as simply a push or a pull on an object. A force is a *vector* quantity, measured in *newtons* (N). A **vector** is any quantity that has both magnitude (size) and spatial direction. *Speed* is the magnitude of velocity. If the velocity of a car is "45 m/s north," its speed is "45 m/s." A force is described, for example, as "10 newtons to the left," or as "45 newtons down."

An object also has *mass*. **Mass** is one of the most difficult concepts in physics, but can be thought of (for our purposes) as a measure of the amount of material in an object. Mass is an example of a **scalar**. A scalar has magnitude but no spatial direction. For example, we do not talk about mass as "12 kg to the left" – it is simply "12 kg." Speed (or the magnitude of any vector quantity) is also a scalar. As the object acquires velocity due to the applied force, it also acquires *momentum*. **Momentum** may be described as a measure of the amount of motion of an object (Figure A.4). It is given the symbol, $\vec{p}$, and is defined as

$$\vec{p} = m\vec{v}, \tag{A.17}$$

where m is the object's mass, measured in kilograms, and $\vec{v}$ is the object's velocity, measured in meters per second. The measurement units of momentum are kg/m/s. Symbols with small arrows above them represent vector quantities. Momentum is a vector with the same direction as the velocity vector to which it is related. Momentum can be increased by increasing either the mass or the velocity of the object. An object with low mass and high velocity may have the same momentum as an object of high mass and low velocity.

Example: Find the momentum of an object of mass 1.5 kg moving at 4.8 m/s.

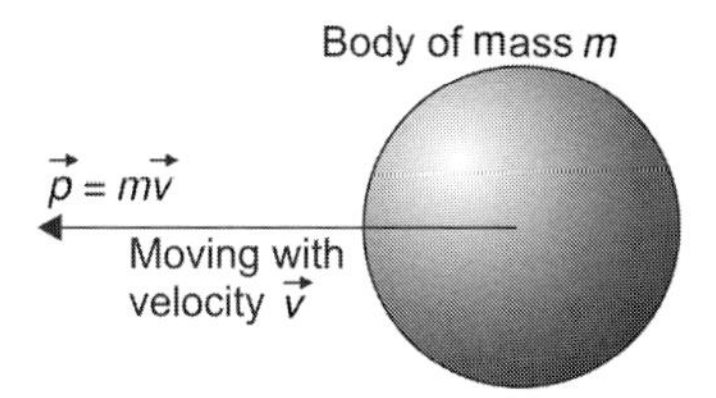

Figure A.4 The linear momentum of any object is equal to the mass of the object multiplied by its velocity.

Using the definition of momentum (A.17),

$$\vec{p} = (1.5)(4.8) = 7.2 \text{ kg/m/s}.$$

If the force creates straight line (linear) motion, the associated momentum is called, **linear momentum**. Force is related to momentum as

$$\vec{F} = \frac{\Delta \vec{p}}{\Delta t}, \tag{A.18}$$

where the Δ symbol means "change in" and t is time. Force is then more strictly defined as the time rate of change of momentum. The faster an object's momentum changes, the greater the force that is being applied to the object. The direction of the change in momentum may be different than the direction of the momentum being changed. This is the case when the *direction* of motion is changing. It is important to understand at this point, that the change in momentum is not necessarily in magnitude. Changing the direction of the momentum requires a force as well.

Using the definition of momentum and using instantaneous (differential) values rather than time average (difference) values, (A.18) becomes

$$\vec{F} = \frac{\mathrm{d}\vec{p}}{\mathrm{d}t} = \frac{\mathrm{d}(m\vec{v})}{\mathrm{d}t}. \tag{A.19}$$

If the mass remains constant while the force is applied, this becomes

$$\vec{F} = m\frac{\mathrm{d}\vec{v}}{\mathrm{d}t} = m\vec{a}, \tag{A.20}$$

where $\vec{a}$ is the acceleration of the object. This is, of course, the more familiar form of Newton's second law of motion.

Probably, the easiest way to understand linear momentum is to think of it in terms of the amount of force it would require to stop a moving object. Momentum is the product of mass and velocity. A low mass object at high velocity would be just as hard to stop as a high mass object at low velocity. A 1000 kg (2200 lb) car traveling at 31.5 m/s (113 k.p.h. or 70 m.p.h.) has as much momentum as a 4570 kg (10 000 lb) truck traveling at 6.3 m/s (22.7 k.p.h. or 14 m.p.h.). These vehicles would be equally difficult to stop – it would require the same amount of force to stop their motion in the same amount of time. Given equal frictional forces in their braking systems, and tires, these two vehicles would require the same stopping distance.

Example: Find the amount of force required to stop a 915 kg car traveling at 25.0 m/s, in 4.00 seconds.

The force required is

$$F = \frac{\Delta p}{\Delta t} = \frac{m(\Delta v)}{\Delta t} = \frac{(915)(25.0)}{4.00} = 5720 \text{ N}.$$

For American readers this is a 2000 pound car traveling at 82 ft/s or 56 miles/h and the required stopping force is 1280 pounds.

Example: Find the acceleration of the car above, as it is stopping.

From Newton's second law (A.20) we have

$$a = \frac{F}{m} = \frac{5720}{915} = 6.25 \text{ m/s}^2.$$

This is equivalent to 20.5 ft/s^2.

Notice the measurement units of acceleration. It measures the rate of change of velocity and thus its units can be stated as "meters per second per second." Because the units of velocity are meters per second, "meters per second per second" tells us by how many meters per second the velocity is changing, per second.

Rotational motion

All objects of any astronomical interest, including, of course, the Earth, spin on an axis. Although we have talked of the sky motions that result from the rotating and revolving Earth, we must understand rotational motion itself, in order to truly understand the seasons and the other Earth–sky motions.

Radian measure

When a particle is in circular motion about an axis, its motion sweeps out an ever increasing angle whose vertex is located at the center of the motion. We have been measuring angles in degrees but this unit will not serve our needs with angular motion. We must now introduce a new unit of angular measure called *radians*. Radians are created by laying the radius of the circle along its circumference. We learn in grade school that the circumference of a circle is equal to the diameter times the number π (the Greek letter pi = 3.1415 . . .).

Mathematicians define π as the ratio of the circumference, C, and the diameter, d,

$$\pi = \frac{C}{d}.$$

The diameter of a circle can be laid along its circumference π times. Using the fact that the diameter, d, is equal to twice the radius, $2r$, we can write,

$$2\pi = 2\frac{C}{2r} = \frac{C}{r}.$$

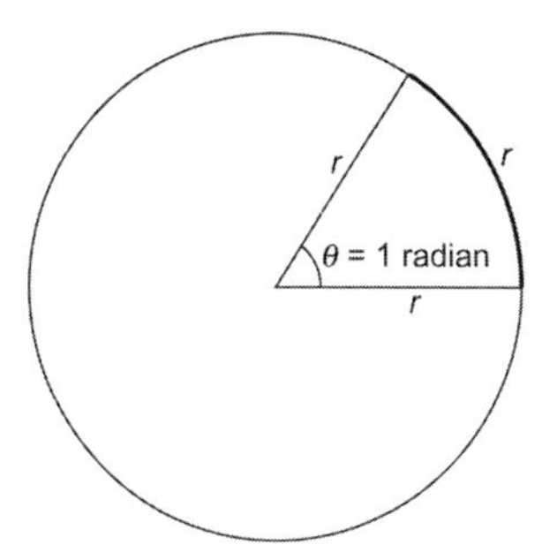

Figure A.5 One radian is the angle created by laying off the radius of a circle along its circumference.

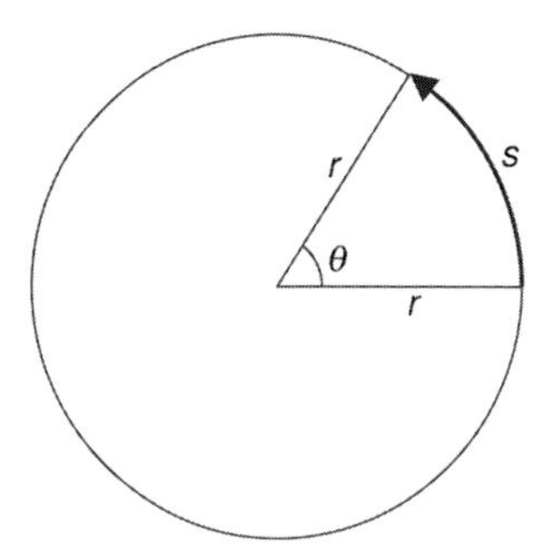

Figure A.6 As a particle moves along the circumference of a circle, its motion produces an angle, θ. By using the radius of the circle, r, we can relate the distance moved along the circumference, s, to the angle, θ, when the angle is measured in radians.

This equation tells us that we can lay the circle's radius along its circumference 2π times. One **radian** is the angle created by laying off the radius of a circle along its circumference. In other words, a complete 360° circle also contains 2π radians (see Figure A.5). One radian is approximately 57°.3.

Radians are the result of a distance (circumference) divided by a distance (radius). The distance units divide out, leaving no units of measure. Radians is an example of measurement with no units or dimensions. (This is a second meaning of the word dimension. It has nothing to do with dimensions of space.) Because radian measure is dimensionless, it has no official symbol (such as ° for degrees); some books use a superscript, r, when it is possible to confuse the two types of measurements. For example, $360^\circ = 2\pi^r = 6.28^r$, or $1^r = 57^\circ.3$. The 57.3 is simply 360 divided by 6.28.

Angular velocity

When a particle is moving along the circumference of a circle, its changing position sweeps out an angle around the center of the circle (Figure A.6). The speed at which the particle moves along the circumference is given by the distance it moves divided by the amount of time it took to move that distance. The distance is equal to the length of the arc of the angle,

$$s = r\theta,$$

where the Greek letter θ (theta), represents the angle generated by the circular motion, of the particle, measured in radians, r is the radius of the circular motion, and s is the distance the particle moved along the circumference.

The speed of a particle is given by the distance moved divided by the time taken to move that distance. So,

$$v = \frac{s}{t} = \frac{r\theta}{t} = r\frac{\theta}{t},$$

where r is assumed constant because the motion is circular. As the rotational motion continues, angle is continuously generated. By letting the time and the angle become differential in size, the more correct version of the above equation is written as

$$v = r\frac{\mathrm{d}\theta}{\mathrm{d}t} = r\omega.$$

This equation introduces a new symbol, the Greek letter ω (omega). Omega represents the time rate of change of the angle associated with the circular motion, and so it is called the *angular speed* of the motion. It is measured in radians per second.

Angular speed is the magnitude of the vector quantity, **angular velocity**. Because the actual direction of motion of a particle in circular motion is constantly changing, defining the direction of angular velocity is a little more complicated than defining the direction of linear velocity.

Imagine wrapping your right hand around the particle's axis of rotation, which is perpendicular to the plane of the circle of its motion. Point your fingers in the direction of the particle's movement around the circle. Lay your thumb along the axis and then your thumb points in the direction of the angular velocity vector. That is, the angular velocity vector lies along the particle's axis of rotation (Figure A.7). This is called the "right-hand rule."

Angular momentum

Angular motion, like linear motion, has an associated momentum. As the particle moves on its circular path around the axis, it has a linear momentum (because it *is* in motion). At any given instant, the linear momentum is tangent to the circular path at the position of the particle. During the circular motion, the linear momentum is constantly changing its direction toward the center of the circle. Hence, there must be a force acting on the particle in order to change the direction of its linear momentum. The force is directed toward the center, the same direction as the change in the linear momentum, as Eqn (A.19) requires.

Earlier, we described momentum as the amount of effort required to stop an object. **Angular momentum** could be thought of as the amount of effort required

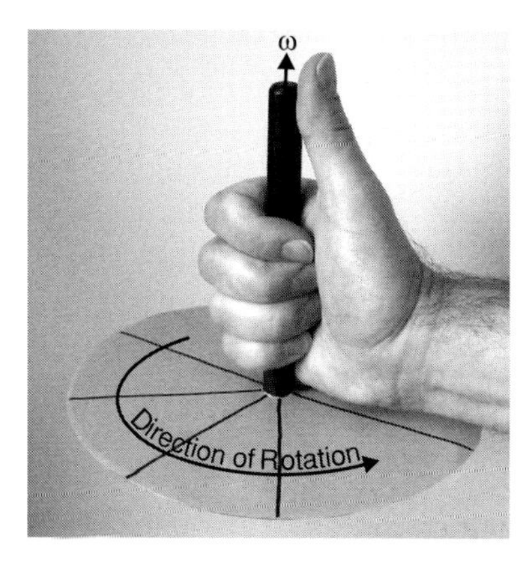

Figure A.7 The right-hand rule gives the direction of the angular velocity and angular momentum vectors.

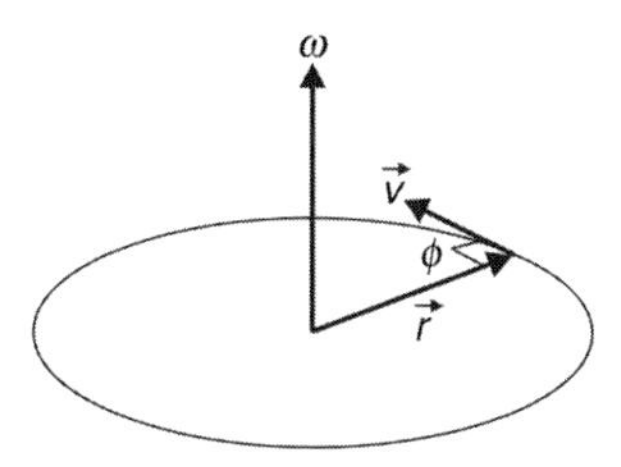

Figure A.8 The magnitude of the angular velocity vector can be found from the particle position vector, $\vec{r}$, and the particle velocity vector, $\vec{v}$. The right-hand rule gives the direction of the angular velocity vector.

to stop the rotational motion of an object. Think for a moment, about a wheel spinning on an axle. If you press your hand to the edge of the wheel, you will create friction and slow the wheel's rotation, eventually stopping it. Another way to imagine this is to think about putting a spinning wheel on the ground. Friction applies a force to the edge of the wheel, which in turn tends to push or pull what ever is holding the wheel's axle.

Angular momentum is ultimately caused by the linear (tangential) momentum of the particle, while in circular motion. Because the linear momentum depends on the speed of the particle, which depends on the particle's distance from the center of motion, the angular momentum must also depend on the distance from the center of motion. By experimental results we can finally conclude that angular momentum depends on three factors – the mass of the particle, its tangential speed (which creates its linear, tangential momentum), and its distance from the axis of rotation. Angular momentum is given the symbol $\vec{L}$. So;

$$\vec{L} = mvr,$$

where m is the mass, v is the tangential speed, and r is the distance between the particle and the axis of rotation. The tangential speed, v, depends on the radius,

r, and the angular speed, ω. So this definition can also be written as

$$\vec{L} = mr^2\omega.$$

Because any object is made of particles (namely atoms), the angular momentum of an object is just the sum of the angular momentum of the particles of the body.

Angular momentum is a vector quantity. The direction of the vector is given by the right-hand rule, with your fingers wrapped around the axis in the direction of the motion and your thumb pointing along the axis in the direction of the angular momentum vector. It is parallel with the angular velocity vector.

Conservation of angular momentum

Momentum is a *conserved* quantity. That means, during any interaction (such as a collision) between two objects (such as billiard balls), the total momentum – the momentum of the two (or more) objects added together – remains constant throughout the interaction. Interactions between the members of a system of objects cannot add momentum to, or remove momentum from, the system. Momentum can be transferred between the members of the system but the total momentum (the sum of the momenta of the system members) remains constant. Because momentum is a vector quantity, interactions cannot change either the magnitude or the direction of the system's total momentum. This concept is called the **conservation of momentum**.

Angular momentum is also conserved. As long as no torque acts on a rotating body, its angular momentum remains constant in magnitude *and* in direction. We may recognize two forms of angular momentum, although they are actually the same. **Rotational angular momentum** is the result of an object's spin, such as the rotation of a gyroscope or the daily rotation of the Earth. **Orbital angular momentum** is when the motion is associated with orbital motion, such as the annual motion of the Earth around the Sun. (They are the same because rotational angular momentum is just the sum of the orbital angular momenta of the individual particles composing the rotating body.) In both cases the angular momentum of the system is conserved.

No interaction between the members of a system of objects can change the total angular momentum of the system. An individual member of the system can experience a change in angular momentum but the angular momentum lost (or gained) by one member of a system is gained (or lost) by another member.

Angular momentum causes the rotational motion (or orbital motion) of a body to define a particular direction in space. By wrapping your right hand around the axis of rotation with your fingers pointing in the direction of the body's angular motion, your thumb will point in the spatial direction associated with the rotational or orbital angular motion (Figure A.7). To fulfill the second requirement of conservation of angular momentum (the direction of the vector cannot change), a rotating body must keep its axis of rotation fixed in space and an orbiting body must orbit (follow a path) confined to a flat plane, like the

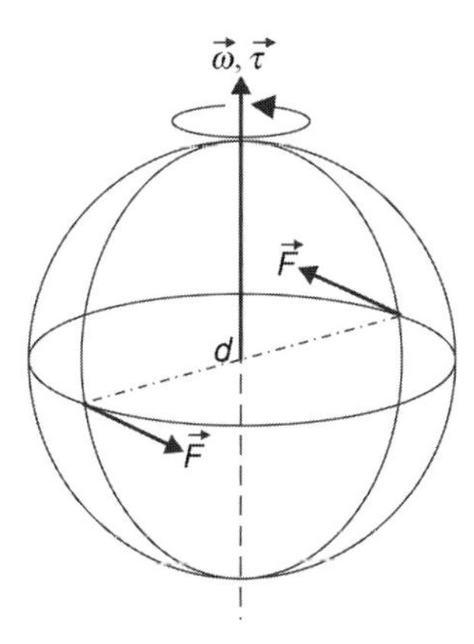

Figure A.9 A pair of forces, $\vec{F}$, acting in opposite directions, held in a common plane and separated by a distance, d, create a torque. As drawn, the plane of the force-pair is the equatorial plane of the sphere. The torque causes the sphere to rotate as shown by the curved arrow at the top. The angular velocity, angular momentum, and torque vectors (all in the same direction) are also shown.

surface of a table. Recall from our discussions on p. 12, the relationship between a line and a plane. With orbital motion, the line is the direction of the orbital angular momentum and the plane is the plane of the body's orbit.

We can now apply the results of this discussion to the Earth. Due to the conservation of angular momentum, the Earth's rotational axis must remain fixed in space. This means the axis points to the same location in space – the celestial poles – at all times. All night long, all year long, the Earth's axis points to the same position of the celestial sphere, i.e. the celestial poles. Conservation of angular momentum also requires the Earth to orbit the Sun on a path that lies in a plane. This is the ecliptic plane. All changes in angular momentum (magnitude or direction) are caused by torques. The Earth's precession and nutation are caused by torques created by the gravitational pull of other objects in the Solar System (mainly the Moon and the Sun).

Torque

Torque results from a force – actually a pair of forces – that changes angular momentum. The symbol for the torque vector is the Greek letter tau $\vec{\tau}$. A torque can be applied to a rotating or non-rotating body. If applied to a non-rotating body, the torque may cause the body to rotate about an axis that is perpendicular to the plane containing the torque's force-pair (Figure A.9). This causes the body to acquire angular velocity and angular momentum. These acquired vectors are parallel to the applied torque vector.

The magnitude of the torque is equal to the magnitude of the applied forces multiplied by the distance between the forces. The unit of measure is N m (newton-meters). The direction of the torque vector is perpendicular to the plane containing the force-pair, with the sense given by the right-hand rule. Strictly

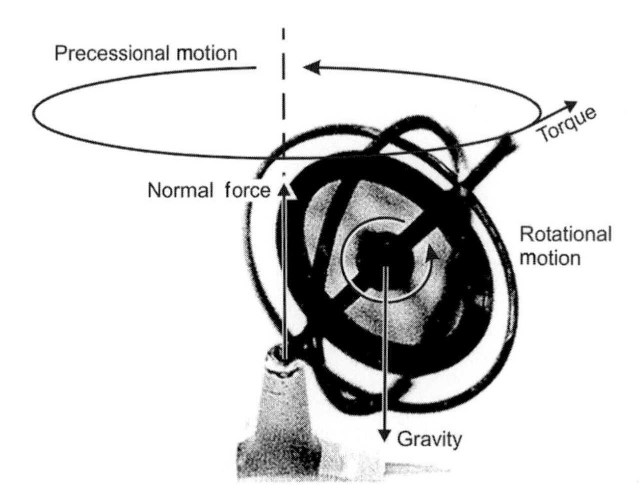

Figure A.10 The normal force between the base and the gyroscope frame couples with the force of gravity acting at the center of mass of the gyroscope to create a torque that causes the gyroscope to precess.

speaking, torque is defined as the time rate of change of angular momentum,

$$\vec{\tau} = \frac{\Delta \vec{L}}{\Delta t},$$

or in differential form,

$$\vec{\tau} = \frac{\mathrm{d}\vec{L}}{\mathrm{d}t}.$$

The torque can be applied in such a way as to cause the body's angular momentum (or angular velocity) to decrease in magnitude. For the example shown in Figure A.9, the most effective way to decrease the angular momentum (after it has been created by the original torque, of course) is to reverse the direction (reverse the arrows) of the force-pair. Notice then, that applying a torque parallel to the angular momentum vector increases the angular momentum magnitude, while applying a torque anti-parallel to the angular momentum vector decreases the angular momentum magnitude.

A torque can also be applied to the rotating body so that the spatial direction of $\vec{L}$ changes, without changing the magnitude of $\vec{L}$. If a torque acts to change the direction of $\vec{L}$, then the *change* in $\vec{L}$ is perpendicular to both $\vec{L}$ and the plane of the forces creating the torque. In the case of the toy gyroscope shown in Figure A.10, the Earth's gravity acting on the center of mass of the gyroscope (paired with the support force between the base and the frame) produces a torque that tries to topple the gyroscope. Instead of toppling the gyroscope, the torque produces a change in the direction of the angular momentum. This persistently changing direction is the precessional motion of the gyroscope.

Appendix B
Astronomical data

Constellations

Note: The right ascension is given in hours. The Age column states whether the constellation is an ancient (Anc., established by Ptolemy) or modern (Mod., sixteenth century or later) constellation.

Name (alternate names) Pronunciation	Representation	Age	Abbr.	Genitive	Location (RA, Dec.)
Andromeda an-drä′meh-deh	The Princess, The Chained Lady	Anc.	And	Andromedae	01^h, $+40^\circ$
Antlia (Antlia Pneumatica) a′nt-lē-eh n(y)ü-ma′ti-keh	The Air Pump	Mod.	Ant	Antliae	10^h, -35°
Apus ā′-pehs	The Bird of Paradise	Mod.	Aps	Apodis	16^h, -75°
Aquarius eh-kwa′r-ĕ-ehs	The Water Bearer	Anc.	Aqr	Aquarii	23^h, -15°
Aquila eh-kwi′l-eh	The Eagle	Anc.	Aql	Aquilae	20^h, $+05^\circ$
Ara a′r-eh	The Altar	Anc.	Ara	Arae	17^h, -55°
Aries a′r-ēz	The Ram	Anc.	Ari	Arietis	03^h, $+20^\circ$

(Cont.)

Name (alternate names) Pronunciation	Representation	Age	Abbr.	Genitive	Location (RA, Dec.)
Auriga or-rī′-geh	The Charioteer	Anc.	Aur	Aurigae	06^h, +40°
Boötes bō -ō′t-ēz	The Herdsman	Anc.	Boo	Boötis	15^h, +30°
Caelum (Caela Sculptoris) sē′-lehm (sē′-leh skehlp-tō′r-ehs)	The Sculptor′s Chisel	Mod.	Cae	Caeli	05^h, −40°
Camelopardus ka′m-eh-lō-pär-deh-lehs	The Giraffe	Mod.	Cam	Camelopardalis	06^h, +70°
Cancer ka′n-sehr	The Crab	Anc.	Cnc	Cancri	09^h, +20°
Canes Venatici kā′-nĕz veh-na′t-eh-sī	The Hunting Dogs	Mod.	CVn	Canum Venaticorum	13^h, +40°
Canis Major kā′-nehs mā′jehr	The Big Dog	Anc.	CMa	Canis Majoris	07^h, −20°
Canis Minor kā′-nehs mī′nehr	The Little Dog	Anc.	CMi	Canis Minoris	08^h, +05°
Capricornus ka′p-reh-kor-nehs	The Sea-Goat, The Goat	Anc.	Cap	Capricorni	21^h, −20°
Carina keh-rī′-neh	The Keel (of Argo Navis, The Ship)	Anc.[a]	Car	Carinae	09^h, −60°
Cassiopeia kas-ē-eh-pē′-eh, kas-ē-o′-pē-eh	The Queen	Anc.	Cas	Cassiopeiae	01^h, +60°
Centaurus sen-to′r-ehs	The Centaur	Anc.	Cen	Centauri	13^h, −50°
Cepheus sē-fyü′s, sē′-fē-ehs	The King, The Ethiopian King	Anc.	Cep	Cephei	22^h, +70°
Cetus sē′-tehs	The Sea Monster, The Whale	Anc.	Cet	Ceti	02^h, −10°
Chamaeleon keh-mē′l-yehn	The Chameleon	Mod.	Cha	Chamaeleontis	11^h, −80°
Circinus seh′rs-ehn-ehs	The (Pair of) Compasses	Mod.	Cir	Circini	15^h, −60°

Name (alternate names) Pronunciation	Representation	Age	Abbr.	Genitive	Location (RA, Dec.)
Columba (Columba Noae) keh-leh′m-beh (nō-ē′)	The Dove, Noah's Dove	Mod.	Col	Columbae	06^h, −35°
Coma Berenices kō-meh ber-eh-nī′-sēz	Berenice's Hair	Mod.	Com	Comae Berenices	13^h, +20°
Corona Australis keh-rō′-neh au-strā′-lehs	The Southern Crown	Anc.	CrA	Coronae Australis	19^h, −40°
Corona Borealis keh-rō′-neh bōr-ē-a′l-ehs, keh-rō′-neh bōr-ē-ā′lehs	The Northern Crown	Anc.	CrB	Coronae Borealis	16^h, +30°
Corvus ko′r-vehs	The Crow	Anc.	Crv	Corvi	12^h, −20°
Crater krä′t-ehr	The Cup	Anc.	Crt	Crateris	11^h, −15°
Crux kreh′ks	The Southern Cross	Mod.	Cru	Crucis	12^h, −60°
Cygnus si′g-nehs	The Swan	Anc.	Cyg	Cygni	21^h, +40°
Delphinus del-fī′-nehs	The Dolphin	Anc.	Del	Delphini	21^h, +15°
Dorado deh-rä′-dō	The Swordfish	Mod.	Dor	Doradus	05^h, −65°
Draco drā′-kō	The Dragon	Anc.	Dra	Draconis	17^h, +65°
Equuleus e-kwü′-lē-ehs	The Little Horse	Anc.	Equ	Equulei	21^h, +10°
Eridanus eh-ri′d-ehn-ehs	The River (Eridanus)	Anc.	Eri	Eridani	03^h, −20°
Fornax fō′r-naks	The Furnace	Mod.	For	Fornacis	03^h, −30°
Gemini je′m-eh-nī	The Twins	Anc.	Gem	Geminorum	07^h, +20°
Grus greh′s grü′s	The Crane	Mod.	Gru	Gruis	22^h, −45°

(*Cont.*)

Name (alternate names) Pronunciation	Representation	Age	Abbr.	Genitive	Location (RA, Dec.)
Hercules heh′r-kyeh-lēz	Hercules, The Warrior	Anc.	Her	Herculis	17^h, +30°
Horologium hor-eh-lō′-jē-ehm	The Clock	Mod.	Hor	Horologii	03^h, −60°
Hydra hī′-dreh	The (Female) Water Snake	Anc.	Hya	Hydrae	10^h, −20°
Hydrus hī′-drehs	The (Male) Water Snake	Mod.	Hyi	Hydri	02^h, −75°
Indus i′n-dehs	The (American) Indian	Mod.	Ind	Indi	21^h, −55°
Lacerta leh-se′rt-eh	The Lizard	Mod.	Lac	Lacertae	22^h, +45°
Leo lē′-ō	The Lion	Anc.	Leo	Leonis	11^h, +15°
Leo Minor lē′-ō mī′-nehr	The Little Lion	Mod.	LMi	Leonis Minoris	10^h, +35°
Lepus lē′-pehs	The Hare	Anc.	Lep	Leporis	06^h, −20°
Libra lī′-breh, lē′-breh	The Scales, The Balance	Anc.	Lib	Librae	15^h, −15°
Lupus lü′-pehs	The Wolf	Anc.	Lup	Lupi	15^h, −45°
Lynx li′nks	The Lynx	Mod.	Lyn	Lyncis	08^h, +45°
Lyra lī′-reh	The Harp, The Lyre	Anc.	Lyr	Lyrae	19^h, +40°
Mensa (Mons Mensae) me′n-seh (mä′nz me′n-sē)	The Table, The Mountain	Mod.	Men	Mensae	05^h, −80°
Microscopium mī′-kreh-skō′-pē-ehm	The Microscope	Mod.	Mic	Microscopii	21^h, −35°
Monoceros meh-nä′s-eh-rehs	The Unicorn	Mod.	Mon	Monocerotis	07^h, −05°
Musca meh′s-keh	The Fly	Mod.	Mus	Muscae	12^h, −70°
Norma nō′r-meh	The Square	Mod.	Nor	Normae	16^h, −50°

Name (alternate names) Pronunciation	Representation	Age	Abbr.	Genitive	Location (RA, Dec.)
Octans ä′k-tanz	The Octant	Mod.	Oct	Octantis	22^h, −85°
Ophiuchus ō′-fē-yü′-kehs	The Serpent Bearer, The Doctor	Anc.	Oph	Ophiuchi	17^h, 0°
Orion or-rī′-ehn	The Hunter	Anc.	Ori	Orionis	05^h, +5°
Pavo pā′-vō	The Peacock	Mod.	Pav	Pavonis	20^h, −65°
Pegasus pe′g-eh-sehs	The Winged Horse, The Flying Horse	Anc.	Peg	Pegasi	22^h, +20°
Perseus purr′-süs, purr′-sē-ehs	Perseus, The Hero	Anc.	Per	Persei	03^h, +45°
Phoenix fē′-niks	The Phoenix	Mod.	Phe	Phoenicis	01^h, −50°
Pictor pi′k-tehr	The Easel, The Painter's Easel	Mod.	Pic	Pictoris	06^h, −55°
Pisces pī′-sēz, pi′s-ēz	The Fishes, The Two Fish	Anc.	Psc	Piscium	01^h, +15°
Piscis Austrinus pi′s-ehs au-strī′-nehs	The Southern Fish	Anc.	PsA	Piscis Astrini	22^h, −30°
Puppis peh′p-ehs	The Stern (of Argo Navis)	Mod.[a]	Pup	Puppis	08^h, −40°
Pyxis pi′k-sehs	The Compass (of Argo Navis)	Mod.[a]	Pyx	Pyxidis	09^h, −30°
Reticulum reh-ti′k-yeh-lehm	The Net	Mod.	Ret	Reticuli	04^h, −60°
Sagitta seh-ji′t-eh	The Arrow	Anc.	Sge	Sagittae	20^h, +10°
Sagittarius saj-eh-ta′r-ē-ehs	The Archer	Anc.	Sgr	Sagittarii	19^h, −25°
Scorpius sko′r-pē-ehs	The Scorpion	Anc.	Sco	Scorpii	17^h, −40°
Sculpter skeh′lp-tehr	The Sculptor	Mod.	Scl	Sculptoris	00^h, −30°
Scutum skü′t-ehm, skyü′t-ehm	The Shield	Mod.	Sct	Scuti	19^h, −10°

(Cont.)

Name (Alternate names) Pronunciation	Representation	Age	Abbr.	Genitive	Location (RA, Dec.)
Serpens[b] seh′r-pehnz (Serpens Caudia) seh′r-pehnz kau′-dē (Serpens Caput) seh′r-pehnz ka′-poŏ	The Serpent (Eastern Serpent) (Western Serpent)	Anc.	Ser	Serpentis	17^h, 0°
Sextans se′k-stanz	The Sextant	Mod.	Sex	Sextantis	10^h, 0°
Taurus to′r-ehs	The Bull	Anc.	Tau	Tauri	04^h, +15°
Telescopium tel-eh-skō′-pē-ehm	The Telescope	Mod.	Tel	Telescopii	19^h, −50°
Triangulum trī-a′n-gyeh-lehm	The Triangle	Mod.	Tri	Trianguli	02^h, +30°
Triangulum Australe trī-a′n-gyeh-lehm au-strā′-lē	The Southern Triangle	Mod.	TrA	Trianguli Australis	16^h, −65°
Tucana tü-ka′n-eh	The Toucan	Mod.	Tuc	Tucanae	00^h, −65°
Ursa Major eh′r-seh mā-jehr	The Big Bear	Anc.	UMa	Ursae Majoris	11^h, +50°
Ursa Minor eh′r-seh mī-nehr	The Little Bear	Anc.	UMi	Ursae Minoris	15^h, +70°
Vela vē′leh	The Sails (of Argo, the ship)	Anc.[a]	Vel	Velorum	09^h, −50°
Virgo veh′r-gō	The Maiden, The Goddess of Justice	Anc.	Vir	Virginis	13^h, 0°
Volans (Piscis Volans) pi′s-ehs vō-lanz	The Flying Fish	Mod.	Vol	Volantis	08^h, −70°
Vulpecula vehl-pe′k-yeh-leh	The Little Fox	Mod.	Vul	Vulpeculae	20^h, +25°

[a] Carina, Puppis, Pyxis, and Vela are parts of the larger constellation of Argo Navis (the ship, Argo, from the myth of Jason and the Argonauts) depicted by Claudius Ptolemy in his book *The Syntaxis*, or its Arabic translation, *Almagest*.

[b] Serpens passes through the constellation Ophiuchus and is therefore sometimes divided into two parts, eastern and western.

Asterisms

Here is a short list of asterisms.

Name[a]	Location
The Arc	Ursa Major
The Arrowhead	Capricornus
The Belt of Orion	Orion
The Big Dipper	Ursa Major
The Circlet	Pisces
The False Cross	Carina, Vela
The Great Square	Pegasus
The Ice Cream Cone	Boötes
The Keystone	Hercules
The Kids	Auriga
The Kite	Boötes
The Little Dipper	Ursa Minor
The Northern Cross	Cygnus
The Pointers	Ursa Major
The Sickle	Leo
The Summer Triangle[b]	Aquila, Cygnus, Lyra
The Teapot	Sagittarius
The Virgo Triangle	Arcturus, Leo, Virgo
The Water Jug	Aquarius
The Wineglass	Virgo
The Winter Hexagon[c]	Gemini, Auriga, Taurus, Orion, Canis Major, Canis Minor
The Winter Triangle[b]	Canis Major, Canis Minor, Orion

[a] Notes are for southern hemisphere observers.
[b] The names of the winter and summer triangles could be reversed.
[c] Or the Summer Hexagon.

Bright stars

The stars on this list should be the easiest stars to find since they are the dominant stars within their constellations. Since this list is intended for naked-eye observations, only the magnitude of the brightest star is given for those that are actually binary or multiple star systems. Use these bright stars as guideposts to find the constellations and learn your way around the sky.

In the following tables, Bayer is the Bayer designation, FN is the Flamsteed number, m is apparent magnitude, and d is distance, measured in light years. If a star is variable, there is a "v" after its apparent magnitude. Variable stars are discussed in Section 3.5.

Bright stars in the northern celestial hemisphere

Proper name	Bayer	FN	RA	Dec. (°)	m	d (ly)
Arcturus	α Boo	16 Boo	14 15.6	+19 12	−0.06	37
Vega	α Lyr	3 Lyr	18 36.8	+38 46	0.03	25
Capella	α Aur	13 Aur	05 16.4	+45 59	0.08	42
Procyon	α CMi	10 CMi	07 39.2	+5 14	0.38	286
Betelgeuse	α Ori	58 Ori	05 55.0	+7 24	0.50v	427
Altair	α Aql	53 Aql	19 50.7	+ 8 51	0.77	17
Aldebaran	α Tau	87 Tau	04 35.7	+16 30	0.85v	65
Pollux	β Gem	78 Gem	07 45.2	+28 02	1.14	34
Deneb	α Cyg	50 Cyg	20 41.3	+45 16	1.25	3229
Regulus	α Leo	32 Leo	10 08.2	+12 00	1.35	77
Castor	α Gem	66 Gem	07 34.4	+31 54	1.52	52
Bellatrix	γ Ori	24 Ori	05 25.0	+6 20	1.64	243
Elnath	β Tau	112 Tau	05 26.2	+28 36	1.65	131
Mirfak	α Per	33 Per	03 24.1	+49 50	1.79	592
Dubhe	α UMa	50 UMa	11 03.6	+61 46	1.79	124
Menkalinan	β Aur	34 Aur	05 59.3	+44 57	1.90v	82
Hamal	α Ari	13 Ari	02 07.0	+23 27	2.00	66
Polaris	α UMi	1 UMi	02 27.9	+89 15	2.00v	431
Mirach	β And	43 And	01 09.5	+35 36	2.06	199
Rasalhague	α Oph	55 Oph	17 34.9	+12 34	2.08	47
Kochab	β UMi	7 UMi	14 50.7	+74 11	2.08	126
Alpheratz	α And	21 And	00 08.2	+29 04	2.10v	97
Algol	β Per	26 Per	03 08.0	+40 56	2.10v	93
Denebola	β Leo	94 Leo	11 48.9	+14 36	2.14	36
Alphecca	α CrB	5 CrB	15 34.6	+26 44	2.20v	75
Sadr	γ Cyg	37 Cyg	20 22.0	+40 14	2.20	1524
Schedar	α Cas	18 Cas	00 40.3	+56 31	2.23	229
Almack	γ And	57 And	02 03.7	+42 19	2.26	355
Mizar	ζ UMa	79 UMa	13 23.7	+54 58	2.27	88
Caph	β Cas	11 Cas	00 09.0	+59 08	2.30v	54
Merak	β UMa	48 UMa	11 01.7	+56 24	2.37	79
Scheat	β Peg	53 Peg	23 03.7	+28 04	2.40v	199
Alderamin	α Cep	5 Cep	21 18.4	+62 34	2.44	49
Markab	α Peg	54 Peg	23 04.6	+15 11	2.49	140

Proper name	Bayer	FN	RA	Dec. (°)	*m*	*d* (ly)
Navi	γ Cas	27 Cas	00 56.6	+60 42	2.50v	613
Menkar	α Cet	92 Cet	03 02.1	+4 04	2.53	220
Sharatan	β Ari	6 Ari	01 54.5	+20 47	2.64	60
Unukalhai	α Ser	24 Ser	15 44.1	+6 26	2.65	73
Kornephoros	β Her	27 Her	16 30.2	+21 30	2.77	148
Cebalrai	β Oph	60 Oph	17 43.3	+4 34	2.77	82
Rastaban	β Dra	23 Dra	17 30.3	+52 19	2.79	382
Cor Caroli	α^2 CVn	12 CVn	12 55.9	+38 20	2.90v	110
Gomeisa	β CMi	3 CMi	07 27.0	+8 18	2.90	170
Abireo	β Cyg	6 Cyg	19 30.5	+27 57	3.08	386
Alfirk	β Cep	8 Cep	21 28.6	+70 33	3.20v	595
Errai	γ Cep	35 Cep	23 39.3	+77 36	3.21	45
Sulafat	γ Lyr	14 Lyr	18 58.7	+32 41	3.24	635
Sheliak	β Lyr	10 Lyr	18 49.9	+33 21	3.40v	882
Metallah	α Tri	2 Tri	01 52.9	+29 34	3.41	64
Kaffaljidhma	γ Cet	86 Cet	02 43.2	+3 14	3.47	82
Rasalgethi	α Her	64 Her	17 14.6	+14 24	3.50v	430
Nekkar	β Boo	42 Boo	15 01.7	+40 25	3.50	219
Altarf	β Cnc	17 Cnc	08 16.3	+9 12	3.52	290
Thuban	α Dra	11 Dra	14 04.5	+64 21	3.67	309
Alshain	β Aql	60 Aql	19 55.5	+06 25	3.71	45

Bright stars in the southern celestial hemisphere

Proper name	Bayer	FN	RA	Dec. (°)	*m*	*d* (ly)
Sirius	α CMa	9 CMa	06 45.1	−16 42	−1.46	9
Canopus	α Car		06 23.8	−52 41	−0.72	312
Rigel Kentaurus	α Cen		14 39.5	−60 49	−0.01	4
Rigel	β Ori	19 Ori	05 14.7	−8 13	0.12	773
Achernar	α Eri		01 37.5	−57 15	0.46	144
Hadar	β Cen		14 03.6	−60 21	0.60v	525
Antares	α Sco	21 Sco	16 29.3	−26 26	0.90v	604
Spica	α Vir	67 Vir	13 25.0	−11 08	1.00v	262
Fomalhaut	α PsA	24 PsA	22 57.5	−29 39	1.16	25
Mimosa	β Cru		12 47.6	−59 40	1.20v	353
Acrux	α^1 Cru		12 26.4	−63 04	1.33	321
Gacrux	γ Cru		12 31.0	−57 05	1.63v	88
Miaplacidus	β Car		09 13.2	−69 41	1.68	111
Alnair	α Gru		22 08.0	−47 00	1.74	101
Regor	γ^2 Vel		08 09.4	−47 19	1.80v	840
Atria	α TrA		16 48.2	−69 02	1.92	415
Peacock	α Pav		20 25.4	−56 45	1.94	183
Alphard	α Hya	30 Hya	09 27.5	−8 39	1.98	177
Murzim	β CMa	2 CMa	06 22.6	−17 57	2.00v	499
Diphda	β Cet	16 Cet	00 43.5	−18 01	2.04	96

(Cont.)

Proper name	Bayer	FN	RA	Dec. (°)	m	d (ly)
Alnitak	ζ Ori	50 Ori	05 40.6	−1 57	2.05	817
Saiph	κ Ori	53 Ori	05 47.7	−9 40	2.06	721
	β Gru		22 42.5	−46 54	2.10v	170
Suhail	λ Vel		09 07.8	−43 24	2.21v	573
Mintaka	δ Ori	34 Ori	05 31.9	−0 18	2.23v	916
	α Lup		14 41.7	−47 23	2.30v	548
Ankaa	α Phe		00 26.2	−42 20	2.39	77
Arneb	α Lep	11 Lep	05 32.7	−17 50	2.58	1284
Geinah	γ Crv	4 Crv	12 15.7	−17 31	2.59	165
Zubeneschamali	β Lib	27 Lib	15 16.8	−9 22	2.61	160
Acrab, Graffias	β Sco	8 Sco	16 05.2	−19 48	2.62	530
Phact	α Col		05 39.5	−34 05	2.64	268
	α Mus		12 37.0	−69 07	2.69v	306
Zubenelgenulbi	α^2 Lib	9 Lib	14 50.7	−16 01	2.75	77
Porrima	γ Vir	29 Vir	12 41.5	−1 26	2.76	32
Cursa	β Eri	67 Eri	05 07.7	−5 06	2.79	89
Nihal	β Lep	9 Lep	05 28.2	−20 46	2.84	159
	β Ara		17 25.0	−55 31	2.85	602
	β TrA		15 54.8	−63 25	2.85	40
	α Tuc		22 18.3	−60 17	2.86	199
	α Hyi		01 58.7	−61 36	2.86	71
	γ Cen		12 41.3	−48 56	2.87	130
	γ TrA		15 18.6	−68 39	2.89	183
Sadal Suud	β Aqr	22 Aqr	21 31.4	−5 35	2.91	1030
Zaurak	γ Eri	34 Eri	03 57.9	−13 31	2.95	221
	α Ara		17 31.6	−49 52	2.95	242
Sadal Melik	α Aqr	34 Aqr	22 05.6	−0 20	2.96	758
Alnasl	γ^2 Sgr	10 Sgr	18 05.6	−30 25	2.99	96
	γ Hya	46 Hya	13 18.7	−23 09	3.00	132
	γ Gru		21 53.8	−37 23	3.01	203
	β Mus		12 46.1	−68 05	3.05	311
Dabih	β^1 Cap	9 Cap	20 20.8	−14 48	3.08	344
	α Ind		20 37.3	−47 19	3.11	101
Wezn	β Col		05 50.8	−35 46	3.12	86
	α Cir		14 42.2	−64 58	3.19	53
	γ Hyi		03 47.3	−74 15	3.24	214
	α Dor		04 33.9	−55 03	3.27	176
	α Pic		06 48.3	−61 56	3.27	99
	β Phe		01 05.9	−46 44	3.31	198
	γ Ara		17 25.1	−56 22	3.34	1136
	α Ret		04 14.4	−62 29	3.35	163
	γ Phe		01 28.3	−43 20	3.40v	234
	β Pav		20 44.6	−66 13	3.42	137
	α Tel		18 26.7	−45 58	3.51	249

Solar eclipses[a] (2005–2025)

Date	Type[b]	Mag	Duration (min : sec)	Location of mid-eclipse (Lat., Long.)
2005, 8 Apr	AT	1.0074	0 : 42	10° 34′ S, 118° 58′ W
2005, 3 Oct	A	0.9576	4 : 31	12° 52′ N, 28° 45′ E
2006, 29 Mar	T	1.0515	4 : 06	23° 08′ N, 16° 46′ E
2006, 22 Sep	A	0.9352	7 : 09	20° 40′ S, 9° 03′ W
2007, 19 Mar	P	0.8742		61° 03′ N, 55° 25′ E
2007, 11 Sep	P	0.7489		61° 00′ S, 90° 17′ W
2008, 7 Feb	A	0.9650	2 : 12	67° 34′ S, 150° 27′ W
2008, 1 Aug	T	1.0394	2 : 27	65° 38′ N, 72° 16′ E
2009, 26 Jan	A	0.9283	7 : 53	34° 05′ S, 70° 16′ E
2009, 22 Jul	T	1.0799	6 : 38	24° 12′ N, 144° 08′ E
2010, 15 Jan	A	0.9190	11 : 08	01° 37′ N, 69° 20′ E
2010, 11 Jul	T	1.0580	5 : 20	19° 47′ S, 121° 51′ W
2011, 4 Jan	P	1.0626		64° 39′ N, 20° 49′ E
2011, 1 Jun	P	0.6016		67° 47′ N, 46° 49′ E
2011, 1 Jul	P	0.0963		65° 10′ S, 28° 39′ E
2011, 25 Nov	P	0.9044		68° 34′ S, 82° 24′ W
2012, 20 May	A	0.9439	5 : 46	49° 05′ N, 176° 19′ E
2012, 13 Nov	T	1.0500	4 : 02	39° 58′ S, 161° 18′ W
2013, 10 May	A	0.9544	6 : 04	02° 12′ N, 175° 31′ E
2013, 3 Nov	AT	1.0159	1 : 40	03° 29′ N, 11° 40′ W
2014, 29 Apr	A*	0.9855		70° 42′ S, 131° 10′ E
2014, 23 Oct	P	0.8112		71° 10′ N, 97° 05′ W
2015, 20 Mar	T	1.0446	2 : 47	64° 25′ N, 6° 34′ W
2015, 13 Sep	P	0.7868		72° 07′ S, 2° 18′ W
2016, 9 Mar	T	1.0450	4 : 10	10° 07′ N, 148° 50′ E
2016, 1 Sep	A	0.9736	3 : 06	10° 41′ S, 37° 48′ E
2017, 26 Feb	A	0.9922	0 : 44	34° 41′ S, 31° 08′ W
2017, 21 Aug	T	1.0306	2 : 40	36° 58′ N, 87° 38′ W
2018, 15 Feb	P	0.5982		71° 02′ S, 0° 45′ W
2018, 13 Jul	P	0.3364		67° 55′ S, 127° 28′ E
2018, 11 Aug	P	0.7363		70° 23′ N, 174° 33′ E
2019, 6 Jan	P	0.7150		67° 26′ N, 153° 37′ E
2019, 2 Jul	T	1.0459	4 : 33	17° 24′ S, 108° 57′ W
2019, 26 Dec	A	0.9701	3 : 39	1° 00′ N, 102° 18′ E
2020, 21 Jun	A	0.9940	0 : 38	30° 31′ N, 79° 43′ E
2020, 14 Dec	T	1.0254	2 : 10	40° 21′ S, 67° 54′ W
2021, 10 Jun	A	0.9435	3 : 51	80° 49′ N, 66° 37′ W
2021, 4 Dec	T	1.0367	1 : 54	76° 46′ S, 46° 18′ W
2022, 30 Apr	P	0.6385		62° 07′ S, 71° 22′ W
2022, 25 Oct	P	0.8613		61° 39′ N, 77° 29′ E
2023, 20 Apr	AT	1.0132	1 : 16	9° 36′ S, 125° 51′ E
2023, 14 Oct	A	0.9520	5 : 17	11° 21′ N, 83° 03′ W
2024, 8 Apr	T	1.0566	4 : 28	25° 17′ N, 104° 06′ W
2024, 2 Oct	A	0.9326	7 : 25	21° 58′ S, 114° 27′ W
2025, 29 Mar	P	0.9365		61° 06′ N, 77° 06′ W
2025, 21 Sep	P	0.8531		60° 54′ S, 153° 29′ E

[a] This data was taken from *Fifty Year Canon of Solar Eclipses: 1986–2035*, NASA Reference Publication 1178 Revised, published in July of 1987, compiled by Fred Espenak of the NASA Goddard Space Flight Center.

[b] Symbols: P, partial eclipse; A, annular eclipse; T, total eclipse; AT, annular–total eclipse; *, no central location – the umbra touches the Earth but the shadow axis passes above or below the Earth.

Lunar eclipses[a] (2005–2025)

Date	Type	Date	Type
2005, 17 Oct	P	2015, 4 Apr	T
2006, 7 Sep	P	2015, 28 Sep	T
2007, 3 Mar	T	2017, 7 Aug	P
2007, 28 Aug	T	2018, 31 Jan	T
2008, 21 Feb	T	2018, 27 Jul	T
2008, 16 Aug	P	2019, 21 Jan	T
2009, 31 Dec	P	2019, 16 Jul	P
2010, 26 Jun	P	2021, 26 May	T
2010, 21 Dec	T	2021, 19 Nov	P
2011, 15 Jan	T	2022, 16 May	T
2011, 10 Dec	T	2022, 8 Nov	T
2012, 4 Jun	P	2023, 28 Oct	P
2013, 25 Apr	P	2024, 18 Sep	P
2014, 15 Apr	T	2025, 14 Mar	T
2014, 8 Oct	T	2025, 7 Sep	T

[a] This list includes only partial and total lunar eclipses. The penumbral eclipses are not listed. Some years, like 2020, contain only penumbral type lunar eclipses. Data obtained from computer software calculations.

Glossary

Italicized words marked with an asterisk, such as **ecliptic line*, are terms that are cross-referenced within this glossary.

A_{CE} The symbol for the **transit altitude* of the **celestial equator*.
A_{LC} The symbol for the **lower culmination* of an object.
A_{UC} The symbol for the **altitude* of the **upper culmination* of an object.
f The symbol for **energy flux*.
f **or** ν The symbol for **frequency*.
$\vec{L}$ The symbol for the **angular momentum* **vector*.
m The symbol for the **apparent magnitude* of an object.
$\vec{p}$ The symbol for **linear momentum*.
Q An alternate symbol for the **transit altitude* of the **celestial equator*.
α The symbol for **right ascension*.
β The symbol for **celestial latitude*.
δ The symbol for **declination*.
λ (1) The symbol for **celestial longitude*.
λ (2) The symbol for the observer's **latitude*.
τ The symbol for an object's **hour angle*.
$\vec{\tau}$ The symbol of the **torque* **vector*.

12-hour circle A circle drawn on the **celestial sphere* from the **NCP* to the eastern **horizon* to the **SCP* to the western horizon and back to the NCP. Any object on the eastern half of the circle requires 12 hours to move to the western half, regardless of its **declination*.

absolute zero The **temperature* where there is no heat **energy*. All motion, down to the smallest scale, stops.

acceleration The rate at which velocity changes with time. The acceleration due to gravity near the surface of a **planet* or **moon* is called it's surface gravity.

active region On the *Sun*, a region of the surface with higher than average **temperature* associated with strong magnetic fields on the surface. Sunspots and other **solar* phenomena are associated with active regions.

albedo Of the light that falls on a surface, albedo is a measure of the portion of light reflected by the surface. Generally expressed as a number between 0 (perfectly absorbing) and 1 (perfectly reflecting) or as a percentage.

Almagest A 13-volume astronomical treatise by the Greek astronomer Claudius Ptolemy published about AD 140. It ranks among the most important works on **astronomy* ever written. It sets out the rules used for calculating the future position of the **planets*, based on an **Earth*-centered (geocentric) **Universe*. *Almagest* is its Arabic name. It's original Greek name is the **Syntaxis*.

almucantar A circle of equal **altitude* in the **horizon coordinate system*.

altitude The angular distance between a celestial object and the observer's **horizon*, measured vertically along the **great circle* passing through the **zenith*. One of the two **coordinates* of the **horizon coordinate system*.

analemma The figure-of-eight shape created by plotting the **Sun*'s position in the sky at the same time of **day* during the **year*.

ancient constellation One of the 48 **constellations* defined by Claudius Ptolemy in his book, **Almagest*.

angular diameter The apparent diameter of an object as measured in units of **radians*, **degrees*, **minutes of arc*, or **seconds of arc*. It is determined by the combination of true diameter and distance.

angular momentum A physical quantity, similar to **linear momentum*, that an object possesses while rotating or while in orbital motion. An object with angular momentum continues to **rotate* at the same rate, with the rotation axis keeping the same spatial orientation, unless acted on by a **torque*. While in orbital motion, its **orbital path* is restricted to a plane.

angular velocity The rate at which an object **rotates*. Generally measured in revolutions per minute or **radians* per second.

annual (Latin) An adjective meaning yearly.

annular solar eclipse A **solar eclipse* where a ring (**annulus*) of the **Sun* is still visible around the **Moon*. Caused by the Moon being too distant from **Earth* to cover the Sun completely.

annulus (Latin) A ring or circle.

anomalistic month The amount of time for the **Moon* to **orbit* the **Earth*, starting and ending at **perigee*, equal to 27.554 **days*.

anomalistic year The amount of time for the **Earth* to **orbit* the **Sun* from **perihelion* to perihelion. Equal to 365.259 **days*. Since the perihelion precesses, it is slightly greater than the **tropical year*.

Antarctic Circle A line drawn on the **Earth*, parallel to the **equator*, at 66°.5 S **latitude*. **South* of this latitude, the **Sun* is above the **horizon* 24 hours on the **December solstice*.

ante meridiem (a.m.) The period of the **solar day* when the **Sun* is in the eastern half of the sky, between the **antimeridian* and the **local meridian*.

antimeridian A line on the *celestial sphere* drawn from the *south* point of the *horizon,* through the *nadir* to the *north* point on the horizon. The line opposite to the *local meridian.*

aphelion The point farthest from the *Sun* in the *orbit* of any object orbiting the Sun.

apogee The point farthest from the *Earth* in the *orbit* of the *Moon* or a satellite.

***Apollo* spacecraft** The third manned space program of the USA. Carried three men to the *Moon,* allowing two to land on the Moon in the *lunar* module while the third remained in *orbit* in the command module.

apparent magnitude (symbol *m*) A number describing the relative *brightness* of a *star* (or other celestial object) as seen from *Earth.* Apparent *magnitude* depends on the *luminosity* and distance of the object. Lower values correspond to greater brightness. It usually applies only to the visible portion of the spectrum and is thus called visual apparent magnitude.

apparent solar time The measure of time based on the true motion of the *Sun,* which is not uniform because the *Earth*'s *orbit* is slightly elliptical. The difference between apparent *solar time* and mean solar time is called the *equation of time.*

apparition The period of time during which an astronomical object is visible. Usually applied to an object that appears only once or only rarely, as opposed to those that are seen every day such as the *stars.*

arc That portion of a circle subtending an angle where the circle is centered on the *vertex* of the angle.

arc minute A unit of measure for small angles. Equal to one-sixtieth of a *degree.*

arc second A unit of measure for very small angles. Equal to one-sixtieth of an *arc minute* or 1/3600 of a *degree.*

Arctic Circle A line drawn on the *Earth,* parallel to the *equator,* at 66°.5 N *latitude.* *North* of this latitude, the *Sun* is above the *horizon* for 24 hours on the *June solstice.*

asterism An easily recognized pattern of *stars* that is not one of the official 88 *IAU* *constellations.* For example, the Big Dipper, the *Summer Triangle,* or the *Winter Triangle.*

astrology The tradition that claims to connect human traits and future events with the positions of the *Sun,* *Moon,* and *planets* in relation to the *vernal equinox.*

astrometry The measurement of the position and motions of celestial objects.

Astronomer Royal Formerly the title of the director of the Royal Greenwich Observatory in the United Kingdom (England), but since 1972, an honorary title bestowed on a distinguished astronomer who is not necessarily the director of the Royal Observatory.

astronomical horizon The circle on the *celestial sphere* where all points on the circle are 90° away from the *zenith.*

astronomical twilight The *period* before sunrise or after sunset where sunlight is scattered by the upper layers of *atmosphere.* Astronomical *twilight* begins or ends when the center of the Sun is 108° from the *zenith* or 18° below the *horizon.* At this *altitude* the scattered sunlight is on the same level as the average level of starlight.

astronomical unit (symbol AU) The radius of an unperturbed circular *orbit* around the **Sun* with a **period* of one Gaussian **year* (365.257 **days*). Equal to 149 597 870 km or 92 955 807 miles. Used for measuring distances within **planetary* systems. There are 63 240 AU in one light year.

astronomy The study of the **Universe* outside of the **Earth*'s **atmosphere*. Until recently, a branch of **astrology* concerned with measuring the position and motion of celestial objects.

atmosphere The layer or layers of gases surrounding a **star*, **planet*, or **moon* with sufficient gravity to retain the gases from expanding into space.

atmospheric refraction A small change in the path of light from astronomical objects as the light passes through the **atmosphere*. The amount of change increases as the object comes closer to the **horizon* causing the light to pass through more atmosphere.

atom The smallest particle of a chemical **element*, typically about 0.1 nanometers in diameter.

AU Abbreviation for **astronomical unit*.

aurora A display of rapidly changing, luminous patterns of various colors, observed in the night sky, usually near the polar **latitudes*.

Aurora Australis The southern lights. See **aurora*.

Aurora Borealis The northern lights. See **aurora*.

autumnal equinox In the northern hemisphere, when the **Sun* crosses the **celestial equator*, moving from positive to negative **declination* (September). In the southern hemisphere, when the Sun crosses the celestial equator, moving from negative to positive declination (March). In either case, generally considered to be the beginning of the season of Autumn or Fall.

azimuth The position of an object measured as an angle, eastward along the **horizon*, usually starting from due **north*. There are situations where due **south* is used as the starting point. One of the two **coordinates* of the **horizon coordinate system*.

base The number that a **power* is multiplying. In the expression, 10^4, 10 is the base.

Bayer designation The combination of a Greek letter with the Latin genative of the **constellation* name, used to designate a bright **star* within the constellation. The Greek letters are assigned to the **stars* within each constellation in order of their **apparent magnitude*. The bright star Betelgeuse is also known as α Orionis.

binary star A pair of **stars* held in **orbit* around each other by their mutual gravitational attraction. There are different types of binary star systems: apparent binaries, visual binaries, **eclipsing binaries*, and spectroscopic binaries.

brightness The intensity of the **radiation* emitted by an object. Apparent brightness is the intensity of the radiation received from an object.

Calendar of Sothis The Egyptian calendar based on the **heliacal rising* of **Sothis*. The first calendar to recognize 365.25 **days* in a **year*.

calendar year Equivalent to a **tropical year*, but containing an integer number of **days*.

cardinal point The four compass directions: **north*, **east*, **south*, and **west*.

Cartesian coordinate system A **coordinate system* with the **reference lines* constructed of straight lines at **right angles* to each other. Attributed to French mathametician, René Descartes. Also known as a rectangular coordinate system.

celestial equator The **great circle* on the **celestial sphere* 90° away from the **celestial poles*. It forms the boundary between the northern and southern celestial hemispheres.

celestial latitude (symbol β) The angular distance measured on the **celestial sphere*, **north* and **south* of the **ecliptic line*, along a **great circle* passing through the **ecliptic poles* and an object. Also known as ecliptic **latitude*.

celestial longitude (symbol λ) The angular distance on the **celestial sphere* measured eastward along the **ecliptic line* from the **vernal equinox* to a **great circle* passing through the **ecliptic poles* and the object. Also known as ecliptic **longitude*.

celestial pole The two points on the **celestial sphere* directly above the **Earth*'s **poles* or the points on the celestial sphere pierced by the Earth's axis of rotation.

celestial sphere The ancient idea of the **Earth* centered inside a giant sphere on which the **stars* and **planets* are positioned and are moving. In **modern astronomy*, a model of the sky used for visualizing the motion of celestial objects as seen from Earth and for positioning celestial objects with **coordinate systems*.

centi (symbol c) A metric prefix meaning one-hundredth or 10^{-2}. There are 100 centimeters (cm) in a meter (m).

centigrade (symbol °C) The standard, international unit of measure for **temperature*. Water freezes at 0 °C and boils at 100 °C.

cepheid variable The class of **variable star* that varies in **apparent magnitude* by about one magnitude in about 5 to 50 **days*. Named for their prototype, δ Cephei. The **period* of their variation depends on their **luminosity*, thus they can be used for measuring distance.

chromosphere The layer of the **Sun*'s **atmosphere* just above the **photosphere*. It has a slightly pink glow, surrounding the **Moon* during a **total solar eclipse*.

chronometer A device used to measure time, but the more general use of the term is in reference to a high precision clock.

circumpolar constellation **Constellations* that never go below the **horizon* at the observer's **latitude*.

circumpolar star A **star* that never goes below the **horizon* at the observer's **latitude*.

civil twilight The **period* before sunrise or after sunset where sunlight is scattered by the upper layers of **atmosphere*. Civil **twilight* begins or ends when the center of the Sun is 96° from the **zenith* or 6° below the **horizon*. At this **altitude* the level of scattered sunlight allows the brightest **stars* to be seen as well as a clearly defined horizon.

conservation of angular momentum The physical principle stating that neither the **magnitude* nor the spatial direction of **angular momentum* contained in a system of objects can change with events or time.

conservation of momentum The physical principle stating that neither the **magnitude* nor the spatial direction of **linear momentum* contained in a system of objects can change with events or time.

constellation One of the 88 regions of the sky whose borders were designated by the **IAU*, or the easily recognized pattern of **stars* within such a region. The boundary lines lie along lines of **right ascension* and **declination*.

coordinate One of the numbers or values within a **coordinate system*.

coordinate system A system or method of specifying the position of an object in terms of linear or angular distance from a specified starting point or **reference line* based on a starting point.

Coordinated Universal Time (UTC) A time scale available from broadcast signals. UTC differs from international atomic time by an integer number of seconds. It is maintained within +0.90 seconds of UT1 by the introduction of **leap seconds* on 30 June or 31 December.

corona The outermost layer of the **Sun*, extending for millions of kilometers above the **photosphere*. It becomes visible as a white halo surrounding the **Moon* during a **total solar eclipse*.

coronagraph An instrument used to study the **Sun*'s **corona* or upper **atmosphere* by creating an artificial **total solar eclipse*.

cosmology The study of the possible **origin*, structure, and evolution of the **Universe* as a whole.

crescent A **phase* of the **Moon* where the face is less than half lit. Venus and Mercury also have crescent **phases*.

culmination The point where an object crosses or **transits* the (local or anti) **meridian*. Circumpolar objects transit the **local meridian* twice each **day*.

cycle One repetition of a **periodic motion*.

cycle of Sothis A 1456-**year* period marked by the rising of **Sothis* with the **Sun*.

day One rotation of the **Earth* on its axis. One day is 24 hours, 1440 minutes, or 86 400 seconds. A **solar day* is measured with reference to the **Sun*. A **sidereal day* is measured with reference to the **stars*.

Dec. An abbreviation for **declination*. As is also, the Greek letter δ (delta).

deca The metric prefix equal to 10^1.

December solstice The **day* or date in the **month* of December when the **Sun* reverse its **annual* motion in **declination* from **south* to **north*.

deci The metric prefix equal to 10^{-1}.

decimal form A manner of writing numbers in which there is no power of ten as a multiplier. The number consists of only a **mantissa*.

declination (abbr. Dec., symbol δ) One of the **coordinates* of the equatorial **coordinate system* used to specify a location on the **celestial sphere*. Declination is equivalent to **latitude* in the **terrestrial coordinate system*. It is measured in **degrees* from the **celestial equator* along a **great circle* containing the **celestial poles* and the object. Northerly declinations are positive and southerly ones are negative.

degree (1) (symbol $^\circ$) A unit of angular measurement. There are 360° in a full circle.

degree (2) A unit of **temperature* measurement. See **Fahrenheit*, **centigrade*, and **kelvin*.

diurnal (Latin) An adjective meaning daily.

diurnal motion The apparent daily motion of any sky object, caused by the rotational motion of the **Earth*.

double star See *binary star.*

Earth The third **planet* from the **Sun*.

east The point where the **celestial equator* crosses the **ante meridiem *horizon*.

eccentricity (symbol e) A number from zero to one describing the oblateness of an **ellipse*. An ellipse with zero eccentricity is a circle. When the value is one the curve becomes a parabola and a value greater than one decribes a hyperbola. All of these curves are used to describe the possible **orbit* of one body around another.

eclipse A phenomenon in which one celestial body moves through the shadow of another.

eclipse season One of two (possibly three) times each **calendar year* when a **solar* or **lunar eclipse* is likely. They occur when the **Moon's *line of nodes* is nearly lined-up with the **Sun*.

eclipse wind A cool wind that blows through the regions in the path of the **Moon's* shadow during **totality* of a **total solar eclipse*.

eclipsing binary A **binary star* system whose total **brightness* varies with a regular, periodic **cycle* because the orbital motion of the **stars* causes them to pass in front of one another, one eclipsing the other.

ecliptic limit The maximum distance (measured as an angle) between a **lunar node* and the **Sun*, or the **Earth*'s shadow, in which a **solar* or **lunar eclipse* may occur.

ecliptic line The apparent **annual* path of the **Sun* on the **celestial sphere*, caused by the orbital motion of the **Earth*. It passes through the **zodiac *constellations*. It is also the result of the intersection of the **ecliptic plane* with the celestial sphere.

ecliptic plane The plane of the **Earth*'s **orbit* around the **Sun*. It is perturbed by the gravitational interactions between the Earth and the other **planets*, so the plane is seen as the average plane.

ecliptic pole The point directly above or below the **ecliptic plane*. It is found at the center of the **orbit* of the **Earth*'s **pole*, created by the Earth's **precessional* motion.

electromagnetic Of or pertaining to **electromagnetic energy*.

electromagnetic energy A form of **energy* transported by **photons* or by waves passing through linked electric and magnetic fields. Also known as **electromagnetic radiation*.

electromagnetic radiation A form of **energy* that passes through a vacuum at the **speed* of 3×10^8 m/s, or 186 000 miles/s (known as the "speed of light," c). In classical terms it consists of linked, rapidly varying electric and magnetic fields. In modern physics, it is carried by particles called **photons*. There are many forms of **electromagnetic *radiation* including: radio waves, microwaves, infrared, visible light, ultraviolet, X-rays, and gamma rays.

electromagnetic spectrum A list of the forms of **electromagnetic radiation*. Each form consists of a range of wave **frequencies* or **wavelengths*.

electromagnetic wave In classical physics, the mechanism by which **electromagnetic energy* is transported from one region of space to another. The waves travel as linked variations in electric and magnetic fields, at the **speed of light*. Light is an electromagnetic wave.

electron An elementary particle of small *mass* and negative electrical charge.

element A substance which cannot be reduced or simplified by chemical reactions. Each of the elements has its own uniquely structured *atom*. Elements are identified by the number of *protons* in the *nucleus* of their *atoms*.

ellipse A closed curve, with two foci, symmetrical about two perpendicular axes, with one (the major) axis longer than the other (the minor) axis. The ratio of length of the axes is related to the *eccentricity*. The sum of the distances between any point on the curve and the foci is a constant for the ellipse.

energy Typically described as the ability to do work or to create change.

energy flux (symbol f) A measure of the amount of *energy* (usually *electromagnetic*) passing through a unit of surface area (usually one square meter) per second.

engineering notation A form of *power of ten* notation where the *exponent* on the ten is restricted to multiples of three.

ephemeride See *ephemeris*.

ephemeris A table giving the celestial *coordinates* and other data for astronomical bodies.

epoch The moment in time at which the given celestial *coordinates* (generally on a *star* chart or in a star catalog) are strictly correct. Or a period of time during the history of the *Universe* or any celestial object.

equation of time The *day*-to-day variation in the time of the *Sun* crossing the *local meridian*. Created by the difference between the mean *solar time* (shown by a mechanical clock) and *apparent solar time* (shown by a sundial).

equator The line (a *great circle*) drawn 90° from either *pole* of a rotating body. The line represents the intersection of a plane with the body's surface, where the plane passes through the center of the body and is perpendicular to the axis of rotation.

equatorial chart A *star* or *constellation* chart showing only the region around the *celestial equator*, excluding the *celestial sphere*'s polar regions.

equatorial coordinate system A system of celestial *coordinates* based on the *celestial equator* and the *equinoctial colure*. Its two coordinates are called *declination* and *right ascension*. The *origin* is the *vernal equinox*.

equinoctial colure The *great circle* on the *celestial sphere* passing through the *celestial poles* and the *equinoxes*.

equinox Either of two points where the *ecliptic line* crosses the *celestial equator* or the time of *year* when the *Sun* passes through those points. The Sun passes from *south* to *north* at the *March equinox* (*vernal equinox*), and north to south at the *September equinox*. On the equinox *days*, all *latitudes* (except the *poles*) receive equal length daylight and night.

ET *Ephemeris* time. An older time scale used for the equations of motion of bodies in the *Solar System*. Based on the motion of the *Sun*.

evening star The *planet* Venus as seen in the evening (western) sky. The name comes from the fact that it is the first "star" visible as the *Sun* sets.

exponent The number in a *power* which indicates the number of times the *base* is multiplied by itself. In the expression 10^4, the 4 is the exponent.

Fahrenheit The old English (now American) unit of **temperature* measurement. No longer used in science. Water freezes at 32 °F and boils at 212 °F.

fall equinox In the northern hemisphere, the **September equinox*. In the southern hemisphere, the **March equinox*. The point where the **Sun* crosses the **celestial equator*, moving from positive **declination* to negative.

far side The hemisphere of the **Moon* that is permanently turned away from the **Earth*.

first contact The event during a total or **annular solar eclipse* when the disk of the **new Moon* first touches the disk of the **Sun*. The event during a **total lunar eclipse* when the Moon first touches the **Earth*'s **umbra*. In any eclipse, the beginning of the eclipse event.

first point of Aries An alternate name for the **vernal equinox*.

first quarter The **phase* of the **Moon* when half the visible disk (the western half) is lit and the Moon is **waxing* toward full.

Flamsteed number Numbers assigned to **stars* by order of the stars' **right ascension*. Used in combination with the **constellation*'s Latin genitive to form a designation for the star. For example, Betelgeuse (*α Orionis) is also 58 Orionis. First used (implicitly) in the star catalog *Historia coelestis Britannica* created by the first **Astronomer Royal*, John Flamsteed, and published posthumously in 1725.

force A push or a pull on an object. Any physical phenomenon which may cause an object to **accelerate*.

fourth contact The event during a **total solar eclipse* or an **annular solar eclipse* when the **Moon* is no longer in front of the **Sun*. The event during a **lunar eclipse* when the Moon is no longer in contact with the **Earth*'s **umbra*. In any eclipse, the end of the eclipse.

frequency (symbol f, sometimes ν) The number of times a repetitive event occurs in a unit of time, usually one second.

full Moon The **phase* of the **Moon* when it is located on the opposite side of the sky from the **Sun*. The entire surface facing the **Earth* is sunlit.

geocentric parallax A **parallax* effect caused by the observer being on the surface of the **Earth* rather than at its center and, thus, not at the center of the **celestial sphere*.

gibbous The **phases* of the **Moon* (**waxing* gibbous and **waning* gibbous) when it is more than half lit, but not full.

GMT Greenwich Mean Time. The **mean solar day* based on the **local meridian* of Greenwich, England.

gravitational constant (symbol $\mathcal{G}$) The constant of Newton's law of gravity, describing the relative strength of gravity between two **masses*.

great circle A circle on the surface of a sphere where the plane passing through the sphere to create the circle also passes through the center of the sphere.

Gregorian calendar The calendar used in most of the world. Named for Pope Gregory XIII who oversaw its production and introduction in 1582, replacing the **Julian calendar*.

Harvest Moon The **full Moon* nearest the time of the **equinox* during harvest season. The northern hemisphere's Harvest Moon is nearest the **September equinox* and the southern hemisphere's is nearest the **March equinox*.

Harvest Moon effect The time of **year* when the **day*-to-day change in rise time of the **Moon* is shortest, because of the low inclination of the Moon's **orbital path* with the **horizon*. It occurs at approximately the time of the **Harvest Moon* for the particular hemisphere.

hecto The metric prefix equal to 10^2.

heliacal rising When an object rises with the **Sun* on the eastern **horizon*.

high noon That (local) time of **day* when the **Sun* is on the **local meridian*, independent of the **time zone*.

horizon The circle on the **celestial sphere* found 90° away from the observer's **zenith*, marking the line between the visible and non-visible halves of the sphere. Used as one of the **reference lines* for the **horizon coordinate system*.

horizon coordinate system A **coordinate system* for specifying the location of celestial objects, based on the observer's **horizon*. Its two **coordinates* are called **altitude* and **azimuth*.

hour angle (symbol $*\tau$) An angle measured along the **celestial sphere* from the intersection of the **local meridian* with the **celestial equator* to the intersection of an object's **hour circle* with the celestial equator. Equal to the amount of **sidereal time* passed since the object transited the local meridian.

hour circle A circle on the **celestial sphere* passing from the north to the south **celestial pole* and containing the object of interest. This circle represents an object's **right ascension*. On the sphere, this circle represents locations of same right ascension.

IAU The abbreviation for the International Astronomical Association.

inertia The tendancy of an object or body to oppose changes in its motion. Inertia is considered proportional to **mass*.

infinity The largest of any possible value.

intercalary month See **leap months*.

International Date Line The line located on the opposite side of the **Earth* from the **prime meridian*. When it is **high noon* on the prime meridian, it is midnight on the International Date Line and the **day*/date advances. Used for making proper corrections for date and time while traveling **east* or **west* around the world.

intrinsic brightness See **luminosity*.

intrinsic variable A **star* that varies in **energy* output or **luminosity* because of periodic, physical changes within the body of the star.

JD Julian Date or **day*. The **Julian day* is a measure of time since noon of 1 January, 4713 BC at Greenwich, England.

joule (symbol J) The standard international unit of measurement of **energy*.

Julian calendar A calendar introduced by Julius Caesar to the Roman Empire starting from 46 BC, based on the **calendar of Sothis*. It was replaced by the **Gregorian calendar* in 1582.

Julian Day A number equal to the number of **days* since the date of 1 January 4713 BC.

June solstice The point along the **ecliptic line* where the **Sun*'s **declination* stops increasing and begins to decrease.

kelvin (symbol K) The standard international metric unit of **temperature*, referenced at **absolute zero*. Water freezes at 273 K and boils at 373 K.

Kepler's laws See **laws of planetary motion*.
kilo (symbol k) A metric prefix meaning one thousand, 10^3.
last quarter See **third quarter*.
latitude One of the two **coordinates* of the **terrestrial coordinate system*. It is the angular distance of a location as measured directly **north* or **south* of the **Earth*'s **equator*.
laws of planetary motion Also known as **Kepler's laws*. Three mathematical laws describing a **planet*'s **orbit* shape, size, and the planet's motion along that orbit.
leap day A **day* added to the calendar every four **years* (except in most century years) to keep the calendar in sync with the motion of the **Earth* around the **Sun* and therefore the seasons.
leap month A **month* added to a **lunar* based calendar to keep the calendar in closer step with the motion of the **Earth* around the **Sun*.
leap second Used to keep **UTC* time to within ±0.90 seconds of UT1 (or universal time). A leap second is introduced, as needed, on either 31 December or 30 June of a given **year*.
LHA Local **Hour Angle*.
libration Any of the effects which cause a slight variation of the hemisphere of the **Moon's* surface visible from the **Earth*. In combination, these variations allow us to see approximately 59% of the Moon's surface from Earth.
light curve A graph of the light (**apparent magnitude*) from a **variable star* plotted against time.
light minute Equal to the distance traveled by light in one minute. (300 000 km/s × 60 s/min). About 18 million km or 11 million miles.
light second The distance light travels in one second. Equal to 300 000 km or 186 000 miles.
light year (symbol ly) The distance light travels in one **year*. Approximately 9.46×10^{12} km or 5.9×10^{12} miles. Calculated from: 300 000 km/s × 60 s/min × 60 min/h × 24 h/day × 365.25**days*/year.
limb The edge of the visible disk of an object such as the **Sun* or **Moon*.
limiting magnitude The **apparent magnitude* of the faintest object visible under the current observation conditions or with a particular instrument.
line of nodes A line drawn between the **Moon's* orbital **nodes*. It passes through the center of the **Earth*. It is created by the intersection of the Moon's **orbital plane* with the **ecliptic plane*.
linear momentum (symbol p) The **momentum* associated with straight line motion. $p = mv$, where m is the **mass* of the object and v is its velocity.
local meridian The sky line drawn from the observer's due **north* **horizon*, through the **zenith*, to due **south* on the horizon. The visible half of the **great circle* which, generally, is simply called the **meridian*. Also see **antimeridian*.
longitude One of the two **coordinates* of the **terrestrial coordinate system*. The angular distance along the **equator* (or a **parallel*) from the **prime meridian* to a location on the **Earth*'s surface.
lower culmination The **altitude* of an object when it crosses the **local meridian* below the **pole*.

lower meridian That half of the **meridian* containg the **nadir*.

LST Local **Sidereal Time*.

luminosity (symbol *L*) The actual amount of **energy* radiated (per unit of time, usually one second) by a luminous body such as a **star*.

lunar Of or pertaining to the **Moon*. From "Luna," the Latin name of the Moon.

lunar eclipse An event where the **Moon* passes through the **Earth*'s shadow, causing it to "disappear." There are three types: total, partial, and prenumbral.

lunar ecliptic limit The distance between a **lunar node* and the farthest point along the ecliptic at which a **lunar eclipse* is possible.

lunar node A point where the **Moon's* **orbital path* crosses the **ecliptic line*.

lunar phase One of the **phases* of the **Moon*.

lunation One **cycle* of **lunar phases*, which takes the time equal to one synodic **period* of the **Moon's* **orbit*: 29.531 **days*.

M The prefix given to the 110 objects listed in the **Messier Catalog*.

magnitude The relative measurement scale representing the **brightness* of a **star* or other celestial object. This logarithmic compares the brightness of an object with the brightness of one or more objects used as a reference standard.

major lunar eclipse period A longest **period* of **days* during an **eclipse season*, during which a **lunar eclipse* is possible.

major solar eclipse period A longest **period* of **days* during an **eclipse season*, during which a **solar eclipse* is possible.

mantissa The number multiplying the power of ten in engineering or **scientific notation*.

March equinox The sky point where the ecliptic crosses the **celestial equator* or the date when the **Sun* crosses the celestial equator when moving from negative to positive **declination*. Also known as the **vernal equinox*.

mare (pl. maria) A Latin word meaning *sea*, representing the vast dark areas on the face of the **Moon*. It was once thought that these dark areas were seas of liquid water on the Moon's surface.

mass In elementary terms, a measure of the amount of material in an object. Or a measure of an object's **inertia*. Actually, probably the most difficult physical term (or quantity) to define or understand.

mean local solar time A clock keeping time with the **Sun* based on the **local meridian*, ignoring **time zones*.

mean solar day The length of a **solar day* averaged over a **year*.

mean sun An imaginary **sun* that travels along the **celestial equator* at a constant rate, representing the average position of the apparent sun during the **year*.

mega (symbol M) A metric prefix meaning one million, 10^6.

meridian The **great circle* on the **celestial sphere* containing the **north celestial pole*, the **south celestial pole*, and the observer's **zenith* and **nadir*.

Messier Catalog A catalog of 110 bright galaxies, star clusters, and nebulae. Created by French astronomer, Charles Messier (1730–1817). The objects in the catalog are numbered, with an M prefix.

micro (symbol μ) The metric prefix meaning one-millionth, 10^{-6}.

milli (symbol m) The metric prefix for one-thousandth, 10^{-3}.
minor solar eclipse period A shortest **period* of **days* during an **eclipse season*, during which a **solar eclipse* is possible.
minute of arc See **arc minute*.
Mira variable A member of a class of long **period *variable stars* of which the star Mira (Omicron Ceti) is the prototype.
MJD Modified Julian Day (date). Equal to the **Julian Day* minus 2 400 000.5.
modern astronomy **Astronomy* since 1850, beginning with the use of spectroscopy to analyze the light emitted or reflected by celestial objects.
modern constellation Those **constellations* created after the Greeks, mainly by sixteenth and seventeenth century celestial map makers and not appearing in Ptolemy's **Almagest*. Of the 88 **IAU* defined constellations, 40 are modern. They are mainly in the southern half of the **celestial sphere*.
modified Julian Day A version of the **Julian Day* where 2 400 000.5 has been subtracted from the Julian Day. This makes the Julian Day number an easier number to work with.
molecule A combination of two or more **atoms*.
momentum Loosely, a measure of the motion of a body. Described in two forms: **linear momentum* and **angular momentum*. Momentum is conserved when there is no net **force* or **torques* acting on the body.
month Approximately equal to the amount of time taken for the **Moon* to **orbit* the **Earth* with reference to the **Sun* (one synodic **period*). The word is derived from moon.
moon (1) Any natural satellite (generally, also of sufficient size) of a **planet*.
Moon (2) The **Earth*'s natural satellite. Its Latin name is Luna.
morning star The **planet* Venus seen in the morning (eastern) sky. The name comes from the fact that it is the last "star" visible as the **Sun* rises.
nadir The point on the **celestial sphere* opposite to the observer's **zenith*.
naked-eye astronomy Observing astronomical objects or phenomenon without the aid of a telescope or other instruments.
nano (symbol n) The metric prefix meaning one-billionth, 10^{-9}.
nautical twilight The period before sunrise or after sunset where sunlight is scattered by the upper layers of **atmosphere*. Nautical **twilight* begins or ends when the center of the Sun is 102° from the **zenith* or 12° below the **horizon*. At this **altitude* there is no longer enough scattered sunlight to make the horizon visible (at sea) and altitude measurements made with reference to the horizon are no longer possible.
NCP The **north celestial pole*.
near side The hemisphere of the **Moon* that is permanently turned toward the **Earth*.
neutron Electrically neutral subatomic particles found in the **nucleus* of **atoms*.
neutron star A dead **star* that has collapsed under gravity consisting almost entirely of **neutrons*, having a **mass* between 1.5 and 3 **solar* masses. They are formed by supernova explosions and are observed as pulsars.

new Moon The *Moon's *phase* when it is located between the **Earth* and the **Sun*. Since we are observing the unlit half of the Moon's surface and also trying to observe it while looking into the Sun, we cannot observe a new Moon except during.

node A point where the *Moon's *orbital path* crosses the **ecliptic line*. A point where the orbital path of a moon of any **planet* crosses the planet's **orbital plane*. A point where the orbital path of any object crosses a reference plane.

north The point on the **horizon* directly below or closest to the **north celestial pole*, or that point on the horizon, the *zenith *meridian* of which contains the north celestial pole.

north celestial pole (NCP) The point on the **celestial sphere* located directly above the **Earth*'s **North Pole*. Its **declination* is $+90°$.

north ecliptic pole The point of intersection between the northern half of the *celestial sphere* and the line perpendicular to the **ecliptic plane*.

North Pole The location in the northern hemisphere where the **Earth*'s rotational axis passes through its surface.

North Star A common name for the **star* Polaris, which is located very close to the **north celestial pole*.

nucleus The central **mass* of an **atom*, containing its **protons* and **neutrons*.

nutation A short **period* motion happening on top of the **precessional* motion of a rotating object or body. Generally caused by gravitational interactions with nearby bodies such as **moons*.

obliquity of the ecliptic The angle between the **ecliptic plane* and the plane of the **equator* of the **celestial sphere*. Also, the tilt of the **Earth*'s axis of rotation.

orbit The path (usually closed and elliptical) traveled by a body moving under the gravitational influence of a larger object such as the **Sun* or a **planet*.

orbital angular momentum The **angular momentum* associated with the orbital motion of an object (**planet* or **moon*).

orbital element A set of values determined by observation that define the shape, orientation, and timing of a body's orbital motion.

orbital path See **orbit*.

orbital period The time it takes to complete one revolution around the parent body in a closed **orbit*.

orbital plane A mathematical plane surface that holds the **orbital path* of a celestial object. Its existence is required by **conservation of angular momentum*.

origin The starting point of any **coordinate system*. The point where the **reference lines* intersect and all coordinate values are (usually) zero.

parallax (symbol π) A change in the observed position of an object when it is viewed from different locations. Generally measured as an angle, which is larger valued for nearby objects and perhaps unmeasurable for distant objects. Measuring parallax is a primary distance measuring tool in **modern astronomy*.

parallel A small circle on the surface of the **Earth*, parallel to the **equator*, marking locations of equal **latitude*.

partial lunar eclipse An **eclipse* of the **Moon* where only part of the Moon is immersed into the **Earth*'s **umbra*.

partial solar eclipse An *eclipse of the *Sun where only part of the Sun is covered by the *Moon.

penumbra The region of partial shadow outside the *umbra region. During a *solar eclipse, observers located in the penumbra region of the *Moon's shadow see a *partial solar eclipse.

penumbral lunar eclipse An *eclipse of the *Moon where the Moon passes through only the prenumbral portion of the *Earth's shadow – generally ignored by observers.

perigee The point of closest approach to the *Earth in the *orbital path of the *Moon or a satellite.

perihelion The point of closest approach to the *Sun in the *orbital path of the *Earth or any other object orbiting the Sun.

period An interval of time, or the time taken to repeat a cyclical event.

periodic motion A motion requiring a constant interval of time between repetitions, known as the *period of the motion.

phase The ratio of the observable lit area of the disk of a body with the entire disk. The apparent shape of an object due to the position of the observer relative to the lit half of the object.

photometer An electronic device used to measure the intensity of light received from an object or *star.

photometric magnitude The *magnitude of an object as measured by photographic or photometric techniques.

photon In modern physics, a particle that carries *electromagnetic energy or *electromagnetic radiation. Each photon carries a specific amount of energy, related to the *frequency of the photon as $E = hf$, where E is the energy, h is Planck's constant, and f is the frequency of the photon.

photosphere The "surface" of the *Sun. The layer of gas where the Sun's material changes from being opaque to being transparent. The layer of gas which creates the *photons which expose and create an image on our photographic plate.

pico (symbol p) A metric prefix equal to 10^{-12}.

planet A body considered to be of sufficient size, in *orbit about a *star.

planetary Of or pertaining to the *planets (except in the case of planetary nebula).

planetary precession The motion of the *ecliptic plane caused by the gravitational influence of the other *planets in the *solar system.

polar chart A *star or *constellation chart showing only a polar region of the *celestial sphere.

pole A point on the surface of a rotating sphere (or any other shaped object) where the axis of rotation passes through the body of the object. Any rotating object has two poles.

post meridiem (p.m.) The period of the *solar day when the *Sun is in the western half of the sky, between the *local meridian and the *antimeridian.

power (1) A form of writing numbers involving a *base and an *exponent. For example 10^4: the 10 is the base and the 4 is the exponent. The combination is a power.

power (2) The rate at which work is done or *energy is expended.

power of ten The number ten, raised to the **power* of an **exponent*. Examples: $10^2 = 100$, $10^5 = 100\,000$. It is a shorthand method of writing large multiples of the number ten.

precession The slow **periodic motion* (wobbling motion) of the rotation axis of the **Earth* or other rotating body. Caused by the presence of a **torque* acting on the body.

precession of the equinoxes The observation of the motion of the **equinoxes* along the ecliptic, caused by the **precessional* motion of the **Earth*.

precessional An adjective meaning to precess, or to have **precession*.

prime meridian The dividing line between **east* and **west* in the **terrestrial coordinate system*. Drawn from the **North Pole* to the **South Pole* through the Old Royal Observatory in Greenwich, England, by international agreement.

proper motion The motion of a **star* on the **celestial sphere*, caused by its motion through space, relative to our **Sun*. One of the reasons for our need to update the **right ascension* and **declination *coordinates* of celestial objects.

proton Positively electrically charged subatomic particles found in the **nucleus* of an atom.

quadrant An instrument for measuring the **altitude* of objects from the **horizon*. It is made using a graduated quarter circle and sighting arm with another (possibly weighted) arm which moves along the angle markings on the quarter circle.

RA The abbreviation for **right ascension*. Also abbreviated as the Greek letter α (alpha).

radian A unit of angular measure related to the radius of the circle. One radian is approximately $57^\circ.3$. (2π radians is 360°).

radiation The emission of **electromagnetic energy* by any object.

Rayleigh scattering The scattering of light by the particles of a gas or liquid. The maximum **wavelength* of the scattered light related to the size as the particles.

reference lines The lines passing through the **origin* of a **coordinate system* along which the coordinate values are measured.

revolve To **orbit* or circle around a central point or object.

right angle An angle of 90°.

right ascension (RA) One of the **coordinates* in the **equatorial coordinate system*. Measured as hours, minutes, and seconds, along the **celestial equator* from the **vernal equinox* to the hour line of the object. The right ascension of an object is the reading of a local **sidereal clock* when that object is on the observer's **local meridian*.

rotate The motion of a body as it spins around an axis.

rotational angular momentum The **angular momentum* associated with the rotation of an object.

RR Lyrae variable A class of pulsating or **variable stars*. Similar to (but much dimmer than) **cepheid* variables, they can be used to measure distance.

Saros The **period* after which the sequence of **lunar eclipses* and **solar eclipses* repeat. Equal to 6585.32 **days*. Discovered by ancient astronomers.

Saros cycle See **Saros*.

scalar A quantity having only **magnitude* and no associated direction (compare to **vector*).

scientific notation A method of writing very large or very small numbers using a **mantissa* with **power of ten* notation. The mantissa must be greater than or equal to one, but less than 10.

SCP See **south celestial pole.*

second contact The event during a **total solar eclipse* when the **Moon* has completely covered the **Sun*, or during an **annular solar eclipse* when the Moon is completely surrounded by the **annulus*. The event during a **total lunar eclipse* when the Moon has moved completely within the **Earth*'s **umbra*. In any eclipse, the beginning of **totality* or annularity.

second of arc See **arc second.*

seeing A judgement of conditions while observing through a telescope. "Good seeing" is the condition of being able to disern details in a telescopic image. "Poor seeing" occurs with greater atmospheric turbulence.

semi-major axis (symbol a) Half the length of the longest axis of an **ellipse.*

semi-minor axis (symbol b) Half the length of the shorter of the two axes of an **ellipse.*

September equinox The sky point where the ecliptic crosses the **celestial equator* or the date when the **Sun* crosses the celestial equator when moving from positive to negative **declination.* In the northern hemisphere, also known as the fall or **autumnal equinox*. In the southern hemisphere, also known as the **spring equinox*.

SI second The standard metric second, defined by a type of **radiation* emitted by cesium-133 **atoms.*

sidereal (Latin) Of or pertaining to the **stars.*

sidereal clock A clock that keeps local time according to the **stars* (**sidereal time*).

sidereal day The rotational **period* of the **Earth* with respect to the **stars*. The time it takes for the same star to return to the **local meridian* each day.

sidereal month The length of time required for the **Moon* to **orbit* the **Earth* referenced to the **stars*. The time required for the Moon to line-up with the same star for one revolution around the Earth.

sidereal period The time it takes for a **planet* or satellite to complete one **orbit*, measured with respect to the **stars.*

sidereal time Time kept with reference to the apparent **diurnal motion* of the **stars.*

sidereal year The amount of time required for the **Earth* to **orbit* the **Sun* relative to the **stars.*

sky brightness Caused by **aurora*, moonlight, and light pollution. It reduces our ability to see the dimmer **stars* or those extent objects such as galaxies or nebulae that we view through a telescope.

small circle A circle on the surface of a sphere created by the intersection of the sphere's surface with a plane, where the plane does not pass through the center of the sphere.

Sol The Latin proper name of our **star*, the **Sun.*

solar (Latin) Of or pertaining to the **Sun*, **Sol.*

solar day The time between two successive crossings of the **local meridian* by the **Sun.* The same as an apparent solar day.

solar eclipse An event where the **Sun* passes behind the **Moon*, causing the Sun to "disappear." There are three types: total, partial, and **annular*.

solar ecliptic limit The distance between a **lunar node* and the farthest point from that node at which a **solar eclipse* may occur.

Solar System The **Sun* and everything in **orbit* around it.

solar time Time kept with reference to the **Sun*, **Sol*.

solstice A point on the ecliptic where the **Sun* reaches its maximum and minimum **declination*. The times at which the Sun is at these points.

solstitial colure The **great circle* passing through the **celestial poles* and the **solstices*.

Sothis The Egyptian name of the **star* Sirius.

south The point on the **horizon* directly below or closest to the **south celestial pole*, or that point on the horizon, the **zenith *meridian* of which contains the south celestial pole.

south celestial pole (SCP) The **celestial pole* located directly above the **Earth*'s **South Pole*. Its **declination* is $-90°$

.

south ecliptic pole The point of intersection between the southern half of the **celestial sphere* and the line perpendicular to the **ecliptic plane*.

South Pole The location in the southern hemisphere where the **Earth*'s rotational axis passes through its surface.

speed The rate of change of the position of an object.

speed of light (symbol c) The **speed* at which **electromagnetic waves* travel. In empty space it is equal to 300 000 km/s or 186 000 m/s.

spring equinox In the northern hemisphere, the **March equinox*. In the southern hemisphere, the **September equinox*. Usually associated with the **vernal equinox*.

star An intrinsically luminous sphere of ionized gas (plasma). A living star is generating **energy* in its core through nuclear fusion. A dead star may still be giving off energy by virtue of stored heat, but it is no longer producing energy by nuclear fusion.

star clock An ancient Egyptian method of keeping time by the motion of the **stars*.

summer solstice In the northern hemisphere, the **June solstice*. In the southern hemisphere, the **December solstice*.

summer triangle An **asterism* made of the three **stars*: Altair in Aquila, Deneb in Cygnus, and Vega in Lyra. It is high in the northern hemisphere's summer skies and is used to find other stars and **constellations*. In the southern hemisphere it may be called the **Winter Triangle*.

Sun The **star* at the center of our **Solar System*.

synchronous orbit The motion of a body such that its rotational and **orbital periods* are the same. Such a body, like the **Moon*, keeps the same hemisphere toward its parent **planet* or **star* during its orbit. This is caused by tidal interactions acting over a long period of time.

synodic month (period) The time between two successive **new Moon *phases* (or any other phase). Equal to 29.531 **days*.

Syntaxis The original name of the **Almagest*.

TAI International Atomic Time. A time standard based on the statistical average of 200 **frequency* standards located around the World. Based on the **SI second*.

TCB Barycentric **Coordinate* Time. Time scale for a **coordinate system* with its **origin* at the center of **mass* of the **Solar System*.

TCG Geocentric **Coordinate* Time. Time scale for a **coordinate system* with its **origin* at the center of **mass* of the **Earth*.

TDB Barycentric Dynamical Time. Used as the time scale of ephemerides based on the center of **mass* of the **Solar System*.

TDT **Terrestrial* Dynamical Time. Used as the time scale of geocentric ephemerides.

temperature A measure of heat **energy*.

terra The highlands areas of the **Moon*. By naked eye, the bright regions on its face.

terrestrial Of or pertaining to the **Earth*.

terrestrial coordinate system The system of **latitude* and **longitude* used to specify locations on the surface of the **Earth*.

third contact The event during a **total solar eclipse* at which the **Moon* no longer covers the **Sun*, or during an **annular solar eclipse* when the Moon is no longer surrounded by the **annulus*. The event during a **total lunar eclipse* when the Moon is no longer completely immersed in the **Earth*'s **umbra*. In any eclipse, the end of **totality* or annularity.

third quarter The **phase* of the **waning* **Moon* when (the eastern) half the face is lit.

time zone One of the regions on the **Earth* (established by political agreements) where the **solar* clocks are all set to the same time, partly independent of the actual location of the **Sun* (or local **solar time*).

torque A twisting or turning **force* that causes a change in **angular momentum*.

total lunar eclipse An **eclipse* of the **Moon* where the entire Moon passes into the **umbra* of the **Earth*'s shadow.

total solar eclipse An **eclipse* of the **Sun* where the Sun is completely covered by the **Moon*.

totality The **period* in the course of a **solar eclipse* during which the **Sun* is totally obscured by the **Moon* or during a **lunar eclipse* when the Moon is totally immersed in the **Earth*'s **umbra*.

transit The passage of any celestial object through the **local meridian*.

transit altitude The **altitude* of an object as it crosses (**transits*) the **local meridian*.

transparency The optical clarity of the **atmosphere*. It is affected by clouds, air moisture content, and particulate matter created by industrial pollutants or natural causes such as volcanos.

Tropic of Cancer A small circle, parallel to the **Earth*'s **equator*, located at 23°.5 N. It is the northern most **latitude* where the **Sun* may pass through the **zenith* at least once each **year*.

Tropic of Capricorn A small circle, parallel to the **Earth*'s **equator*, located at 23°.5 S. It is the southern most **latitude* where the **Sun* may pass through the **zenith* at least once each **year*.

tropical year Equivalent to a **calendar year.*

TT **Terrestrial* Time. Considered equivalent to Terrestrial Dynamical Time (**TDT*).

twilight A condition just after sunset or just before sunrise where sunlight is scattered by the upper layers of the **atmosphere.*

twinkling An effect caused by turbulent air refracting starlight as it passes through.

umbra A region of complete darkness in a body's shadow.

universal gravitation The concept of applying Newton's law of gravity to every object in the **Universe.*

Universe All that exists.

upper culmination The **altitude* of an object when it crosses the **local meridian* above the **pole.*

upper meridian That half of the **meridian* containing the **zenith.*

UT Universal Time. A measure of time closely related to apparent **solar* motion and is the basis of civil time keeping. It is mathematically related to and thus derived from **sidereal time.* This derived time is labeled UT0. When adjusted for variations in **longitude* caused by the **Earth*'s polar motion, it is labeled UT1. Civil time is based on UT1 and UT generally refers to UT1.

UTC See **coordinated universal time.*

variable star A **star* whose **energy* output varies either regularly or irregularly.

vector A quantity having both **magnitude* and direction (compare to **scalar*).

vernal equinox The point on the **celestial sphere* where the **Sun* crosses the **celestial equator* moving from negative to positive **declination.* Also called the **March equinox.*

vertex (pl. vertices) The point of intersection of the two lines forming an angle.

visual magnitude See **apparent magnitude.*

waning The part of the **cycle* of the **Moon's *phases* when the sunlit portion of the face is becoming smaller.

watt (symbol W) The standard unit of measurement of **power.*

wavelength (symbol λ) The shortest distance between two similar points on a wave. For visible light, wavelength is related to color.

waxing The part of the **cycle* of the **Moon's *phases* when the sunlit portion of the face is becoming larger.

weight The **force* of gravity acting on an object. The force is proportional to the **mass* of the object.

west The point where the **celestial equator* crosses the **post meridiem *horizon.*

white dwarf A small, dense, dead **star* whose **mass* is supported against gravity by **electron* degenerate matter.

winter solstice In the northern hemisphere, the **December solstice.* In the southern hemisphere, the **June solstice.*

winter triangle The **asterism* formed by the three **stars*: Betelgeuse in Orion, Sirius in Canis Major and Procyon in Canis Minor. In the southern hemisphere it may be called the **Summer Triangle.*

year The *period* of time for the *Earth* to *orbit* the *Sun* once. There is more than one way to determine this period.

zenith The point on the *celestial sphere* directly over the observer's head.

zodiac The set of *constellations* through which the ecliptic passes. The set of astrological signs related to the position of the *Sun*.

zodiac constellation One of 13 *constellations* through which the *Sun* moves during its *annual* motion.

References

Allen, Richard Hinckley (1963) *Star Names. Their Lore, and Meaning*. New York: Dover Publications.
Barnett, Jo Ellen (1999) *Time's Pendulum*. New York: Harcourt Brace & Company.
Berry, Arthur (1961) *A Short History of Astronomy*. New York: Dover Publications.
Burgess, Eric. (1982) *Celestial Basic: Astronomy on Your Computer*. Berkeley, CA: Sybex Inc.
Comins, Neil F. (1993) *What if the Moon didn't Exist?* New York: HarperCollins.
Comins, Neil F. (1999) *Heavenly Errors*. New York: Columbia University Press.
Davis, Kenneth C. (2001) *Don't Know Much About the Universe*. New York: HarperCollins.
Duffett-Smith, Peter (1988) *Practical Astronomy with Your Personal Computer*, third edition. Cambridge: Cambridge University Press.
Duffett-Smith, Peter (1997) *Easy PC Astronomy*. Cambridge: Cambridge University Press.
Ehmann, James (1999) Lonesome in a crowd. *Sky and Telescope* **98**(5): 10.
Espenak, Fred (1987) *Fifty Year Canon of Solar Eclipses: 1986–2035*, NASA reference publication 1178 edition. Washington, DC: National Aeronautics and Space Administration.
Heifitz, Milton D. and Tirion, Wil. (2004) *A Walk Through the Heavens: A Guide to Stars and Constellations and their Legends*. Cambridge: Cambridge University Press.
Hiscock, Philip (1999) Once in a blue moon. *Sky and Telescope* **97**(3): 52–55.
Jesperson, James and Fitz-Randolph, Jane (1999) *From Sundials to Atomic Clocks*, second revised edition. New York: Dover Publications.
Johnson, H. L. and Morgan, W. W. (1953) Fundamental stellar photometry for standards of spectral type on the revised system of the *Yerkes Spectral Atlas*. *The Astrophysical Journal* **117**(3): 313–352.
Kaler, James B. (1994) *Astronomy!* New York: HarperCollins College Publishers.
Kaler, James B. (1996) *The Ever Changing Sky*. Cambridge: Cambridge University Press.
Krupp, Edwin C. (1999) Bear country. *Sky and Telescope* **97**(5): 94–96.
Lang, Kenneth R. (1980) *Astrophysical Formulae*. New York: Springer-Verlag.
Lang, Kenneth R. (1992) *Astrophysical Data: Planets and Stars*. New York: Springer-Verlag.
Mayor, Michel and Queloz, Didier (1995) A Jupiter-mass companion to a solar-type star. *Nature* **378**(23): 355–359.
Meeus, Jean (1991) *Astronomical Algorithms*. Richmond, VA: Willmann-Bell Inc.

Meeus, Jean (2000) Where eclipses come thrice. *Sky and Telescope* **99**(4): 63–65.
Mills, H. Robert (1994) *Practical Astronomy: A User-Friendly Handbook for Sky Watchers*. Chichester: Albion Publishing.
Montenbruck, Oliver and Pfleger, Thomas (1994) *Astronomy on the Personal Computer*, second edition. Berlin: Springer-Verlag.
Moore, Patrick and Tirion, Wil. (1997) *Cambridge Guide to Stars and Planets*. Cambridge: Cambridge University Press.
North, John (1995) *The Norton History of Astronomy and Cosmology*. New York: W. W. Norton & Company.
Olcott, William Tyler (2004) *Star Lore. Myths, Legends, and Facts*. New York: Dover Publications.
Olson, Donald W. and Sinnott, Roger W. (1999) Blue-moon mystery solved? *Sky and Telescope* **97**(3): 55.
Pogson, Norman (1856) Magnitude of 36 of the minor planets. *Monthly Notices of the Royal Astronomical Society*, **17**(12).
Rey, H. A. (1988) *The Stars*. Boston, MA: Houghton Mifflin.
Ridpath, Ian and Tirion, Wil. (2003) *The Monthly Sky Guide*, sixth edition. Cambridge: Cambridge University Press.
Sagan, Carl (1980) *Cosmos*. New York: Random House.
Seidelmann, P. Kenneth, ed. (1992) *Explanatory Supplement to the Astronomical Almanac*. Mill Valley, CA: University Science Books.
Sobel, Dava and Andrewes, William J. H. (1995) *Longitude*. New York: Walker & Company.
Teresi, Dick (2002) *Lost Discoveries*. New York: Simon & Schuster.
Tirion, Wil and Sinnott, Roger W. (1998) *Sky Atlas 2000.0*, second edition. Cambridge: Cambridge University Press.
Tirion, Wil. (1991) *Cambridge Star Atlas 2000.0*. Cambridge: Cambridge University Press.
Willmoth, Frances, ed. (1997) *Flamsteed's Stars: New Perspectives on the Life and Work of the First Astronomer Royal, 1646–1719*. Woodbridge: Boydell & Brewer.

Bibliography

Abel, George O., Morrison, David, and Wolff, Sidney C. *Realm of the Universe*, fifth edition. New York: Saunders College Publishing, 1994.

Arny, Thomas T. *Explorations: An Introduction to Astronomy*, third edition. New York: McGraw-Hill, 2002.

Bakich, Michael E. *The Cambridge Guide to the Constellations*. Cambridge: Cambridge University Press, 1995.

Bennet, Jeffery, Donahue, Megan, Schneider, Nicholas, and Voit, Mark. *The Cosmic Perspective*, third edition. San Francisco, CA: Pearson, Addison Wesley, 2004.

Berry, Richard. *Discover the Stars*. New York: Harmony Books, 1987.

Bierhorst, John. *The Mythology of Mexico and Central America*. New York: William Morrow and Company, 1990.

Burnham, Robert, Jr. *Burnham's Celestial Handbook*, volume one. New York: Dover Publications, 1978a.

Burnham, Robert, Jr. *Burnham's Celestial Handbook*, volume two. New York: Dover Publications, 1978b.

Burnham, Robert, Jr. *Burnham's Celestial Handbook*, volume three. New York: Dover Publications, 1978c.

Chaisson, Eric and McMillian, Steve. *Astronomy: A Beginner's Guide to the Universe*, fourth edition. Englewood Cliffs, NJ: Prentice-Hall, 2003.

Chartrand, Mark R. *Night Sky: A Field Guide to the Heavens*. New York: St. Martin's Press, 1990.

Cornelius, Geoffrey. *The Starlore Handbook*. San Francisco, CA: Chronicle Books, 1997.

Crowe, Michael J. *Theories of the World from Antiquity to the Copernican Revolution*. New York: Dover Publications, 1990.

Culver, Roger B. and Ianna, Philip A. *Astrology: True or False?* Buffalo, NY: Prometheus Books, 1988.

Cutnell, John D. and Johnson, Kenneth W. *Physics*, third edition. New York: John Wiley & Sons, 1995.

Davidson, Norman. *Sky Phenomena: A Guide to Naked-Eye Observation of the Stars*. Hudson, NY: Lindisfarne Press, 1993.

Dickinson, Terence. *Night Watch, An Equinox Guide to Viewing the Universe*, third edition. Willowdale, ON: Firefly Books, 2001.
Dickinson, Terence and Dyer, Alan. *The Backyard Astronomer's Guide*. Camden East, ON: Camden House Publishing, 1991.
Dixon, Robert T. *Dynamic Astronomy*, fifth edition. Englewood Cliffs, NJ: Prentice-Hall, 1989.
Dunlop, Storm, ed. *Atlas of the Night Sky*. New York: Crescent Books, 1984.
Ebbighausen, E. G. *Astronomy*, fifth edition. Columbus, OH: Charles E. Merrel Publishing, 1984.
Engelbrektson, Sune. *Astronomy Through Space and Time*. Dubuque, IA: William C. Brown, 1994.
Evans Martin, Martha and Menzel, Donald Howard. *The Friendly Stars, How to Locate and Identify Them*. New York: Dover Publications, 1966.
Fix, John D. *Astronomy. Journey to the Cosmic Frontier*, third edition. New York: McGraw-Hill, 2002.
Goldman, Stuart J. Where are you, really? *Sky and Telescope* **99**(5): 72, 2000.
Gutsch, William A., Jr. *1001 Things Everyone Should Know about the Universe*. New York: Doubleday, 1998.
Haddad, Leila and Cirou, Alain. *Mapping the Sky: The Essential Guide to Astronomy*. San Francisco, CA: Chronicle Books LLC, 2003.
Hamburg, Michael. *Astronomy Made Simple*. New York: Doubleday, 1993.
Harrington, Philip S. *Star Watch*. Hoboken, NJ: John Wiley & Sons, 2003.
Harris, Joel and Talcott, Richard. *Chasing the Shadow: An Observer's Guide to Eclipses*. Waukesha, WI: Kalmbach Books, 1994.
Heath, Sir Thomas L. *Greek Astronomy*. New York: Dover Publications, 1991.
Hemenway, Mary Kay and Robbins, R. Robert. *Modern Astronomy, an Activities Approach*, revised edition. Austin, TX: University of Texas Press, 1991.
Jones, Brian. *The Practical Astronomer*. New York: Fireside, Simon & Schuster, 1990.
Karttunen, H., Kröger, P., Oja, H., Poutanen, M., and Donner, K. J., eds. *Fundamental Astronomy*, second edition. Berlin: Springer-Verlag, 1993.
Kaufmann, William J., III. *Discovering the Universe*, third edition. New York: W. H. Freeman & Company, 1993.
Kepple, George Robert and Sanner, Glen W. *The Night Sky Observer's Guide*, volume one. Richmond, VA: Willmann-Bell, 1998a.
Kepple, George Robert and Sanner, Glen W. *The Night Sky Observer's Guide*, volume two. Richmond, VA: Willmann-Bell, 1998b.
Kerrod, Robin. *The Book of Constellations*. Hauppauge, NY: Barron's Educational Series Inc., 2002.
Kring, David A. Calamity at meteor crater. *Sky and Telescope* **98**(5): 49–53, 1999.
Krupp, Edwin C. Moonlit highways. *Sky and Telescope* **96**(3): 94–96, 1998.
Krupp, Edwin C. The stellar ties that bind . . . *Sky and Telescope* **97**(1): 101–103, 1999.
Krupp, Edward C. Noticing the eclipse. *Sky and Telescope* **100**(6): 94–96, 2000a.
Krupp, Edwin C. The big room. *Sky and Telescope* **99**(1): 94–96, 2000b.
Kuhn, Karl F. *In Quest of the Universe*, second edition. Minneapolis/St. Paul, MN: West Publishing Co., 1994.
Kyselka, Will and Lanterman, Ray. *North Star to Southern Cross*. Honolulu, HI: University of Hawaii Press, 1976.

Link, F. *Eclipse Phenomena in Astronomy*. New York: Springer-Verlag, 1969.
Littmann, Mark and Willcox, Ken. *Eclipses of the Sun*. Honolulu, HI: University of Hawaii Press, 1991.
McCready, Stuart, ed. *The Discovery of Time*. Naperville, IL: Sourcebooks, Inc., 2001.
Meyers, Robert A., ed. *Encyclopedia of Astronomy and Astrophysics*. San Diego, CA: Academic Press, 1989.
Mitton, Jacqueline. *A Concise Dictionary of Astronomy*. New York: Oxford University Press, 1991.
Moche, Dinah L. *Astronomy: A Self-Teaching Guide*, sixth edition. New York: John Wiley & Sons, 2004.
Moore, Patrick. *Exploring the Night Sky with Binoculars*, fourth edition. Cambridge: Cambridge University Press, 2000.
Moore, Patrick. *Patrick Moore's A–Z of Astronomy*. New York: W. W. Norton & Company, 1987.
Motz, Lloyd and Nathanson, Carol. *The Constellations*. New York: Doubleday, 1988.
Muir, Hazel, ed. *Larousse Dictionary of Scientists*. Edinburgh: Larousse Kingfisher Chambers Inc., 1994.
North, Gerald. *Advanced Amateur Astronomy*. Edinburgh: Edinburgh University Press, 1991.
O'Meara, Jane. Strange eclipses. *Sky and Telescope* **98**(2): 116–118, 1999.
Olsen, Donald W., Fienberg, Richard Tresh, and Sinnott, Roger W. What's a blue moon? *Sky and Telescope* **97**(5): 36–38, 1999.
Ottewell, Guy. *Astronomical Calendar 2004*. Greenville, SC: Astronomical Workshop, Department of Physics, Furman University, 2004.
Parker, Sybil P. and Pasachoff, Jay M., eds. *McGraw-Hill Encyclopedia of Astronomy*, second edition. New York: McGraw-Hill, 1992.
Pasachoff, Jay M. and Covington, Michael A. *The Cambridge Eclipse Photography Guide*. Cambridge: Cambridge University Press, 1993.
Pasachoff, Jay M. and Kutner, Marc L. *University Astronomy*. Philadelphia, PA: W. B. Saunders, 1978.
Raymo, Chet. *365 Starry Nights*. New York: Prentice-Hall, 1982.
Rey, H. A. *Find the Constellations*, revised edition. Boston, MA: Houghton Mifflin, 1988.
Ridpath, Ian, ed. *Norton's Star Altas and Reference Handbook*, twentieth edition. New York: Pi Press, 2004.
Sanford, John. *Observing the Constellations*. New York: Simon & Schuster, 1989.
Seeds, Michael A. *Horizons: Exploring the Universe*, eighth edition. Belmont, CA: Wadsworth Publishing, 2003.
Shaff, Fred. *Wonders of the Sky*. New York: Dover Publications, 1983.
Sherrod, P. Clay and Koed, Thomas L. *A Complete Manual of Amateur Astronomy*. New York: Prentice-Hall, 1981.
Shu, Frank. *The Physical Universe*. Mill Valley, CA: University Science Books, 1982.
Sinnott, Roger W. The best star catalog ever. *Sky and Telescope* **102**(1): 22, 2001.
Snow, Theodore P. *Essentials of the Dynamic Universe*, fourth edition. New York: West Publishing Company, 1993.
Steel, Duncan. *Eclipse*. London: Headline Book Publishing, 1999.

Stephenson, F. Richard. *Historical Eclipses and Earth's Rotation*. Cambridge: Cambridge University Press, 1997.

Stephenson, F. Richard. Early Chinese observations and modern astronomy. *Sky and Telescope* **97**(2): 48–55, 1999.

Thompson, C. J. S. *The Mystery and Romance of Astrology*. New York: Barnes & Noble Books, 1993.

Toomer, G. J. *Ptolemy's Almagest*. Princeton, NJ: Princeton University Press, 1998.

Verdet, Jean-Pierre. *The Sky: Mystery, Magic, and Myth*. New York: Harry N. Abrams, 1992.

Wagman, Morton. Who numbered Flamsteed's stars? *Sky and Telescope* **81**(4): 380–381, 1991.

Westrheim, Margo. *Calendars of the World*. Rockport, MA: Oneworld Publications, 1993.

Zeilik, Michael. *Astronomy: The Evolving Universe*, ninth edition. Cambridge: Cambridge University Press, 2002.

Zirker, Jack B. *Total Eclipses of the Sun*. New York: Van Nostrand Reinhold Company, 1984.

Zombeck, Martin V. *Handbook of Space Astronomy and Astrophysics*. Cambridge: Cambridge University Press, 1982.

Website interest

General interest websites

http://www.usno.navy.mil – the United States Naval Observatory
http://www.nist.gov – the National Institute of Standards and Technology
http://www.nasa.gov – the National Aeronautics and Space Administration
http://www.iau.org – the International Astronomical Union
http://www.astronomy.com – the website of *Astronomy* magazine
http://www.skyandtelescope.com – the website for *Sky and Telescope* magazine
http://www.astrosociety.org – the Astronomical Society of the Pacific
http://www.aavso.org – the Amateur Astronomer Variable Star Observers Website

Astronomy software websites

http://www.bisque.com – the home page for "The Sky" astronomy software (PC)

http://www.starrynight.com – the home page for "Starry Nights" astronomy software (PC and Mac)

http://www.viva-media.com – the home page for "Redshift" astronomy software (PC)

http://www.carinasoft.com – the home page for "*Voyager*" astronomy software (PC and Mac)

http://seds.lpl.arizona.edu/billa/astrosoftware.html – an extensive list of planetarium software for all operating system platforms – commercial, shareware, and public domain

Index

Page numbers in **bold** indicate pages with term definitions.